...nced
...patial
...alysis

...he CASA book of GIS

...ey and Michael Batty, editors
...nced Spatial Analysis

ESRI
Advanced Spatial Analysis: The CASA Book of GIS
ISBN 1-58948-073-2

First printing June 2003.

Printed in the United States of America.

Library of Congress Cataloging-in-Publication Data
Advanced spatial analysis : the CASA book of GIS / Paul A. Longley and Michael Batty, editors.
p. cm.
ISBN 1-58948-073-2 (pbk. : alk. paper)
1. Geographic information systems. I. Longley, Paul. II. Batty, Michael.
III. University College, London. Centre for Advanced Spatial Analysis.
G70.212.A394 2003
910'.285—dc21 003012142

Published by ESRI, 380 New York Street, Redlands, California 92373-8100.

Books from ESRI Press are available to resellers worldwide through Independent Publishers Group (IPG). For information on volume discounts, or to place an order, call IPG at 1-800-888-4741 in the United States, or at 312-337-0747 outside the United States.

Contents

Foreword

GIS means different things for different people. Some people use it simply to document geographic reality. Some people use it for visualisation and presentation. But for me the real heart of GIS is the analytical part, where you actually explore at the scientific level the spatial relationships, patterns and processes of geographic phenomena, cultural phenomena, biological phenomena and physical phenomena. I think this is the area that holds the greatest promise for creating insight into how our world works and how it is evolving, connecting and changing.

This book focuses on advanced spatial analysis. The material is presented in such a way that the concepts of spatial analysis are easy to understand and are anchored within a human context. Heretofore much of the advanced GIS literature concentrated on the mathematics and algorithms that people would apply. What I like about this book is that non-domain specialized people (laymen) can read the case studies and understand what spatial analysis is and how it could apply to their own work.

Spatial analysis is actually not a new concept, as pointed out in the case study showing Booth's poverty maps of London from the 1890s. People have always wanted to document and analyse patterns and interpret them for decision-making purposes. And now the very concepts of spatial decision making are expanding far beyond a simple visual of a map.

Spatial analysis is really a kind of data-mining technology. With it you can document different layers of separate information sets and use the spatial analytic tools in the same way that businesses currently mine tabular databases for business intelligence. Spatial analysis is the basis for spatial data mining, which will increasingly emerge as the basis for decision support systems across a diverse range of industries and jurisdictions to answer such questions as: Where should I locate? What areas should be targeted? What areas are biologically productive? What areas are at high risk due to natural or manmade hazards? What is the best strategy to take in the context of a more detailed and accurate understanding of place and location? These are the types of questions that are pervasive in the human world as our society becomes more spatially literate. The examples in this book serve as footsteps that will show the way and demonstrate what will become commonplace in the near future. In other words, this book provides evidence of how important geographic information systems are and illuminates their potential.

A book like this allows us to become more conscious of this shift towards spatial literacy, illustrating the costs of urban sprawl, the benefits of using advanced location-allocation algorithms for site location, the selection of optimum transportation routes, and many other examples. In all of these, we see the evidence, and it is powerful.

Jack Dangermond, President, ESRI

Chapter 1

[Prologue]

Advanced spatial analysis: extending GIS

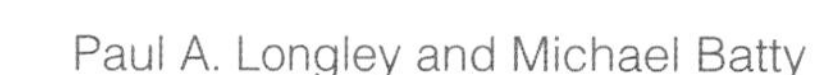

Paul A. Longley and Michael Batty

To introduce this edited collection of contributions from members of CASA at University College London, we begin by setting them within the general historical context of GIS as it has developed over the last 40 years. We adopt a broad definition of GIS and define our purpose as showing how GIS has been and continues to be extended to encompass a wide range of theories and methods that are applicable to processes that require spatial analysis. We identify three main themes that recur again and again throughout this book; these themes are about ways in which time and space should be represented within GIS, ways in which agents and institutions communicate and function, and ways in which networks are changing how we communicate, process and visualise data across the full range of geographic scales. From these themes we identify seven issues that are elaborated through the various contributions to this book. These are scale in mapping form and representing process; confidentiality and ethical issues in geodemographics and spatial behaviour; incompleteness of representations across space and time; visualisation in communication and user interaction; representation in the space of the third dimension and beyond; representation of dynamics and spatial processes; and policy applications through planning and design.

1 How we got here: a short history of GIS

Geographic information systems (GIS) are integral to a wide, applications-led field, with diverse roots in many spatial sciences ranging from geography and the earth sciences to architecture, urban planning and ecology. It is widely accepted that the term GIS was first coined by Roger Tomlinson in 1963 (figure 1), and Tomlinson is also accredited with developing the 'first GIS'—the national natural resource inventory for Canada created under his directorship (Tomlinson 1998). However, GIS as a cognate area and as a significant domain of commercial activity developed only in the late 1970s and 1980s after the costs of computer processing had made hardware, particularly display devices, affordable with the commercial costs of GIS software development then becoming shared amongst a wide range of users.

Popular reviews of the field (for example, Burrough and McDonnell 1998; Longley et al 2001) trace the principal roots to at least four distinct activities. First, the early innovations in computer graphics in the late 1940s, when various engineers working with machine diagnostics used oscilloscopes to experiment with graphics in order to simulate bouncing balls and various other entertainments. This ultimately led to the development of computer cartography in the mid-1960s and the creation of rudimentary maps using line plotters and printers. The micro-revolution, which began in earnest after the invention of the microprocessor in 1971, dramatically changed these possibilities as computer memory itself came to be associated with pictorial representation.

The second motivation arose through mainstream developments in database technologies and information systems. Representing spatial units required the development of algorithms to search spatial data structures and to solve problems such as the 'point in polygon', and these did not evolve until the late 1960s and 1970s. A third force was generated by the development of environmental remote sensing, initially to service the military satellites of the 1950s and subsequently the civilian systems of the 1960s. The fourth motivation arose out of the development of techniques to merge data representing different layers of activity such as terrain and land use. A variety of associated simple but effective overlay techniques, particularly developed for problems of landscape planning and best illustrated in the work of McHarg (1969), were first automated using early ideas of map algebra with their rudimentary display on line printers. All of these ideas came together in the 1960s at places like the Harvard Laboratory for Computer Graphics that served as a focal point for conceptual developments and for the basic software. It is from these beginnings that public domain and commercial software subsequently evolved (figure 1).

Figure 1 Roger Tomlinson who coined the term GIS, and Jack Dangermond who developed the first commercial software

Parallel to these developments, spatial analysis emerged largely from the advancement of quantitative methods in geography, geology and the earth sciences, as well as from regional science and macro-economics. Spatial statistics and locational analysis dominated geography in the 1960s, and the protagonists of geography's 'Quantitative Revolution' were responsible for the development and application of techniques that together laid the foundations of a new scientific geography (Johnston 1999). *Spatial Analysis* (Berry and Marble 1968) illustrated the rapid growth of the field, and symbolised the quantitative revolution in geography. The approach retained a core following in geography and set the agenda for the next generation. GIS emerged in parallel to developments in spatial statistics and locational analysis during the 1970s and 1980s, with a number of researchers moving between quantification and computer application. In this way, some of the more general of these techniques became slowly but systematically incorporated into GIS software. By the mid-1990s, GIS had broadened its remit sufficiently to embrace spatial analysis, while modelling and simulation were being developed in complementary fashion to GIS. Our own edited book *Spatial Analysis: Modelling in a GIS Environment* (Longley and Batty 1996) summarised the field at the time. This current book attempts to take our definition much further based on our perception that GIS is now extending into many fields of spatial representation, analysis, modelling, policy and design. This, then, is what we mean by the slightly precocious term 'Advanced Spatial Analysis'.

Spatial analysis takes GIS a major step forward, from a restricted range of 'off-the-shelf' representational forms to custom-made depictions of the world and critical interpretation of the assumptions that are invoked to represent it. It extends GIS to the concept of the model and beyond, to ideas about policy and design, to how we might change as well as simply understand the geographic dimension. Moreover, because of this, spatial analysis also points the way to advanced applications. The standard functionality of GIS provides a framework for data manipulation and visualisation, and this provides a supportive environment for structuring routine applications in operational management. However, harnessed to spatial analysis techniques, GIS also provides powerful ways of generating and evaluating strategic policies and plans.

As with most other software applications, the cumulative development of GIS software has been accompanied by transitions from the mainframe to workstation to desktop, with the packaging of particular GIS functions in software (and sometimes also data) products designed for particular applications. Today's commercial GIS provide a family of products ranging from light Internet browsers through to analytical engines encompassing a comprehensive range of functions for spatial data manipulation. It is not within the scope of this introductory chapter to attempt a broad-ranging history of the technical challenges and changes in computer architecture that have characterised the recent development of GIS (see Longley et al 2001: 9-25), although before introducing the research areas that make up this book, it is relevant to sketch out some of them. First, it is clear that the universality and generality of GIS has increased as hardware has become more pervasive. Second, until the mid-1990s, most of the emphasis in GIS was on developing fast, reliable software that made conversion between different data structures routine and safe, and thus diversified the range of data sources that could be combined within GIS. Third, the 1980s mission of commercial GIS to 'satisfy 90 percent of applications needs 90 percent of the time' has broadened as applications themselves have become more demanding. This has led to the adaptation of methods and models of spatial analysis to GIS software, and a range of new functions has been incorporated into desktop software in particular (for an early review, see Maguire 1995).

Fourth, modularity has become the dominant procedure for adding new functionality. Fifth, developments in 3-D representation have also provided a cutting edge to research. Sixth, the miniaturisation of computer devices, allied to ongoing developments in computer networking, is taking some of academic geography back to its roots in field measurement and environmental cognition, and this poses important new challenges for relaying real-time applications to the field. In fact, the move from fixed hardware installations to networks on the one hand, to small mobile devices on the other hand represents a new cutting edge of GIS development. This is part of the massive decentralisation of computing that is still ongoing, where data and software and even applications themselves are being distributed across networks and access is increasingly being effected through wireless devices. We will point the way to some of these applications in this book, but in our final chapter, the epilogue, we will also argue that these are likely to be the most important directions in which this field will evolve in the next decade.

2 Themes in this book

All GIS-based representations are selective abstractions, simplifications of reality. What we choose, or are forced, to leave out can be as important as the aspects of reality that we choose to retain. Fifty years ago, the idea of the model as a simplification of reality entered our vocabulary (Chorley and Haggett 1967), and in the social sciences in particular, the idea that models could be based on mathematical relationships rather than physical materials fitted very well a world in which simulation on computers was becoming ever more important. When we classify models, we usually fall back on the idea of simplification in that the essence of a model is to gain understanding through simplification. But in order to retain realism as well as simplification, we can embody models through very different media.

In general terms, we may distinguish between iconic and symbolic models, where iconic models tend to be scaled-down versions of the real thing often with working parts which are similarly scaled down. In their most superficial form, such physical models are often thought of as 'toys' and the way young children learn is through such media. In practical terms of some relevance here, the architect's block model, where buildings are made from balsa wood or cardboard or the scientist's wind tunnel or sediment tray, perform the same function. In contrast, symbolic models are based on logical relationships, usually mathematical and statistical where less obvious attributes of the system in question are modelled or simulated. Spatial analysis is part of this latter symbolic domain, while paper maps are best thought of as iconic. Digital maps that embody topological relations are best thought of as symbolic, as the notion of scale has no physical meaning within the confines of the computer. In passing, we should also mention the halfway house of analogue modelling where the model is generated in analogy to some other system, for example, simulating the interactions within a local economy using electrical networks whose various components represent different flows, resistances and potentials.

The computer revolution has taken us one step further by enabling us to not only model symbolic representations of systems but to model the physical representations within which such symbolic interactions usually exist. This has largely happened through computer graphics, but in terms of GIS, this means that the physical map is now the object of simulation. Indeed, GIS is a very clear example of a medium in which physical models of the system—its geography—merge with symbolic models of this geography through spatial relationships. This is what makes it so powerful. An excellent illustration of this merging, which we will exploit extensively in this book, is the move from the 2-D map to the 3-D scene, but the superficiality of this development masks the real power of such developments. Extending geographic representation from the icon of the map to the icon of the building or landscape gives us the possibility of extending symbolic modelling from 2-D to 3-D.

The picture of GIS that emerges today is of powerful ideas embodied in software and capable of immediate visual communication and that have important implications for many aspects of our day-to-day activities. Today's GIS:

- selectively integrate time and space across a far wider range of spatial and temporal scales than has been the case hitherto. Effective management of the granularity, or detail, of time and space can preserve the unique attributes of places and particular time periods, whilst retaining the power of generalisation between places and times.
- promote use by a wide range of agents and institutions in a wide range of casual as well as specialist settings. GIS has become not only a widely used tool for scientists to generalise and theorise about human activity patterns, but also an important visual communications medium for improving and promoting wider understanding of such generalisations. As such, it improves not just our scientific understanding of the public domain but also the public

understanding of our science, providing, that is, that the GIS is efficient, effective and safe to use (Rhind 1999).

- link the smallest handheld devices into a global network of computing, for exploration, simulation and analysis of geographic phenomena. Computer networks thus connect ideas in 'organized activity by which people measure and represent geographic phenomena, then transform these representations into other forms while interacting with social structures' (Chrisman 1997).

Several pertinent issues arise from these themes. In particular, there are issues of:

- scale
- representing human behaviour
- understanding the implications of ethics for spatial analysis
- incompleteness of spatio-temporal representations
- visualisation and our limited abilities to communicate amongst ourselves and to others
- representation in the third dimension and beyond
- understanding dynamics and spatial process
- policy application, planning and design

These issues emanating from our themes are clearly evident in the research at CASA contributed here, and we will elaborate in much of the rest of this chapter.

3 Some issues arising out of these themes

Scale in mapping form and representing process

At the finest scale of granularity, GIS depicts humans as mobile point georeferenced 'events' and human agency and activity patterns as changes in their relative locations. It has long been recognised that the best ways to represent such events are grounded at the level of the individual, epitomised, for example, in microeconomic models of travel choice (Hensher and Johnson 1981) and the behavioural geography of cognition and mental maps (Golledge and Stimson 1987). However, in the early days of spatial analysis, the restricted processing power of computers made analysis of multiple individual activity patterns impossible, and such data were, in any case, simply not available prior to the development of digital data capture technologies. The pragmatic response that has ensued is familiar to any geographer—representation through zonal aggregation with a strong emphasis on searching for patterns through choropleth mapping. But a fundamental paradox in geography as a scientific discipline is that aggregation is fundamental to achieve generalisation across space, yet the aggregated zones that are created to facilitate such generalisation usually have no validity independent of particular applications (or, indeed, any validity at all). Geography is unlike many sciences in that the basic atoms of information—points located in time and space—can rarely be aggregated into units that are obviously 'natural'. Moreover, the scale at which aggregation occurs and the configurations of areal units that are adopted often exert a critical influence upon the outcomes of the spatial analysis with contradictory conclusions being generated for the same data at different scales (Openshaw 1984).

The outcomes of such spatial pattern analysis may be misleading if zones are configured or aggregated in unusual ways—the whole notion of gerrymandering (Schietzelt and Densham, this volume) emerged as a consequence. The core problem is that there are few general guidelines to inform us whether a zonal scheme is unusual. These problems are compounded when the focus is upon spatial interaction and spatial process or the representation of temporal dynamics. Early digital representations of spatial interactions were usually framed within very coarse zonal geographies. Typical were intraregional representations of shopping patterns in the 1960s and 1970s where critical interactions were present between only a handful of zones. Over time, these constraints were relaxed to some degree so that by the mid-1990s retailers, for example, were building much more partial models of systems around finer-scale zonal geographies based on much more restricted domains of socio-economic behaviour.

Constraints on data availability, the depiction of smaller-scale objects, and their model-based processing, have also relaxed as data and simulation technologies have become more powerful. The consequences of this can be seen in many contributions to this book. Most obviously, the models of individual pedestrian flows presented by Batty (this volume) represent human beings as individual agents and appraise behaviour through a series of scenarios at scales from the architectural to the neighbourhood. The spirit of this approach is to develop plausible models of individual movement based around sensory attraction to features (sound systems in the case of the Notting Hill Carnival or exhibits in the case of the Tate Gallery) within the constraints of built form and street geometry. In these examples, the geographic units of analysis are individual human agents, and, as such, the application is the logical end point of the trend to successively finer levels of granularity. Although still foremost a research application, there is clear interest in this approach from a number of applications areas, such as retailers who have followed the trend down-scale from aggregate to small area data and now consider models of pedestrian flows as the ultimate level of disaggregation in devising performance measures for all store formats.

Although this approach has methodological strengths beyond representing the true unique individual, similar approaches can be taken with respect to other aggregations, as illustrated by Torrens (this volume) in his appraisal of the broad swathe of cellular automata (CA) and related agent-based approaches to urban development. In its simplest form, CA can be thought of as a form of digital analogue modelling, whereby the properties of cells (which may represent individual land parcels of developed space or individual buildings) are deemed to interact according to prespecified rules. With regard to urban development, most research to date has related to cities in the developed world, and thus the wider applicability of agent-based approaches has received little attention. Barros and Alves-Junior (this volume) go some way towards resolving this deficiency with a far-reaching study of peripherisation dynamics in Latin America and empirical reference to São Paulo and Belo Horizonte. Although these chapters all involve a move down-scale from the aggregate to the individual, such applications are temporally dynamic as well as spatially disaggregate, a trend that is evident in all GIS as we come to embrace more detailed data at ever finer scales.

Confidentiality and ethical issues in geodemographics and spatial behaviour

It is clear that developments in software, linked to vast increases in the capacity and processing power of hardware, now make it possible to simulate the movements of large numbers of individual points and to conduct spatial analysis at very fine levels of granularity. The case studies reported here by Smith (this volume) illustrate the recent development of digital 'framework' data by national mapping agencies (Rhind 1997), and there have also been improvements in the accuracy and precision with which conventional socio-economic data sources can be anchored to such sources. It is unfortunate, but perhaps inevitable, that the creation of socio-economic digital data infrastructures to support these developments has not sustained a similar pace of development. Indeed, fine-grained public sector datasets, such as national censuses, today make up a smaller real share of available data, given the rise of private sector data warehouses.

The maturation of global positioning system (GPS) technologies, coupled with the advent of portable GIS devices, allows much more detailed monitoring of individual activity patterns, as well as tagging of individual characteristics and actions (Li and Maguire, this volume). In important respects, these are some of the developments that take geography back to its roots in the measurement of conditions in the field, potentially reinvigorating the field of behavioural geography (Golledge and Stimson 1987). However, although this is undoubtedly the case in the research domain, there are few individuals who are likely to allow computers to monitor the intricacies of their daily activity patterns—and those that do are unlikely to be representative of the population at large. Thus, most GIS applications that require multi-attribute socio-economic data that are representative of populations will remain constrained to a more limited range of data sources, aggregated in order to preserve respondent confidentiality. Such are the datasets used by Evans and Steadman (this volume) to develop urban indicators consistent with the more aggregate land-use transport modelling that is being adapted to GIS. The technical niceties of scale and aggregation involve constraints on socio-economic data availability arising out of ethical considerations to protect privacy. The chapters in this book that describe coarse zonal geographies are those that require multiple attributes of individuals to be measured.

Other contributions to this book suggest more adaptive solutions to the available range of socio-economic data. The field of geodemographics has become accustomed to fusing records pertaining to individuals with aggregations, as in the use of lifestyles data to freshen up and extend the remit of census-based geodemographic classifiers. Much time in analytical geography has been spent trying to bridge scales, and the late 1990s generation of geodemographic classifiers to some extent epitomise the problem of how to relate rich, pertinent indicators of social conditions with the local scales that concern many business and service planning applications. The related approach taken by Webber and Longley (this volume) seeks to exploit the properties of social similarity and locational proximity by examining the area effects that are not specified in geodemographic classifiers.

Incompleteness of representations across space and time

The contribution by Webber and Longley (this volume) also raises a point that strikes back at the heart of a representation—namely, how to specify and, hence, generalise about 'place' effects of present or past environments that are either unmeasured or even immeasurable (in practical terms). In applications to the contemporary world, it is important to recognise that the effects of place are not wholly encapsulated in individual agents, that we cannot capture the effects of all individuals in all places and all time periods, and, even if we could, it would be wasteful to do so. It is not just the agents existing today that create place effects, and Webber and Longley demonstrate that GIS can be used to identify place effects that encapsulate different event histories and simplify historic chronologies.

In other applications settings, representations are simply incomplete. Sometimes, as in the cellular models described by Torrens (this volume) and the agent-based movement models developed by Batty (this volume), it is not possible to observe and record the data on individual decision making that embody the rules for action and interaction. For example, data on where people originate and where they are destined are notoriously difficult to collect for large crowds and although advances in laser scanning and remote sensing are able to produce aggregate volume data, there is no fix other than direct interaction with the objects and subjects of the simulation to generate the data required. Representations may also be incomplete because relevant information has been destroyed. This is the case in using GIS to reconstruct past environments as illustrated by Grajetzki and Shiode (this volume) where information about the past is incomplete and fragmented. In such circumstances, GIS do enable information to be assembled from different sites and different time periods, in order to create as complete and as consistent a representation as possible.

Visualisation in communication and interaction

The illustrations and scenarios presented by Grajetzki and Shiode (this volume) not only illustrate the advances that have been made in visualisation, but the persuasive power of the visual medium in conveying the interpretations of the past that particular researchers may wish to impart. They also illustrate how changes in computer architectures enable interrogation of databases from a distance across networks, and how GIS can create a better sense of immediacy in remote locations. Today, the ubiquity of GIS in developed countries means that more people than ever can use it to gain an impression of different places and different times.

The implication of this is that the desire, even of rather novice users, to perform advanced spatial analysis implies the need for guidance. Improved methods of visualisation and user interaction are key to this requirement. Tobón and Haklay (this volume) and Lloyd et al (this volume) each illustrate how public participation in GIS (PPGIS) can be encouraged through new techniques and procedures that foster clearer perception, improved interpretation, and more thorough interrogation of spatial data. The techniques presented in this contribution make it possible to visualise the effects of outlying points: the medium does not dominate the message, but rather is sensitised to revealing it in a data-led way. The important point here is that ‘advanced’ spatial analysis does not necessarily imply 'complicated': in the spirit of other developments at the research frontier (for example, Fotheringham, Brunsdon and Charlton 2001), advanced tools of spatial data exploration can be readily assimilated by an increasingly broad user base.

Some of these approaches are already used in full-fledged applications of GIS to scenario development as illustrated by Evans and Steadman (this volume) in the context of the urban sustainability debate. One of the problems with scenario analysis is the length of time required to refresh a simulation. Schietzelt and Densham's (this volume) research on location-allocation can similarly result in generating a wide range of scenarios for emergency management under any conceivable system state (for example, preprogramming the daily commuting rush), and this can reduce the processing needed for real-time analysis. This creation of a scenario base then also creates demands on human-computer interaction to manage the scenarios—that is, to make the selection of the most appropriate scenario from meta scenario categories as easy as possible.

Hudson-Smith et al (this volume) provide a wide-ranging example of how GIS and related multimedia information is being used to communicate ideas about the spatial urban environment. Visualising built environments in effective and easy to understand ways requires the development of multimedia techniques that need to communicate in direct and immediate ways. PPGIS and its extensions into multimedia are increasingly being aided by developments in 3-D GIS and CAD that are delivered to the public across networks rather than residing on the desktop. Visualisation can be thought of as the domain of GIS, but effective communication also requires good networking. It is in the general area of visualisation that extensions to 3-D and to the networked and wireless worlds are all coming together.

Representation in the space of the third dimension and beyond

Extending GIS to the third dimension has relied heavily on developments in hardware speeds, rendering and computer graphics software. It is now possible to display 3-D environments almost as quickly as the 2-D flat map, and this is providing new insights from visualisation that translate more abstract information into a form that many non-expert users can immediately understand. Sometimes these extensions are referred to as 3-D models or even urban models, but these are very different from the symbolic models that have dominated quantitative geography and spatial analysis. In fact, the extension of iconic but digital modelling from 2-D to 3-D also enables us to extend symbolic modelling of spatial relationships. The third dimension remains much underplayed in geographic and geometric analysis. For example, urban land use is largely collected at the ground floor level and in dense cities, much of the variety of land use and urban activity locations is lost because of this simplification. GIS in 3-D makes it possible to extend GIS functionality to analyse such diversity while providing a much more complete picture of how cities and landscapes work.

Hudson-Smith and Evans (this volume) illustrate how these extensions are being developed through the idea of the virtual city. They show how 3-D GIS is converging towards computer-aided architectural design (CAD) and providing a new lease of life for CAD, which to date has often remained preoccupied with issues of aesthetics and rendering. The media that they introduce is used in the application of Web-based public participation, and Hudson-Smith et al (this volume) also use these media in an application of urban regeneration in Woodberry Down. Some of the more architectural techniques presented in both of these applications are taken further by Grajetzki and Shiode (this volume) in their reconstruction of temple sites in ancient Egypt, while the kinds of spatial infrastructures needed to sustain such representation are considered by Smith (this volume). For example, thin clients, which perform little processing locally, are now widely used to support the kind of Internet browsing that is being developed for applications described by Hudson-Smith and Evans (this volume).

A related question involves the way in which such 3-D representations scale to smaller handheld devices. The frontier question is how such devices, already used in the field and associated with field data capture and field location of static objects, can be refreshed to contribute to the challenges of navigating urban environments and wayfinding. This is part and parcel of bringing GIS closer to our direct experience and our need to understand how location-based service (LBS) technologies become incorporated into behaviour. The historic demarcation between direct and indirect experience blurs when handheld devices are used as an adjunct to reality in the field. In this context, Li and Maguire (this volume) reflect that it is spatial knowledge acquired via the survey method, such as map reading, that is usually assumed to be the most advanced level of spatial knowledge. This implies that GIS are now more directly about spatial knowledge acquisition than just data input and the translation of data into information. In this sense, extensions of 3-D GIS into other forms of multimedia based on virtual reality (VR) systems are likely to dominate the future. Thus, the methods used in constructing multimedia Web sites that link photogrammetric with 3-D graphics and other forms of multimedia delivered across the Internet will be central to future developments in GIS.

Representation in time: dynamics and spatial processes

Like the third dimension, ideas about time provide a clear challenge to new ways of representation within GIS. GIS are manifestly spatial and, hitherto, have largely treated questions of time as simply static spatial snapshots. Dynamic processes are hard to represent within GIS, and typical GIS structures focus largely on spatial representation as the starting point. Yet throughout this book, we develop tools in which dynamics figure strongly. Agent-based, CA and the more macro-population models developed, for example, by Batty and Shiode (this volume) all involve dynamics which cannot be represented in conventional GIS software, and this poses a challenge to researchers in the field. In fact, fully functional temporal GIS software of a proprietary nature is long overdue, for there are a plethora of problems that require explicit dynamic representations. A primary reason why such systems have not yet emerged is that there is no single large niche area of application that is forcing the pace. It is easier to structure GIS around spatial variation and representation because this provides a much more common base, while dynamics, insofar as they can be represented, are still regarded as an add-on to be represented as part of the modularity that is a feature of contemporary GIS software. Indeed, some of the techniques developed by de Smith (this volume) involve transformations of distance mapped directly onto the kinds of dynamics that are needed in terms of temporal GIS, and he shows that the challenges that need to be addressed in the future will take us back to fundamentals despite the advances that have been made during the last half century.

Shifting spatial analysis from statics to dynamics or, in other words, from product and structure to process and behaviour, is another challenge for the future. In particular, most applications require at least an implicit form of dynamics through thinking about changes to the systems that are represented, or more formally through forecasting or even policy and design. Besussi and Chin (this volume), for example, develop implicit dynamics to their analysis of urban sprawl, although the project that they describe uses GIS extensively in the form of snapshots of growth and change. In the same way, Barros and Alves (this volume) have a different take on what in the United States and Europe is termed ‘sprawl’ but in Latin America is considered to be ‘peripherisation’. In fact, as in many applications reported here, this analysis of sprawl uses a variety of models and techniques to represent urban development, illustrating the obvious but often overlooked feature of GIS usage where many related methods and software are used for any application. GIS is employed here in its very widest sense, almost as a shorthand for spatial analysis, modelling, design and policy. This suggests how different conceptual frameworks for GIS can be formulated into different models to drive different scenarios, thus adding dynamics to GIS through its wider usage.

Finer scale dynamics still constitute a challenge for GIS, especially where models based on 'far-from-equilibrium' concepts are important. Batty (this volume) illustrates some of these ideas in the context of pedestrian modelling, but in most of this book, there is a strong sense in which applications of GIS require relatively stable problem structures and well-defined forecasting methods. The idea that systems are never in equilibrium is one which is hard to embrace within an applications-oriented paradigm, although it is very likely that one of the challenges of the next decades in GIS will be the notion that the world is never in equilibrium, that change is central to any problem that we care to consider, and that our methods must be adapted accordingly. This will require new ways of exploration, simulation and optimisation that we will address at the end of this book in our epilogue.

Policy applications through planning and design

Policy is woven throughout the applications described here, for many of the projects reported are motivated by problems resulting from urban change and which require planning and design. Even the new and speculative methods and models that we report on, such as those that pick up in agent-based technologies, are influenced by policy issues such as crowding in shopping malls, problems of containing sprawl, urban regeneration, and problems in developing appropriate information infrastructures. Here we present a range of policy applications with different motivations. For example, Thurstain-Goodwin (this volume) develops techniques of representing data in a form which is immediately relevant to retail policy applications. His techniques aid communication, while Evans and Steadman (this volume) adapt GIS technologies for the same kinds of purpose—to give decision makers and users of these technologies more user-friendly interfaces and more organised ways of representing data and forecasting for problems of strategic land use-transportation planning.

Longley et al (this volume), in their chapter on the complexity of the retail system and the need for new ways of dealing with retail geodemographcis in terms of location, show that it is no longer satisfactory to represent retail space in a uniform way. Activity must be differentiated according to more general patterns and objectives which, for aggregates of individuals, bear an identifiable correspondence with time of day, type of land use, mode of transport, and a whole range of much more individualistic factors than have been considered by retail planners and developers hitherto. This poses enormous challenges in terms of how we can represent retail centres as discussed by Thurstain-Goodwin (this volume), involving issues such as secondary retail centres and the emergence of new forms of organisation through the concentration of retail capital.

The policy and design processes within which the new technologies of advanced spatial analysis might be set are explored by Alexiou and Zamenopolous (this volume) in their analysis of planning processes in terms of distributed learning and control coordination. They develop models of the planning process itself. This process is somewhat different to others in this book for it deals with ways in which we might plan and design, although the models that are used (based on neural nets) are close to many of the techniques currently being developed in the spatial analysis of geographic systems. In a sense, these methods relate to those developed by Schietzelt and Densham (this volume) in their work on location-allocation models where the choice of optimal solutions within formalised solutions spaces is central to their application. Once again, policy, although important to our survey of advanced applications, is an area that we feel will dominate GIS in the next decades as we develop and improve our applications through experiences such as those reported here.

4 Ways forward: a connected world

When we compiled our book *Spatial analysis: modelling in a GIS environment* (Longley and Batty 1996) nearly a decade ago, the idea that computing would move so quickly from the desktop to the network was still fanciful despite the facts that the Web had been invented in the early 1990s and that most of us were using e-mail routinely. What we will see here in the many contributions that constitute much of 'advanced spatial analysis' is the use of very diverse software, albeit much of it associated with proprietary GIS but much of it also based on wide linkages between different types of software on the desktop. In wider context, many of the applications that we include also use the Web not only to retrieve and store data but also to operate software. However, the biggest advantage that the Web offers is the ability to communicate not only amongst ourselves as researchers and professionals but also to those who provide our mandates and those who we hope will benefit from the kinds of data and information that we work with. Visualisation is thus a major theme and issue throughout this book, and every chapter translates GIS technologies into a form that must be communicated visually. The various themes and issues identified here recur throughout the chapters that follow, and in an epilogue at the end of this book we will draw these together and point to the future.

Section *I*

Virtual cities and visual simulation

GIS is transforming the way that we represent cities, urban areas and built forms. The contributions to this part of the book illustrate that this, in turn, has profound repercussions on the way that we think about the urbanisation phenomenon. Naru Shiode and Wolfram Grajetzki illustrate this with reference to the nascent urban system of ancient Egypt, in a convincing exposition of the way that GIS allows us to assemble digital shards of evidence from different spaces and different times so we can speculate about the environments of some of the first urban dwellers. Andrew Hudson-Smith and Steve Evans then illustrate the richness and detail of 3-D representations of today's cities. They present a wide-ranging review and interpretation of recent international experience and set the scene for developing a geographically extensive representation of London. The real and the virtual are then located in time as well as space, by Paul Torrens and then by Michael Batty, in contributions that examine, respectively, the way that cities grow and change over time, and the ways in which the micro-scale movements and activities of individual human beings are guided and constrained by built form and the movement patterns of other individuals.

Chapter

Digital Egypt: reconstructions from Egypt on the World Wide Web

Wolfram Grajetzki and Narushige Shiode

This chapter focuses on a joint project, Digital Egypt for Universities, between CASA and the Petrie Museum of Egyptian Archaeology. Taking copyrighted images, we assemble an online resource on Egyptian archaeology, which offers a wide range of audio-visual representations of this ancient culture and its technology. In particular, we create 3-D reconstructions of a series of historic sites, which has never been created on such a scale in the field of Egyptology, for the purposes of learning and teaching. Although the subjects of many of our reconstructions are monuments and buildings that are now almost completely destroyed, we produce several models of the same sites based on different interpretations as a stimulus to scholarly debate and discussion. Three such exemplars are described in detail here: Hemamieh, a prehistoric settlement; Naqada, one of the earliest towns in Egypt; and Koptos, a temple from which several important finds reside in the Petrie Museum, including some of the earliest monumental sculptures.

1 An online virtual archaeological site

One very effective way of delivering information resources on archaeological contents is to create an online virtual archaeological site that features images of found artefacts, maps, 2-D graphics, audio, 3-D visualisation utilities and other forms of learning and teaching materials. This chapter explores the media of virtual encounters as vehicles for learning about the cultural, social, technological and various other aspects of the past via an online resource that can be used by a wide audience. Reproducing a destroyed historic landscape using excavation records and surviving artefacts provides us with the opportunity to visualise, explore and present ancient sites in their probable forms and to compare them with alternative scenarios that have been suggested by archaeologists in the past.

Some recent novel attempts have been made in the fields of virtual presentation and reproduction of historical or heritage sites (see, for instance, Forte and Siliotti 1997 for a collection of various reconstructions). However, most projects are restricted to visualising an existing site or constructing a realistic model where the original plan is known. In our own project at CASA, we propose a generic method for visualising the possible forms of historic sites whose precise forms are unknown, and thus have to be reconstructed from the few surviving artefacts. Our 3-D reconstructions of these places seek to incorporate the known archaeological and geographical data, including landscape, other contemporary edifices at the same site, the past and present colouring of individual buildings, and the likely phasing of settlement construction. We hope to reconstruct the most likely appearance of a given site at a certain time as a way of encouraging viewers to evaluate the evidence for themselves, for example, by comparing different paths that might have been taken in reconstruction. This represents a quite radical departure from the idealised versions that have often hitherto been produced for other sites. In broader terms, this approach can be considered as consistent with a shift in archaeology from presentation of a single confirmatory representation that is one person or group's interpretation of the evidence to a more exploratory and participatory approach of sifting evidence and encouraging wider participation in the evaluation of scenarios. As such, it is similar in spirit to other research that is being undertaken in different applications fields at CASA (see Tobón and Haklay, this volume, Hudson-Smith et al, this volume).

This is a major breakthrough not only in the field of Egyptian archaeology but also for other archaeological applications. Many Egyptian sites have never previously been visualised using any form of 3-D representation, let alone delivered with the options of walking through the models or comparing different interpretations side by side. The examples that we draw in this chapter are all adopted from our ongoing three-year Digital Egypt for Universities project, hereafter simply referred to as Digital Egypt. We are in a privileged position in that we have ready access to a huge range of artefacts and the archaeological records of W. M. Flinders Petrie and other excavators of the late nineteenth and twentieth centuries. These are preserved at the Petrie Museum of Egyptian Archaeology, University College London, in one of the largest collections of its kind in the world. W. M. Flinders Petrie (1853–1942), after whom the museum is named, was one of the main figures in Egyptian archaeology. He excavated over several decades (from about 1881 to 1938) on many sites in Egypt (and Palestine). His records and the material that he excavated are of central importance for reconstructing Egyptian culture. The collection comprises some 80 000 objects covering almost all periods of Egyptian history from the Palaeolithic to the Islamic period. By summer 2002 a project had completed a photographic inventory of all of these objects. One of the aims of our project is to deliver these images across the Web. Using these copyrighted images, Digital Egypt will assemble an important resource for university courses, providing background information on many of these artefacts. The connection with related sources such as excavation notebooks, tomb cards and the publications of the British Institute of Archaeology enhances this unique asset.

The methods that we propose and apply to each case study here are generic, and they potentially pertain to a vast range of historic environments. The contents are offered through a range of media including 3-D VR models, audio files and digital images of archaeological finds, as well as information provided under several different cultural categories or based on temporal and locational classification. We adopt VRML97 as a primary means to visualise texture and create 3-D models. This is complemented by other forms of online material such as movies and still images, allowing low-end users to access the contents thus delivering to a wider audience.

Digital Egypt is now in its third year; in 2000 a first version of several hundred pages was launched on the Web (www.petrie.ucl.ac.uk/digital_egypt/Welcome.html). The first part of the project has been principally concerned with Egypt's prehistory and Pharaonic Egypt. The Petrie Museum is a particularly important source of information concerning Egyptian prehistory and has many very significant finds from all of Petrie's excavations. These include Naqada, the key site of prehistoric Egypt, after which the whole period is called; Tarkhan, a cemetery with about 2000 tombs; Abydos, the site of the earliest royal tombs; and

Hierakonpolis, the main cult centre of the period. The collection also includes several thousand objects from Guy Brunton's expeditions, including those to the Qau-Badari region. Together the objects from these historically pivotal sites provide an excellent background for recreating important aspects of ancient Egypt.

The focus of CASA's Digital Egypt project is upon the archaeological artefacts, but not to the exclusion of written sources. Indeed the Petrie Museum has a huge collection of papyri, including unpublished fragments of important funerary manuscripts and Lahun papyri (mainly administrative documents and letters) which are also incorporated into our project. Also included is a new translation (including philological and historical comments) of the *Tale of Sinuhe* (a masterpiece of ancient Egyptian literature) and of its contemporary *Teaching of Amenemhat* (both translated by Stephen Quirke). Audio fragments allow users to experience the ancient Egyptian language, insofar as it is possible to reconstruct its sounds. This is very important for students, not only of Egyptology in particular, but also of archaeology and, more broadly still, in cultural studies. The ancient Egyptian language contained many sounds not known in European languages; only audio can bring these to modern speakers of those languages.

In terms of Egyptology, Digital Egypt will also be a useful resource for archaeologists working in the field. Its typologies of pottery and everyday objects are essential for dating finds on archaeological sites, and this knowledge can be served anywhere in the world through the Web. Egyptological research often focuses on single aspects of material or written culture. Although one role of the project is to concentrate on particular artefacts, the broader objective is to combine the knowledge of different research areas to provide a full picture of this important culture. Single objects are related to other objects and to the architectural context in which they were found, thus establishing a basis for comparison across space and time.

The rest of this chapter comprises three sections. First, we address the methodology and techniques we applied to develop the online learning and teaching resource. Next, we showcase examples by focusing on some of the individual sites that we have reconstructed in our study. We conclude by summarising some of the insights obtained and discuss some of the issues and problems that we have encountered whilst creating the online resource.

2 Methodologies used to construct the online resource

Digital Egypt for Universities as a learning resource

The central aim of Digital Egypt is to provide a learning resource on the Web. Several informal learning courses at different levels will be provided. The site also provides reading lists for the subjects covered in Digital Egypt, some of which are accompanied with summaries and reviews. A large number of high-resolution digital photographs will provide a valuable resource for studying objects, especially for students in distant locations. The publication of artefacts using high-definition black-and-white and colour plates has often posed problems in archaeology because the print runs of research publications are often small numbers and, consequently, the books are very expensive. The budgetary constraints of paper reproductions often dictate that objects be shown only as line drawings. This in turn promotes a need to access the readily available digital images on the Internet (or on CD-ROM). Even recent publications of excavations still use conventional black-and-white pictures and line drawings, although it is very likely that this will change in the near future. Our Web site (www.petrie.ucl.ac.uk/digital_egypt/Welcome.html) also provides short movies to illustrate the various technologies used in ancient Egypt, and elsewhere in the ancient world (figures 1 and 2). Charts and tables provide various data on the subjects covered and clickable maps reveal the locations of important sites in Egypt.

Reconstructing architecture in 3-D

One important aspect of our studies is the 3-D reconstruction of different sites and monuments. Many archaeologically important places have been either completely destroyed or are badly preserved, and this inevitably creates uncertainty in

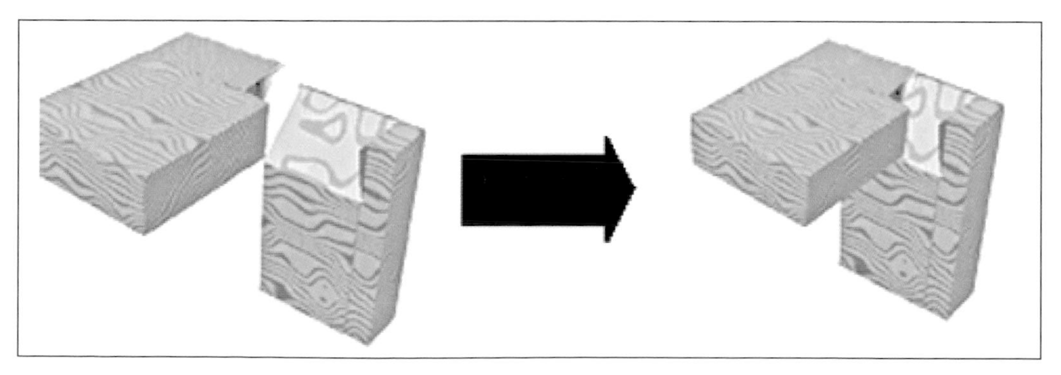

Figure 1 3-D models demonstrating technologies; here, joints in woodworking (used for furniture)

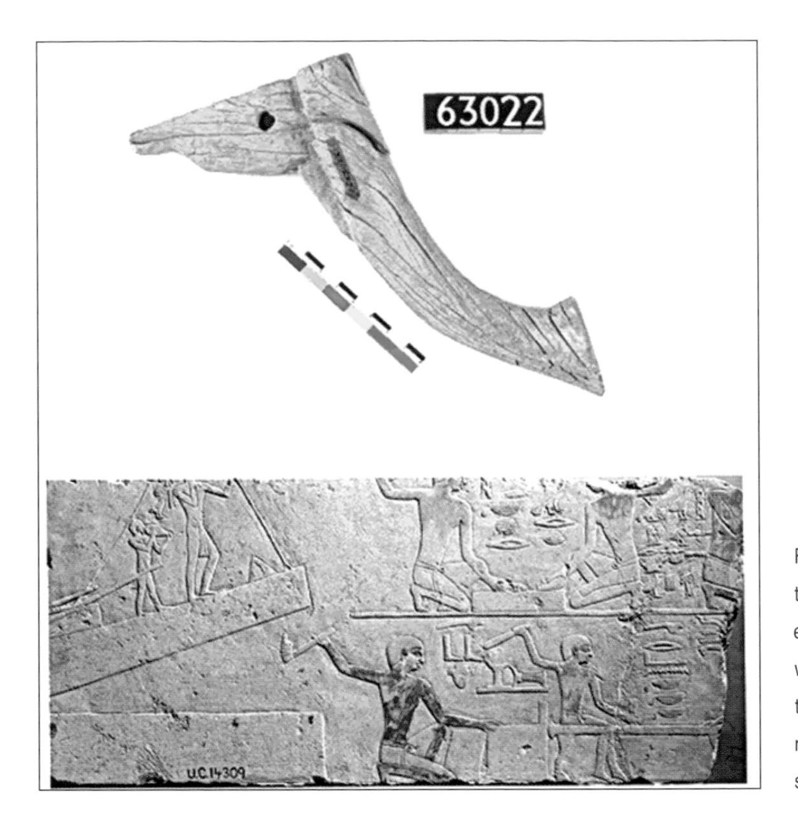

Figure 2 Pictures of objects from the Petrie Museum accompany explanations of technology (top: wooden part of a woodworking tool called *adze*, bottom: tomb relief showing the use of a similar adze)

reconstructing historic forms. In our work, we have used virtual reality to suggest not one but several different possible representations of particular sites in order to present the different forms suggested by ourselves and by other researchers.

Egypt has many well-preserved temples and tombs, although there are only a few good preserved settlements. In Mesopotamia in particular and the Near East in general, only the foundations of buildings have survived. Paper reconstructions have always played an important part in studying the architecture of the Near East (for example, see the series of reconstructions in Heinrich 1982). However, with a few important exceptions, they have never played a very important role in understanding architecture in Egypt. An impressive number of reconstructions is included in A. Badawy's (1954, 1966, 1968) history of Egyptian architecture. The reconstructions by Walter B. Emery (1991) of the Early Dynastic palace tombs at Saqqara (about 3000 B.C.) are an important source, as are the two earlier versions of the reconstruction of the Sun Temple of King Userkaf (c. 2500 B.C.: Ricke 1965). The latter example is unusual in that two different possible reconstructions of one temple are provided, in recognition of the uncertainty arising because this temple structure was destroyed.

Many of the sites excavated by Petrie had been severely damaged (for example, the Min temple at Koptos or the labyrinth at Hawara), and it is therefore very hard to gain even a vague idea of the original appearance of these places in ancient times. While more remains have survived at other sites, Petrie himself never tried to envision the design and layout of the original buildings, and this task has thus been left to others. One example is the governors' tombs at Qau el-Kebir. Petrie excavated here (Petrie 1930), yet he only published the plans of the structures that he found, although Steckeweh (1936), who was also working at the site, published plans and a full 3-D drawing of a reconstruction. Figure 3 shows a 3-D model that we reconstructed after Steckeweh's drawings (1936).

The reconstruction by Steckeweh has been reproduced in many other publications and is persuasive in suggesting that we know the original appearance of the buildings in detail. Yet a closer inspection of the reconstructions and the surviving remains raises many questions. Did the causeway really have a roof? Was there

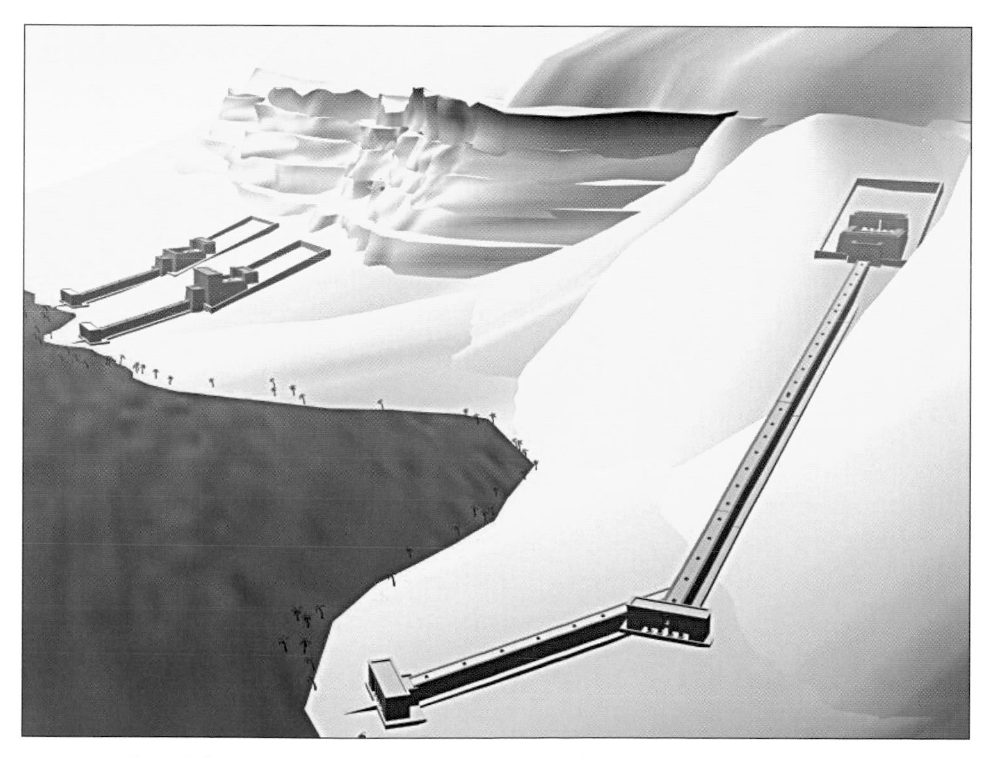

Figure 3 Reconstruction of the tombs of the governors at Qau el-Kebir (after Steckeweh 1936)

really a temple-like building next to the river, or just some kind of small jetty? What kind of structure is the pylon-like building at the end of the last courtyard? Were all of the tombs ever finished? We address these issues when constructing our 3-D model and come up with the most probable form in our representation.

Another example of a reconstruction of an ancient Egyptian building is the mortuary temple of Mentuhotep II in Der el-Bahari. The first excavator Naville published an excavation record that suggested a pyramid on top of the building (Naville 1910: plate XXIV). However, Arnold (1974) proposed a square building, while Stadelmann (1991: 232, figure 74) reconstructed a small mount. This is one of the few cases where Egyptologists seem to be fully aware of the uncertainty surrounding our knowledge (Kemp 1989: 104, figure 38).

In Digital Egypt we will offer several reconstructions for each of the sites excavated by Petrie and others. Several versions of reconstructions will be published on the Web, with the intention of showing what we do not know as much as what we do know. Moreover, because VRML models are relatively low-cost and easy to make, they can be made widely available on the Web. Thus, it is quite straightforward to make changes in the light of new research findings and to disseminate them to many interested parties. VRML models are the perfect media for promoting discussion.

Relevant studies

Digital Egypt is unique in many respects. For instance, there has hitherto been no pedagogic Web site in the field of Egyptology or archaeology. Also, there are a number of 3-D reconstructions of single buildings, but the sites that host them are maintained by private individuals. Some of the architects of these models do not have an academic background and may not have access to the relevant research publications. Their representations frequently suggest a lack of deeper knowledge about the wider ancient landscape and architectural settings. Several noncontemporary building phases are sometimes mixed up; older building phases may be ignored or the colouring of buildings—a very important consideration with respect to visual impact—is not considered. Many 3-D reconstructions have been produced in VRML and other languages and have been published in monographs (more often of a general interest, rather than research, nature) in recent years, but it is more rare for them to be maintained on the Web. It appears that many archaeological institutions, such as universities and museums, remain unaware of the full potential of the Web.

In fact, there are only a handful of important Egyptology, Web-enabled research projects. Among them is the Berliner Wörterbuch (the Berlin dictionary, the main institution for collecting Egyptian vocabularies) that has its whole archive (2.5 million digital pictures of handwritten slips written at the beginning of the twentieth century) on the Web (www.bbaw.de/forschung/altaegyptwb). The Griffith Institute in Oxford, United Kingdom, is in the process of uploading the digitised slips from records made at the discovery of the tomb of Tutankhamun (www.ashmol.ox.ac.uk/gri/4search.html). The Museum of Fine Arts at Boston is also preparing a Web resource of their Reisner archives of handwritten manuscripts and photographs (www.mfa.org/giza/pages/reisner.html). Each of these projects delivers access to firsthand records that have previously been inaccessible to the wider research and lay communities because they were stored in museums and archives that could be visited only with special permits.

Perhaps the most elaborate example of a Web site that seeks to virtually reconstruct an ancient site is that of the Theban Mapping Project, an excavation mission at Luxor (ancient Thebes) and based at the American University in Cairo (www.kv5.com). The Web site includes huge bibliographies, high-resolution pictures of the mission's finds, building inventories and historical notes. The Web site is focussed on Thebes and the New Kingdom (about 1550 to 1070 B.C.) and is usable for teaching, although no online courses are included. There are similar projects based on other excavations, such as the Tomb of Senneferi (www.newton.cam.ac.uk/egypt/tt99/index.html). Other smaller Web projects are often connected with museums and provide a general idea of the objects held in the museum. All such sites provide information that is useful for school education. A summary of relevant Web sites is available at www.newton.cam.ac.uk/egypt.

There is no comprehensive resource of Egyptian archaeology online, and our aim with Digital Egypt is to provide just that. With the aid of 3-D models and other media we will offer information on a variety of social, cultural and technological issues as well as provide data on different time periods. We also offer a series of course modules that can be used as teaching tools in classes. We believe Digital Egypt is a truly unique and valuable resource.

3 Case studies of 3-D reconstruction from 'Digital Egypt for Universities'

In this section we illustrate the resource offered on our Web site by focusing on three excavation sites and the 3-D VR models that we have produced for each.

Hemamieh

Hemamieh is a modern village in Upper Egypt. At the desert edge of this settlement several cemeteries and some domestic remains have been excavated and surveyed. The site also features the remains of a small village, which was excavated in 1924 by Caton-Thompson (Brunton and Caton-Thompson 1928: 69-116). The village is datable to around 4000 to 3500 B.C. and belongs to the so-called Badarian people, the first Chalcolithic (the Copper-Stone Age) culture in this part of Egypt. The main aim of the excavation was not to produce a plan of a Badarian village but to gain information on the relation of the Badarian culture to the Naqada culture. The Naqada was another prehistoric culture in Upper Egypt, whose chronological relation to the Badarian was for a long time unclear. The result of the excavations revealed the Badarian level to be under the levels of the Naqada period. The Naqada is therefore clearly later, although there has hitherto been little discussion of the nature of the excavated village. However, growing interest in the prehistory of ancient Egypt in recent years has focussed on the village

Figure 4 Reconstruction of the village excavated near the modern Hemamieh

because it is one of the few excavated settlement sites of that period in Egypt. The excavated structures create several problems. Most of the structures are round holes in the ground, often just one metre in diameter. It has been assumed that these were storage containers for grain. However, some had been found filled with animal droppings, indicating that they had been used as stables, at least for some periods. The authors' own 3-D reconstruction (based on archaeological research by Grajetzki and VRML modelling by Shiode) suggests that the structures were indeed stables. However, the visualisation in Digital Egypt also allows for the possibility that the structures were initially used as temporary dwelling places, and were subsequently used as stables for animals (figure 4).

Naqada

Naqada was examined in 1895 by W. M. Flinders Petrie (Petrie and Quibell 1896). The site called Naqada actually consists of a series of settlements and cemeteries at the edge of the desert, about 24 km north of modern day Luxor. A temple excavated there by Petrie belonged to the god Seth. The whole site is of special interest because it was here that Petrie first found tombs dateable to before the First Dynasty (earlier than 3000 B.C.). Petrie did not initially recognise the

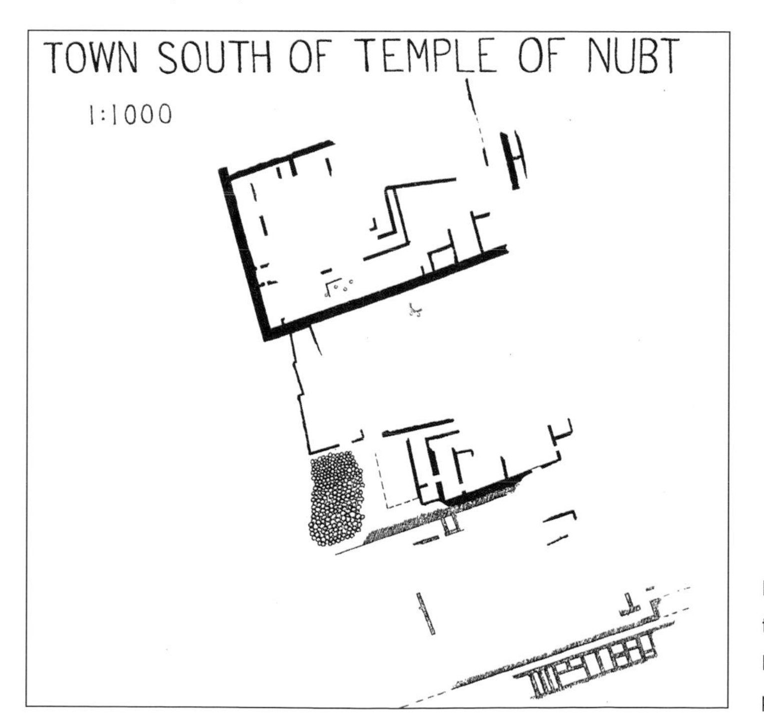

Figure 5 The published plan of the settlement structures found at Naqada (Petrie and Quibell 1896: pl. LXXXV)

extreme old age of the cemetery. He thought that the tombs belonged to a new race that invaded Egypt in the First Intermediate Period (about 2200 to 2050 B.C.), although subsequent research demonstrated that this was incorrect. Naqada must have been one of the major sites in Egypt from about 4000 to 3000 B.C. Petrie concentrated his research on excavating the 2000 tombs but also recorded and excavated structures at the site of the settlement.

The recorded structures (figure 5) are large in scale, and there is some debate about how these remains should be interpreted. In Digital Egypt they are reconstructed as three palace-like residential structures surrounded by a huge town (figure 6). From later sources and from the importance of some of the tombs found at Naqada, it seems very likely that Naqada was the capital of an early kingdom or chiefdom. However, this is only one possible interpretation of the remains of the buildings. The heavy-walled structure in the north could also have been a walled residential area with several buildings inside (that is, some kind of acropolis or fortress). The other structures may be just the scant remains of the town.

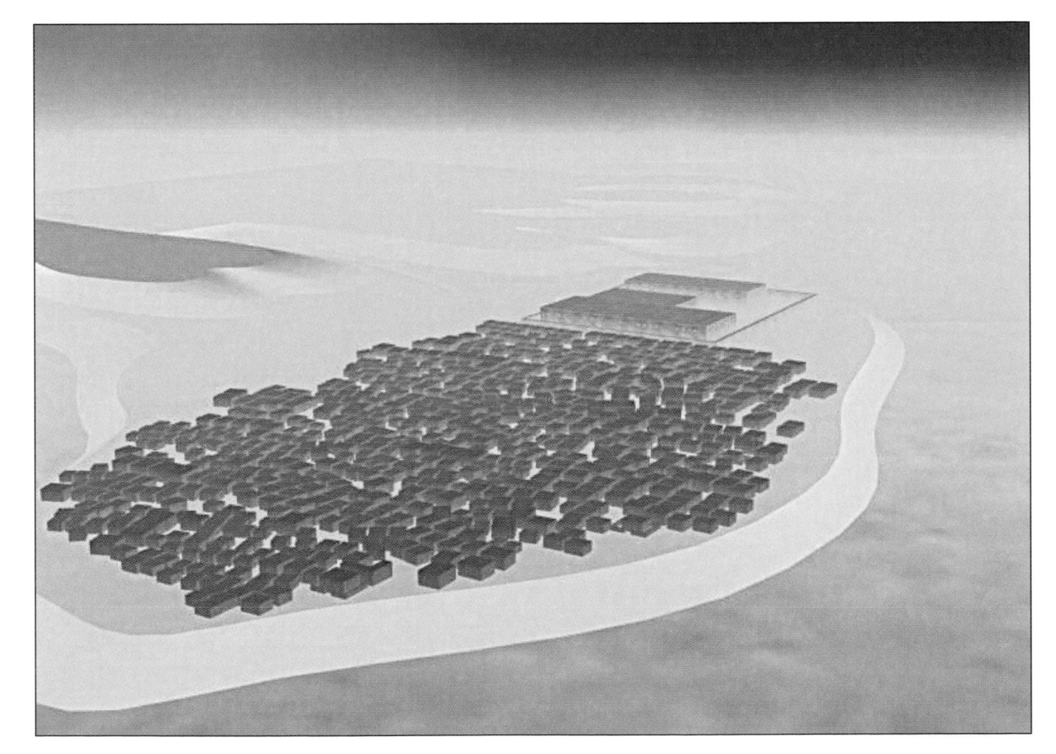

Figure 6 Reconstruction of Naqada

However, despite all the uncertainties concerning the precise shape of this early settlement, the reconstructions in Digital Egypt clearly serve to demonstrate that towns had already developed in fourth millennium B.C. Egypt in a way similar to those traced to the same time period in Mesopotamia. Figure 7 compares the size of certain Egyptian and Mesopotamian towns by showing them side-by-side, demonstrating the larger extent of town developments in the Near East in comparison with the Egyptian settlements.

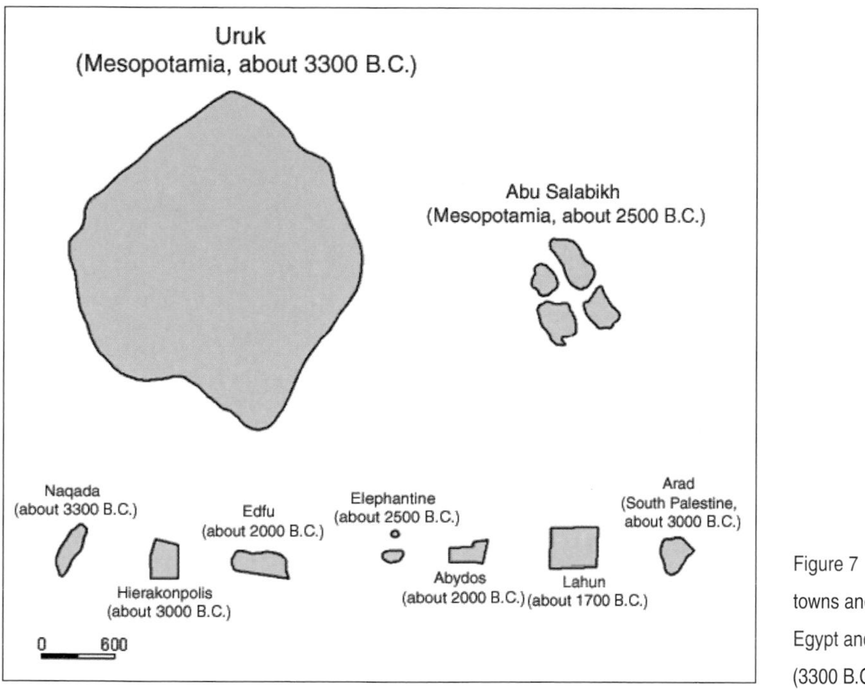

Figure 7 The size of several towns and settlements in Egypt and the Near East (3300 B.C.–1750 B.C.)

Koptos

Koptos is the site of an important town and temple excavated by Petrie in 1893–1894. The main temple, belonging to the fertility god Min, had been largely destroyed prior to Petrie's excavations because the limestone from which it was constructed is perfect for burning lime. Petrie was not even able to draw or reconstruct a detailed plan of the most recent Roman construction, unlike other temple sites in southern Egypt where many temples are built in sandstone rather than limestone. The problems were even greater for the earlier temple buildings at Koptos. Only loose blocks, each decorated on one side only, were found with no clear reference to the original architectural setting. Other artefacts

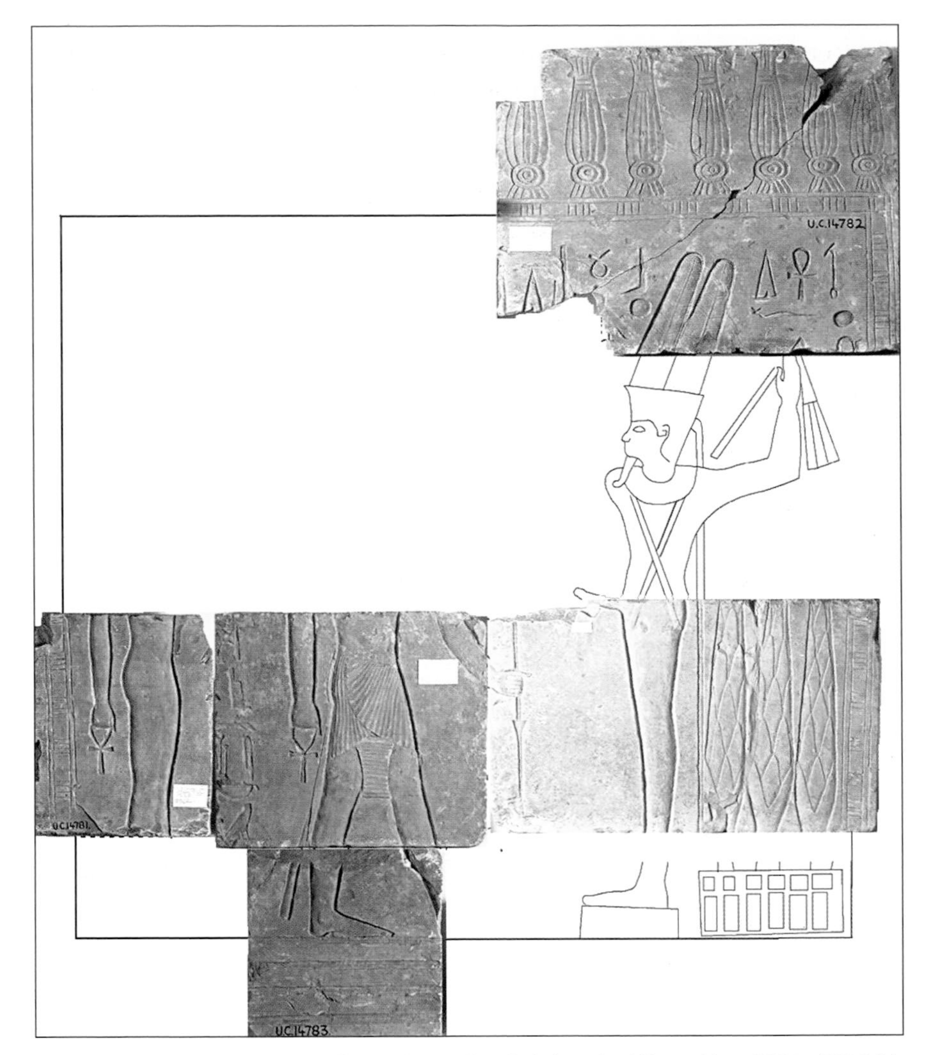

Figure 8 Reconstructed wall from the Second Intermediate Period temple of Koptos (about 1650 to 1550 B.C.)

had been re-used as stones for the pavements of the later temple buildings. Several fragments, including parts of three colossi statues dating to the time of state formation, demonstrate the importance of the temple in Pre- and Early Dynastic times (about 3000 B.C.).

In order to investigate different scenarios, a total of four reconstructions of the temple are offered in Digital Egypt. Two of them depict the temple at the time of state formation (about 3000 B.C.) and the other two show the temple from the Second Intermediate Period (about 1650 to 1550 B.C.). The latter reconstructions are of special interest because several reliefs found at the temple site are now housed in the Petrie Museum, and these make it possible to reconstruct several decorated walls (figure 8) that are the starting points for our reconstructions. These blocks bear the name of the little known King Nubkheperra Intef who reigned at this time in Egypt, and for this reason they are also of some historical importance.

Several sources are available for the temple reconstructions of the Second Intermediate Period. These include other better-preserved temple buildings from more or less the same period and the reliefs in the Petrie Museum (and other collections) that provide a framework. The reliefs come from at least six different walls and several doorways (see figure 9). It is highly likely that they all come from two small chapels.

The two different reconstructions of the Second Intermediate Period temple at Koptos provide a useful illustration and example of the exploratory approach to visualisation that is taken in Digital Egypt. Several reliefs found at Koptos

Figure 9 Koptos in the Second Intermediate Period (about 1650 to 1550 B.C.)

by Petrie and by the French Egyptologist Reinach date from building activity in the reign of King Senusret I (about 1956 to 1911/10 B.C.). It seems that he must have added to a pre-existing temple and may perhaps even have rebuilt the whole temple complex of Min during his reign. In the first reconstruction in Digital Egypt it is assumed that the main temple of the king had been ruined by the onset of the Second Intermediate Period, and was rebuilt under King Nubkheperra Intef. Assuming that the Middle Kingdom temple was destroyed during the Second Intermediate Period, the reliefs of King Nubkheperra Intef would seem to be part of the decoration of the principal sanctuary and would have been entirely rebuilt under that king. The principal sanctuary most likely had three chapels standing next to each other; there is a central sanctuary with raised reliefs and two other sanctuaries that are decorated with both sunken and raised reliefs. The existence of three sanctuaries in one temple was very common in this period. Well-documented examples include a temple in Ezbet Rushi (Lower Egypt) and the Renenutet temple at Medinet Maadi (Fayum, Middle Egypt).

In our second reconstruction, however, we assume that the main building of Senusret I was still standing in the Second Intermediate Period. The reliefs of King Nubkheperra Intef would then have come from two or three small chapels constructed as additions to a pre-existing temple building. They are located inside an open courtyard that had been reconstructed in front of the main temple building. A similar arrangement was revealed by excavations on the island of Elephantine and dates to the reign of King Wahankh Intef (about 2100 B.C.). Here in Elephantine, the king had built an additional chapel in the courtyard of the pre-existing Satet temple.

Some further comments on the reconstruction of the town should also be made. To date, excavations have revealed nothing preserved from the actual town of the Second Intermediate Period. The reconstruction is, therefore, highly speculative. However, some other contemporaneous town sites have been preserved, and these provide a general impression of the towns of that period. In particular, the general appearance of the houses and the street patterns are known from other similar sites.

4 Persuasive visualisations and other conclusions

In this chapter, we have highlighted some of the 3-D VR models that have a strong visual presence, and we have also illustrated the potential of the Web as a medium for visualising and providing information on archaeological sites and ancient architectures that have been long destroyed. This CASA project is a unique collaboration between two different areas of expertise: Grajetzki provides the archaeological background information, while Shiode develops the virtual 3-D models. The 3-D models are extremely persuasive representations of built forms that can give the false impression that the representation is known to be accurate in its precise details. This more certain view of the past, nevertheless, provides a very valuable pedagogic introduction to Egyptology. Where appropriate, we have created several models of the same site and compared them with one another. Nevertheless, many users still find any realistically-rendered 3-D scenario persuasive and visually convincing, even more so when walking inside these models with the aid of immersive virtual environment systems such as the CAVE.

As part of the evaluation of the project, and also to draw the attention of users to the focal point of each excavation site, we have created a list of 3-D impact pages. These elicit user responses to the perceived impact of the 3-D models and seek to gauge the usefulness of the representations in relation to user research questions and requirements. To date, we have reconstructed a total of nine excavation sites from many parts and many time periods of Egypt, and this number will increase as the project progresses. The methodologies described are context-specific in that they can be applied to the development of online content for other archaeological sites. They might also be used to create online teaching resources in other contexts. The combination of various media types such as 3-D models, audio files, excavation records and digital images of the artefacts offer a unique resource for students in Egyptology and archaeology but also in other areas. Furthermore, these resources reproduce historic landscapes at extents not developed previously, and in ways that are only possible using new technologies. Many of the reconstructions offer new solutions for understanding buildings and some even offer a visual reconstruction of buildings for the first time. These reconstructions are supported by our archaeological research, which also provides important background information on many sites.

We have used the services of an external evaluation panel to gauge the usability, perceived validity and usefulness of the site. This has highlighted the various issues and problems that are in need of attention as our work progresses. For instance, while the 3-D models are visually appealing and intuitively plausible, they can on some occasion be overly persuasive when users regard them as a precise reconstruction based on sufficient evidence. In fact, most of the sites that we have studied have been largely destroyed and the representations that we offer are only one of many possible forms. We have, for some sites, suggested multiple choices

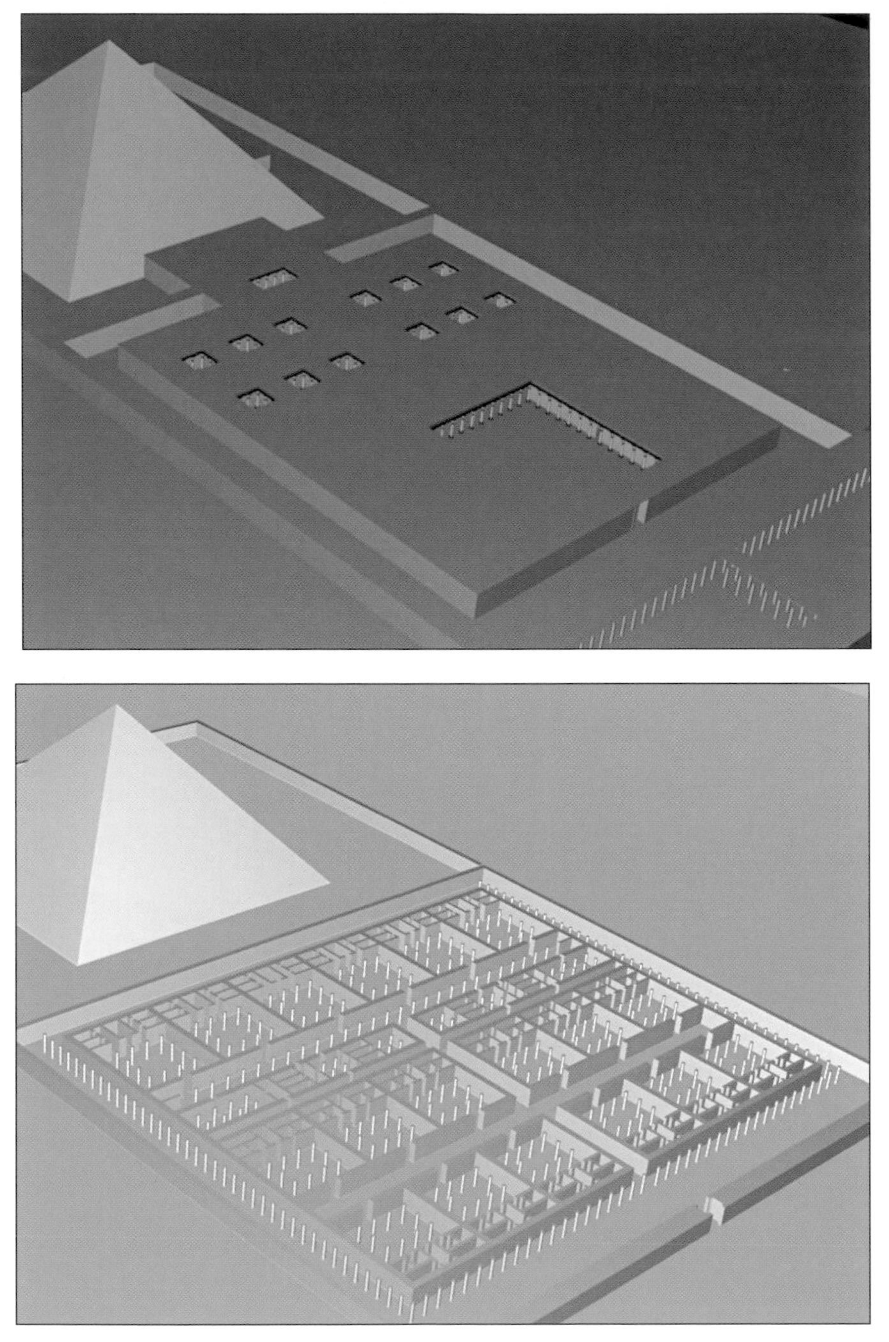

Figure 10 Virtual reconstruction of the great labyrinth and the pyramid complex of Hawara based on different proposals

(as shown, for example, in figure 10), but the external evaluations indicated that their visual impact was still too strong and persuasive for some users. There is no immediate solution to this problem, but if we could offer an interactive environment where the users can move the artefacts and reconstruct their own version of the site by exercising their judgement, then perhaps we could provide a more flexible impression.

In terms of the user interface, results of our user survey suggested that we should keep the structure of the Digital Egypt site as easy to navigate as possible. There is a tension here, for the ability to offer a range of contents sufficient to encourage repeat visits may have the opposite effect of making the site more complex and less intuitive (see Tobón and Haklay, this volume, for a wider discussion of user interaction). We hope to resolve this issue by identifying appropriate trade-offs, and we are also planning to provide a search function to provide users with an overview of the site's contents.

Acknowledgements

This chapter is based on the 'Digital Egypt for Universities' initiative, a three-year project to create an online learning and teaching resource, carried out by Centre for Advanced Spatial Analysis, University College London, and the Petrie Museum of Egyptian Archaeology, University College London. The project is funded by the Joint Information Systems Committee. Digital Egypt is available at www.petrie.ucl.ac.uk/digital_egypt/Welcome.html.

Chapter

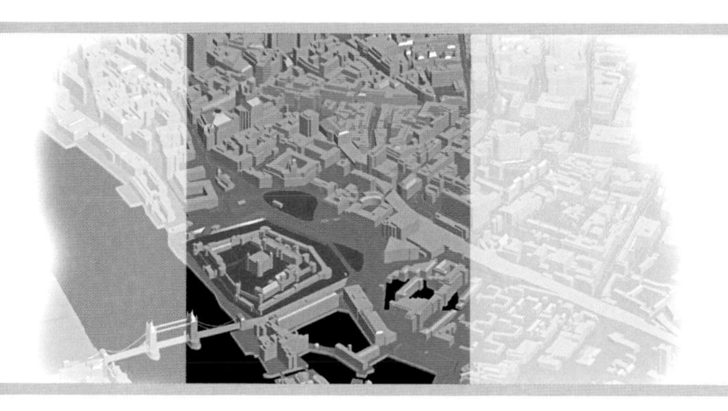

Virtual cities: from CAD to 3-D GIS

Andrew Hudson-Smith and Stephen Evans

We are on the edge of a revolution in how we visualize and query digital data about our environment. To date, computer display of the three dimensions of our environment has been limited to computer-aided design (CAD) packages and the query of related data has been limited to geographical information system (GIS) packages in two dimensions. The current innovation wave across the spatial data information field is based on the development and dissemination of three-dimensional GIS (3-D GIS), which allows data to be visualised and queried on an x, y and z axis. A number of the key players in information visualisation allow conventional two-dimensional data to be viewed and exported in a three-dimensional format, currently using the standard Virtual Reality Modelling Language 2.0 (VRML 2.0). However, such methods of visualisation and data query are limited in their practicality. The move towards 3-D GIS in standard packages has been rather hit and miss, with the third dimension often only used as a substitute for basic CAD-like visualisation. We argue here that 3-D GIS will only become a reality when it is directly linked with CAD models and that the Internet is the most appropriate medium through which this is most likely to occur. We illustrate these arguments in an overview of research into the virtual city in general and our own development of Virtual London in particular.

1 The virtual city

Computer screens are the electronic nodes of the virtual city, just as real cities are nodes in the global settlement system. The term 'virtual city' is now common parlance on the Web and, at the time of writing, a query for the term on the Google search engine returns over 1.6 million links. The term has a wide range of connotations including: digital cities (Mino 2000), city of bits (Mitchell 1995), Web-city (3D Net Productions 2002), telecities (Telecities 2002), wired cities (Dutton, Blumler and Kraemer 1987), infocities (Infocities 1997) and cybercities (Graham and Marvin 1999) to name only a few. The common theme to all of these terms is the use of the city metaphor to describe a network of people or information, information that is digitally communicated with relevance to either a real or imaginary city. In short, the virtual city is a merger of the community and the city, embedding the current functions of the physical city in a digital form. These real/non-real communities have been defined as either grounded or non-grounded digital cities respectively (Aurigi and Graham 1998). Beyond this, however, the characteristics of virtual cities are rather wide-ranging. They can be both grounded and non-grounded, two-dimensional or three-dimensional, service-based or information-based. We will concentrate here on the three-dimensional nature of the virtual city and our abilities to append information directly to 3-D representations. Currently we are in the process of developing Virtual London, and, at this stage in our discussion, it is timely to examine other CAD models that have been developed of London.

2 Three-dimensional models of London

As part of a worldwide survey for the Corporation of the City of London (Batty et al 2001) we appraised the development of digital city models developed for GIS purposes, for CAD visualisations and for the computer games industry. If we focus on London in particular, extents and levels of detail of the models that have been built so far vary considerably, and none has been produced that provides three-dimensional visualisation for the entire central area of London.

Of particular note is the Metropolis Street Racer developed by Bizarre Creations (based in Liverpool, United Kingdom) for the Sega Dreamcast gaming console. Metropolis Street Racer, now renamed as Project Gotham and released on the Microsoft® XBox® console, is a driving simulation based on three cities: London, Tokyo and San Francisco. The XBox version has also added New York,

Figure 1 Metropolis Street Racer on the Dreamcast

with all the cities modelled in high photorealism over a number of fixed racing circuits around each city. The sample screen shot shown in figure 1 illustrates the level of photorealism portrayed in the London model.

Project Gotham features 200 circuits over the four cities, each rendered at 60 frames per second with graphical effects such as real-time weather and lighting changes. The game offers the highest level of realism of all the models we have examined at street level. However, it has been designed purely as a racing game, and, thus, the models do not have consistent three-dimensional geometry. The model is designed to be viewed only from in-car or slightly above-car viewpoints, and so it is not possible to take a bird's eye view or to fly around the city as in other virtual models. Project Gotham does, however, illustrate how the games industry has become a major player in the modelling of built form and demonstrates the clear demand from the consumer for realistic environments as the processing power of the latest gaming consoles improves. With gaming revenues exceeding those of cinema in the United Kingdom, developers clearly have mass-market demand, allowing games to be developed on budgets often exceeding 3 million USD. Such budgets allow realistic city representations to be developed, although it should be noted that the games focus does not render the representations suitable for architectural and planning purposes or for manipulation of the built environment. However, once a city representation has been built, it is possible to port it over to other platforms in order to enhance the geometry by filling in the gaps from a street level view, hence enabling a bird's eye CAD type model to be constructed.

The other five models have been developed along the more traditional CAD route and do not, therefore, represent the state-of-the-art graphically (such as Project Gotham) but do reveal a more accurate geometric approach. One such example is the model developed by Miller-Hare (www.millerhare.com), a small consulting company specialising in the production of high-quality visualisations of architectural design to support planning applications and the marketing of proposed buildings. The development of city models is often ad-hoc with small areas developed over time on a project-by-project basis. This is certainly the case with regard to the approach by Miller-Hare to their three-dimensional model of London, which was developed out of their surveys over the last 15 years. As technology has moved on, so the model has been developed in increasing levels (to A from G) of detail, as shown in table 1.

Level of detail (LoD)	Description
A	Detailed architectural model including fenestration
B	Detail equivalent to 1:100 measured building survey
C	Detailed elevation
D	Major details of building elevation
E	Accurate building volume
F	Roofscape
G	Prismatic block models—coarse massing

Table 1 Level of detail in the Miller-Hare models

The majority of the buildings in the Miller-Hare model are represented as basic blocks from the Ordnance Survey 1:1250 building footprints extruded by height and derived from aerial photogrammetry. Extruded footprints provide a basic prismatic model at level G, while upgrading of the model on a project-by-project

Figure 2 The Miller-Hare model

basis has resulted in a number of sites at levels A and B. Figure 2 illustrates a representative view from the model.

Similar to the Miller-Hare model is a photogrammetric model developed by London's City University (Batty et al 2000). Commissioned in 1996 by a commercial property company (Trafalgar House), the City University model has a spatially extensive coverage of building blocks at a high level of detail and, hence, is probably the most detailed large-scale model of London to date (although at a local scale, the more refined parts of the Miller-Hare model offer a greater detail). The model was produced using 3-D Microstation®, a high-end computer-aided design (CAD) package. The model includes a high level of roof morphology detail and was constructed over three person-months at an estimated cost of 75,000 USD. The model has been used to visualise a number of projects in the city including the proposed Millennium Tower for Foster and Partners. Figure 3 illustrates part of the model with a view from Tower Bridge.

This model is no longer being updated as the researchers that created it have since left City University, although there remains the possibility of licensing the model to third parties to generate revenue for its maintenance. A similar, although coarser scale, model has been produced by the Department of Computer Science

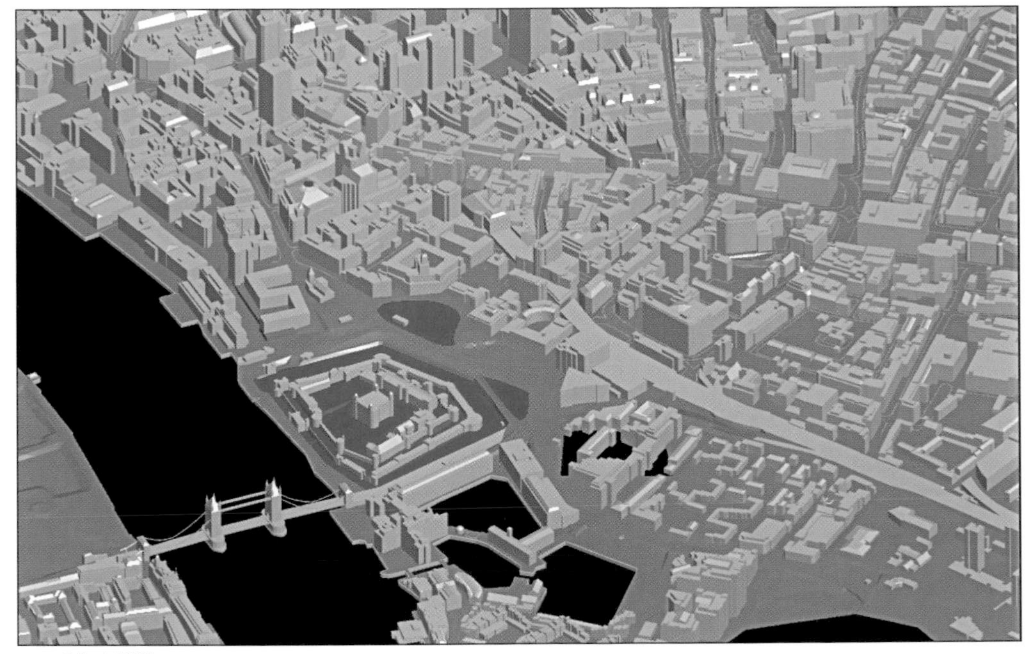

source: Batty et al 2000

Figure 3 The City University model

at University College London. The model was developed using Cities Revealed (www.crworld.co.uk) building outlines and height data to create a simple prismatic model at level G on the Miller-Hare scale. Simple roof structures have been added, although these have been randomly assigned. The model has not been constructed for use with data tagged to the three-dimensional structure but has been designed with the technical objective of ascertaining the speed at which polygons can be rendered into cityscapes using state-of-the-art machine visualisation. Figure 4 presents a screenshot of the model running using the Silicon Graphics® interface.

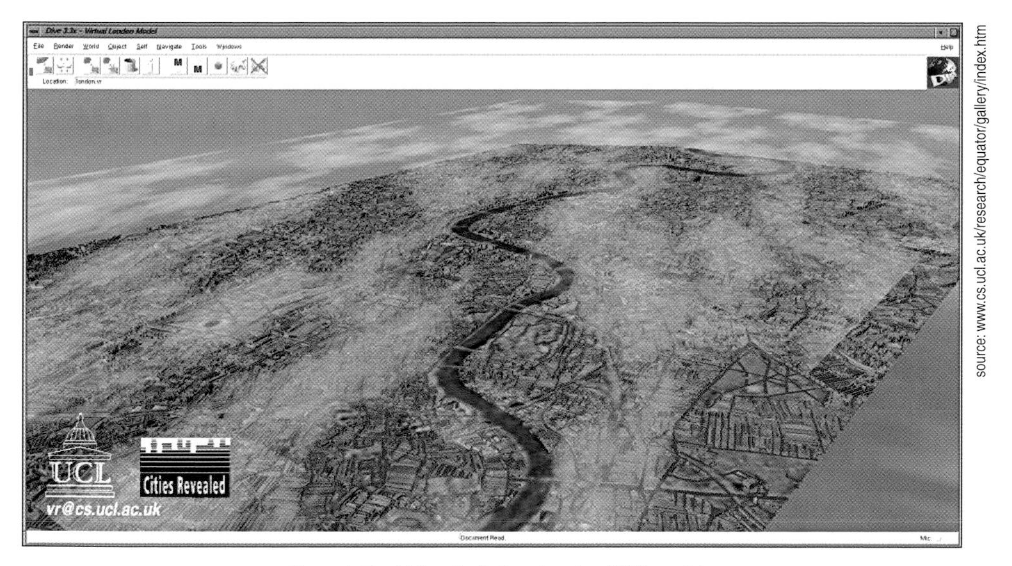

source: www.cs.ucl.ac.uk/research/equator/gallery/index.htm

Figure 4 The University College London (UCL) model

For a more restricted study area, Bath University has developed a prismatic model of the Soho district of Central London in what is the only networked example identified. It is developed in VRML 2.0 and is described on their Web site as 'The Map of the Future'. A representative aerial view is shown in figure 5.

Hayes Davidson (www.hayesdavidson.com/hd/index.html) developed the only non-three-dimensional model of London that we have identified. Developed in association with the Architecture Foundation (www.architecturefoundation.org.uk) and termed 'London Interactive', the system is based on a CD-ROM multimedia presentation and provides information about contemporary urban design and major architectural projects in London. The system is linear in nature and, although two-dimensional, serves to illustrate how multilayered data can

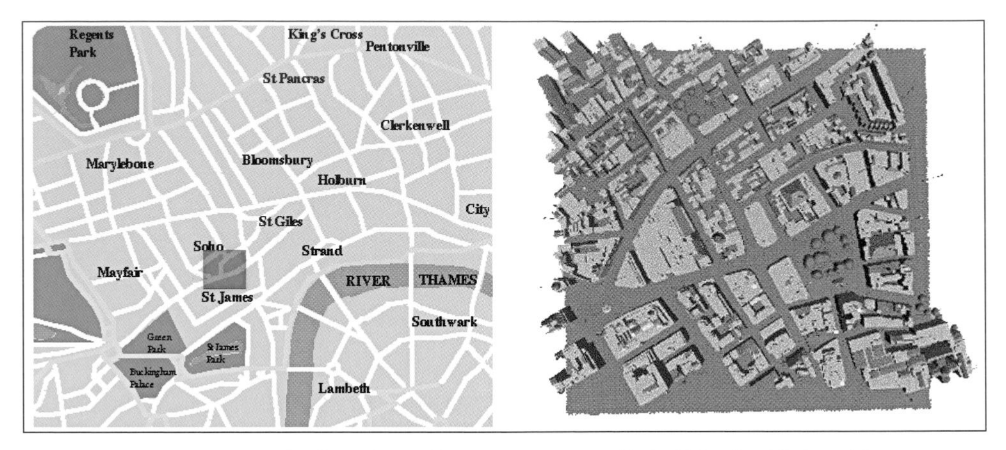

Figure 5 The Soho district of London, as represented in the University of Bath model

be used to provide compelling information about design. The system has made extensive use of panoramic imagery, similar to our example of Wired Whitehall (www.casa.ucl.ac.uk/vuis). Linking panoramic imagery to various levels of data provides a quick and easy way to portray a sense of location and to place the user, and should never be assumed to be inferior to fully three-dimensional models (see also Li and Maguire, this volume). The development of Virtual London will heed the need to offer aspects of panoramic imagery as well as on-the-fly three-dimensional representations where appropriate. This issue is explored further by Hudson-Smith et al (this volume) and can also be explored from the Web site www.casa.ucl. ac.uk/woodberry.

The three-dimensional models discussed thus far do not link to any significant underlying data. They are merely 3-D representations that can be rendered, visualised on high-end machines, or, in the case of the Bath University model, viewed via the Internet. All the models are fixed which restricts their use to 'what is' visualisation rather than application to interactive 'what-if' investigation of scenarios (see Hudson-Smith et al, this volume, for a broader discussion of scenario development). In this broader applications domain, scenarios might also be linked to various types of data and visualised within a collaborative environment. This type of scenario development and data linkage makes up part of the remit of CASA's Virtual London project. To achieve this, we envisage the integration of photographic and textual data, as in the Hayes Davidson example, together with the delivery of non-fixed three-dimensional representations across the network. Thus, three-dimensional representations of a virtual city might be communicated in a collaborative environment.

3 Widespread adoption of 3-D GIS

A recurring theme throughout this book concerns the way in which emerging technologies enable us to query and manipulate our environment at a distance. Applications of 3-D models that can now be effectively streamed over the Internet open up a range of possibilities for development of 3-D GIS for extensive geographic areas. In the United Kingdom the main driver behind the current development of 3-D GIS in general, and the move towards virtual cities in particular, is central government's Electronic Government Initiative with its objective to deliver electronically, and in a customer-focused way, all government procurement services by 2005 (Office of the e-Envoy 2002). This has garnered political support for the vision of a Virtual London and the 3-D GIS research necessary to create it.

Today we are at the beginning of a new era of data delivery and query via the Internet. Such research towards virtual cities can be seen in the wider context in the development of planning support systems (PSS) (see Brail and Klosterman 2001). Snyder (2002) notes that there has been disappointing progress over the last 20 years in the adoption and use of PSS and geographical tools in the field of land-use planning and community design, but there is growing evidence that we are on the cusp of large-scale adoption of these tools. Snyder's views in 2002 are consistent with Klosterman's (1998) prediction that we are on the edge of a new revolution in the use of digital tools for planning and environmental visualisation. But Klosterman's prediction remains as yet unfulfilled, with the 1990s seeing only the development of interesting academic prototypes (Klosterman 1998).

Nevertheless, it is prototypes such as those discussed elsewhere in this book (for example, Hudson-Smith et al, this volume) that can lead to the diffusion of 3-D GIS into practice. Gladwell (2001) reflects how major changes in society happen rather suddenly, and Snyder (2002) develops Gladwell's (2001) key ingredients for the rapid adoption of ideas, practices and products in his discussion of PSS. The research detailed in this book can be seen as developing key prototypes which enable key players to see the value of digital planning, and thus enable the process to gain the backing that academic research needs before these prototypes become widely diffused as tools of policy. Synder (2002) suggests that a tipping point for the widespread adoption of PSS is near at hand. He identifies several reasons that are directly applicable to virtual cities:

- Government departments and communities are increasingly grappling with the growing complexity of land-use planning, resource use and community development.
- Emerging tools for community design and decision making have the potential to dramatically change the planning profession. In particular, computers are becoming more powerful and less expensive (see also Li and Maguire, this volume).
- Data are becoming more readily available and challenges with respect to interoperability of GIS across different platforms and among different models are being resolved.
- Tools are becoming more user-friendly. In addition, the growing number of early adopters of these systems have demonstrated their usefulness and power.
- PSS have the potential to transform decision making in two ways: they can help communities shift land-use decisions from regulatory processes to performance-based strategies, and they can make the community decision-making process more proactive and less reactive.

While Snyder's (2002) arguments are made in the wider context of PSS, we believe that it is the domain of the virtual city that has the most potential to create a dramatic change in how we view, distribute and communicate data. Various projects at CASA such as Woodberry Down Regeneration (see www.casa.ucl.ac.uk/woodberry and Hudson-Smith et al, this volume), funded by the Architecture Foundation, Hackney Building Exploratory and the Woodberry Down Regeneration Team, can be identified amongst Synder's (2002) early adopters. The sponsoring organisations have had the courage and foresight to invest in the series of academic prototypes. Such early adopters provide the all-important working version with which to garner interest amongst other larger interested parties thus creating a snowball effect for wider implementation. It is this snowball effect that is leading to the funding and development of a networked 3-D GIS in the form of Virtual London.

The complete, fully functional model will be developed in different ways for different audiences. There are at least four potential audiences for which the three-dimensional model and its underlying data will be used. The level of usage is likely to be dependent on the delivery method and the method of visualisation used across the network (specifically the Internet). We envisage four broad categories of use:

- Fully professional usage: This is the use of the model by architects, developers, planners and other professionals who are anxious to use its full data query and visualisation capabilities. For example, an architect might place a building within the model in order to assess a variety of issues, from its basic visualisation to the impact it might have on traffic and surrounding land use. This use could potentially be set up with subscription-only access, raising revenue for the continued development of the model. Subscription-only access would also allow potentially delicate data to be distributed and queried via the three-dimensional model, with users requiring a password and user name to view restricted levels of data.
- Concerned citizen usage/public participation: This is use by citizens, and the role of public participation would be both educational and participatory in the sense that interested groups and individuals would use the model to learn about London or to visualise development and traffic proposals. This is seen as the main type of usage of the model with obvious links to the development of e-democracy and the evaluation of 'what if' scenarios to enable digital planning by citizens.
- Virtual tourism: This entails the use of such models for tourist navigation and visualisation, and is a by-product of the development for digital planning (see also Li and Maguire, this volume). A version would be developed in which users of the model on the Web can learn about London as tourists, using it to navigate and view scenes, picking up Web sites of interest, not unlike our Wired Whitehall (www.casa.ucl.ac.uk/vuis). With the appropriate level of underlying data, such a model could also be used for marketing areas in London to encourage outside investment in the city.
- Educational usage: Virtual London will provide a resource for education, linking into the U.K. National Curriculum for high schools through subjects as diverse as geography, civics and history. Education is a key aspect of the virtual city as it reaches an audience that is often not a part of the consultation or data-query process. Educational visualisation would be through a multi-user collaborative system such as that used in our project 30 Days in ActiveWorlds (www.casa.ucl.ac.uk/30days). This would form part of a secure closed educational network in which schools would meet up and discuss issues relating to the National Curriculum in a virtual environment. The model would also be able to link in with university-based education for urban planning, architecture, and virtual environments. In the same way as professional users, students would be able to load their own models in closed

sections of the networked site for visualisation and evaluation of design in the London context.

We will now detail the elements of its construction and identify the various options involved, issues which are central to the development of our 3-D GIS and virtual city concept.

4 Technical development

The development of Virtual London pulls together many if not most of the ideas documented to date by our research team. The goal is to develop a truly virtual city that can be occupied, queried and manipulated by citizens within a collaborative environment. As we have identified, interested citizens will comprise professionals working in London, those using it for educational purposes, and those from outside London using the virtual environment to pursue other interests. As such, the technical development is based not only on the visualisation of London's

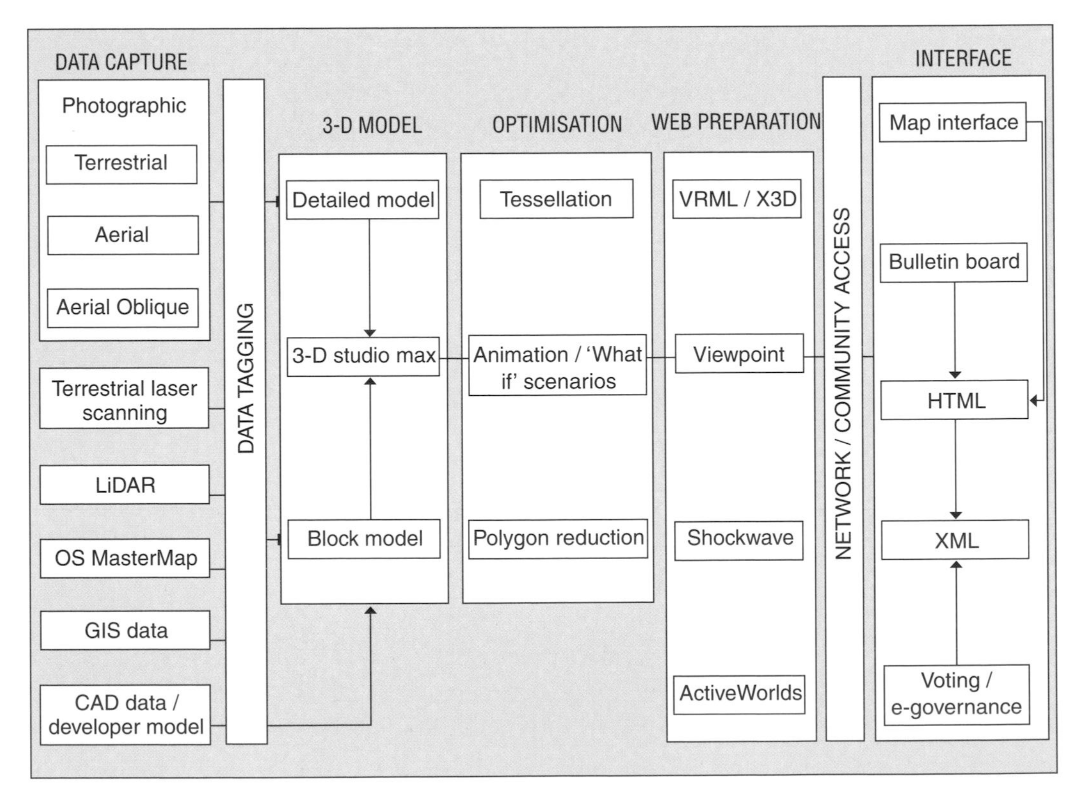

Figure 6 The Virtual London modelling structure

built environment but also on the integration of underlying data with the production of bandwidth-friendly multi-user models. The development route entails a combination of data capture, model development, optimisation and deployment using various technologies and via the Internet. This is illustrated in figure 6.

The current cutting-edge applications in Internet-based three-dimensional visualisation are Viewpoint® (www.viewpoint.com), Shockwave 3-D (www.shockwave.com) and ActiveWorlds (www.activeworlds.com). Each of these technologies allows data to be tagged to its objects, which is central to the development of a 3-D GIS; we explore this in detail later. Examples from each of these applications can be found on our Web site (www.casa.ucl.ac.uk/olp).

Digital map and height data

Acquiring suitable digital data is central to the development of virtual cities. A number of the academic prototypes, developed as part of the Shared Architecture project (www.casa.ucl.ac.uk/public/meta.htm), were assembled using only simple postcard-quality images and had only low-resolution texturing as a consequence. The use of high-quality oblique aerial image data is important if quality is to be maintained in the sections of our model that will be photo-realistically rendered. Oblique aerial data will be acquired from the use of the Ordnance Survey blimp, which is being donated to the data capture section of Virtual London (see Smith, this volume). Blimps are suitable for capturing oblique aerial photography for a number of reasons. First, they are able to fly at a lower level than aircraft and helicopters over urban areas, and second, they are able to carry out a slow and steady flight path suitable for data capture.

To ensure that the model is georeferenced and constructed in a manner that makes it applicable to capture additional data, the Ordnance Survey is contributing its MasterMap™ data for London (see also Smith, this volume). MasterMap data are layered into nine themes: roads, tracks and paths, land, buildings, water, rail, height, heritage, and structures and boundaries (Ordnance Survey 2002). Each feature has a unique TOpographic IDentifier, known as a TOID, with the data supplied in Geographic Mark-up Language (GML). TOIDs provide a unique 16-digit code for each feature, thus enabling easier data analysis and data sharing (Ordnance Survey 2002). Previous applications, such as the Dounreay Nuclear Power Plant model (www.casa.ucl.ac.uk/olp) used Ordnance Survey Landline data (a precursor to the MasterMap data product) merely as topologically unstructured background data. MasterMap is a cleaner, richer, updateable version of this dataset. As a result, the data are presented as a series of closed polygons in which each feature is referenced as part of the TOID system. When linked using GML, data are directly taggable to features using an enquire code which, in terms of three-dimensional modelling, allows integration of a common database feature. As a result of the shared attributes users are able to change the visualisation on-the-fly.

Such tagging makes it possible, for example, to rank commercial rental levels in London. Where data might be shown in a 2-D GIS as a choropleth map with building polygons shaded according to their ranked rental level, in a 3-D GIS the buildings themselves could be shaded or even changed in height according to their rental level. All that would be required would be that the 3-D GIS be developed or modelled with each object having a unique reference number that could always be related back to its parent TOID. In this way, a 3-D extruded building footprint could have the same ID as its parent TOID, but with an extra letter or number prefixed indicating that it is a 3-D object. Likewise children of each basic 3-D block would have unique numbers based upon their parent ID. Hence detailed building morphology (for example, chimneys, porches and so on) would also relate back to each parent block.

This object-oriented approach to the data structure is essential for several reasons. First, it allows management of detail, that is to say, objects in the distance should not be rendered. Hence, a detailed model of London would be delivered in relatively few polygons and yet allow the user to zoom in on an area of interest and see increasing levels of detail as they do so, thus replicating everyday experiences in the physical world. In addition, it would be possible to 'dice and slice' London according to object types. If a user wants to view and calculate the surface area of all roofs in London, it would be possible to pull from the database just the roof morphology to provide an estimate of the roof surface area for London.

One way of achieving an object-oriented approach to the data structure in a Web friendly and open format is by using Extensible Mark-Up Language (XML). XML is the basis for the Viewpoint modelling format (www.viewpoint.com). Each model or collection of models is scripted in a file that can orchestrate the elements of a scene, including animation, interactivity and loading of files. In a similar way, GIS data are increasingly being packaged and delivered in another XML-based format known as Geographic Mark-up Language (GML). GML defines XML schema syntax, mechanisms and conventions that:

- provide an open, vendor-neutral framework for the definition of geospatial application schemas and objects;
- allow profiles that support proper subsets of descriptive capabilities of the GML framework;
- support the description of geospatial application schemas for specialized domains and information communities;
- enable the creation and maintenance of linked geographic application schemas and datasets;
- support the storage and transport of application schemas and datasets; and
- increase the ability of organizations to share geographic application schemas and the information they describe (OGC 2001).

Sharing data is often problematic because of differing standards, but the focus upon GML, linked to the Ordnance Survey TOIDs, allows us to take data supplied and present them in a common form based on a three-dimensional geographic location. Such data provide a level of accuracy to the proposed model which has not been present in many of the applications we have so far documented. With examples such as the Woodberry Down Regeneration (www.hackney.gov.uk/woodberry, Hudson-Smith et al, this volume), it has been argued that the most important factor is the need to develop a sense of location and space in 'cyberplace'. If it is to appeal to a wide audience and range of professional activities, any geographically extensive city model needs also to be accurate for analytical purposes, for example, in line-of-site analysis for the telecoms industry. As such, not only the building outlines and locations need to be georeferenced, but so do the three-dimensional height data. To achieve this aim, a number of routes can be explored using the available data.

For the non-photorealistic sections of Virtual London, simple prismatic models will be extruded from the footprints of MasterMap data. Basic height data are available from a number of sources. Cities Revealed data (www.crworld.co.uk), for example, has average height data for locations in Central London, extending in a circle from the city to the northern edge of Hackney. These data were used to create the UCL Computer Science model (figure 4) through use of a straightforward procedure to link height data to building outlines. Average height data can also be derived from LiDAR provided by InfoTerra (www.infoterra-global.com/laser_spec.html). LiDAR is a relatively new remote sensing technique that is revolutionising topographic terrain mapping. It is a very rich data source, with some laser scanners acquiring 33 000 individually heighted points per second and in typical survey mode, over 700 000 points are measured per square kilometre.

Figure 7 Sample LiDAR model of Finsbury Circus, City of London

In an urban context, these points can be averaged out for individual buildings by overlaying MasterMap data and calculating average height as illustrated in figure 7, which was constructed using OS 1:10 000 scale topographic maps and LiDAR.

With this model we worked around a problem that arises from using surfaces to model urban environments with a GIS. The problem arises because no surface function can have vertical walls (since it cannot have two different z values at the same x, y location). By buffering the building footprints inwards by a small amount (approximately 10 cm) and using this polygon to represent the eaves or roofline of the building, we have managed to create near-vertical walls. These give the building walls the appearance of being vertical, and yet allow us to carry out analysis (such as line-of-sight) and image draping (for example, using orthorectified aerial photography).

The remote-sensing group attached to our team is also extracting roof morphology from LiDAR data. Such research links are important to the success of a data-rich project; the ability to tap into graduate student projects from specialist departments allows a broadening of the research, consistent also with the objectives of digital planning. Satisfactory representation of roof morphology is key to a sense of location and place from the bird's-eye level, yet it proves difficult to obtain if the model is not derived from photographic data. Cities Revealed has categorised its average height data with a range of roof types, and while this will not provide a true representation of London, it will supply a rapid prototype for producing a surrounding context into which more detailed models can be inserted.

Data such as aggregate social deprivation data, property ownership and size, building use and condition, land-use activity types and so on can be tagged to the model using TOIDs. Our first aim is to integrate London-based GIS data with the model for analysis to achieve on-the-fly visualisation of data queries. To achieve a high detail of precision for existing key buildings, the project also has access to terrestrial laser-scanning technology. The system works in a similar fashion to LiDAR, bouncing lasers off the building to build up a point cloud of its external structure. Combined with photographic data captured from the same origin point as the laser scan, a highly accurate model of key buildings can be produced. However, because of the amount of data captured (again similar to LiDAR), the models are not suitable for Internet distribution in their raw format. This is an emerging area of research and although such data may not at first sight be applicable to Internet models, new optimising methods are being developed that may extend our ability to scan local areas and extract basic texture information and land-cover geometry.

Software and hardware

Our research has aimed to develop three-dimensional models for the low-end user (users with standard home and office-based computer systems). We have documented examples of virtual cities that require high-end computers to both produce and render fly-throughs and have located these on the applications development path. However, although the level of available funding makes it possible to utilise such high-end systems (thus straying into the traditional CAD approach), this is not really consistent with our objectives. Software used for the development of such models will be based on the technologies that we have already explored, as well as new ones that are emerging, thus placing the emphasis upon developments that are independent of the tried and tested techniques of CAD development. While this approach may seem contradictory, it is viewed as essential if our ethos of data visualisation is to be maintained. A problem with existing approaches is

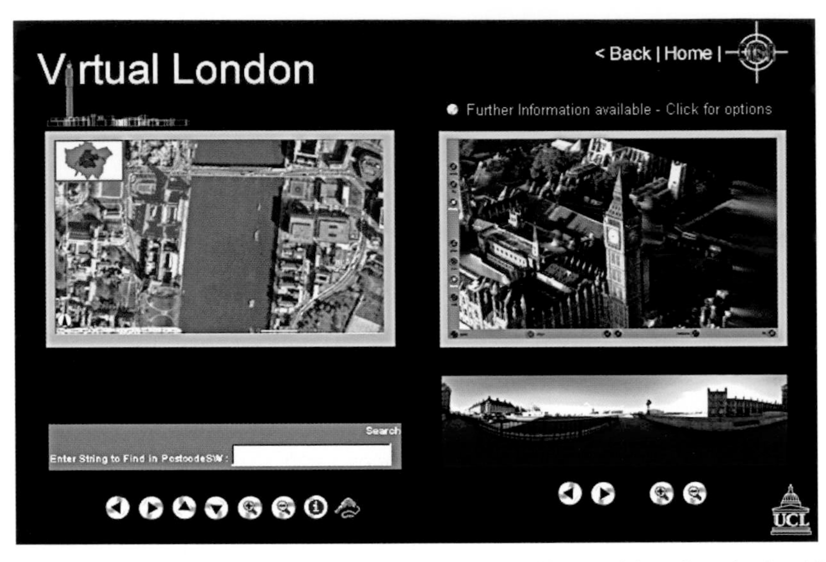

Figure 8 GIS linked to a three-dimensional model and the proposed integrated three-dimensional model

their disregard for the network and the interface constraints of users. Moreover, existing approaches have only used established software, aimed at the planning, design and architecture market, which is dominated by a few software companies. By continuing to research emerging software and finding ways to link them, we can push forward innovative methods for modelling and communication in digital visualisation.

Interfaces

The modelling structure illustrated in figure 6 allows for the development of innovative interfaces. When integrating GIS data with three-dimensional models, the simplest route is to present the two-dimensional map data alongside three-dimensional data. This is how the first prototype for Virtual London was presented, as illustrated in figure 8.

The left side of the screen held the GIS data, which was provided using ESRI® ArcIMS® (Internet Map Server). Data were layered to trigger both a three-dimensional model, in this case of the Houses of Parliament, and an interactive panorama (www.casa.ucl.ac.uk/public/meta.htm). Although this allows the linking of GIS data to a three-dimensional model, the data communication is one-way. Models can be hotlinked to the three-dimensional models but not vice-versa. In effect, the models are only eye-candy and do not contain any data that can be queried. This is typical of current virtual cities. Such a system is simple to develop as

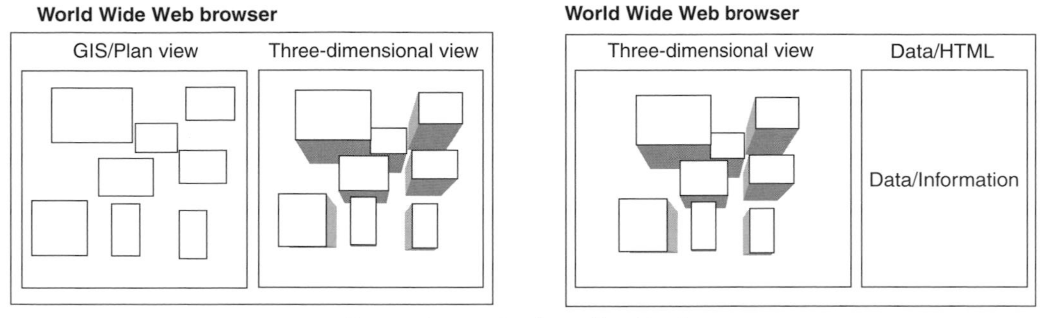

Figure 9 Prototype interface to Virtual London

it utilises off-the-shelf packages but results in running both the GIS and the three-dimensional model independently. This places severe computational demands on the end user's machine, decreases the available space to display data, and increases the amount of bandwidth required to successfully use the system. By contrast, the technical development route presents the opportunity to effectively integrate the functions of the GIS into the three-dimensional model. The implications of this are that it removes the requirement for the traditional two-dimensional plan. We illustrate this move away from the two-dimensional interface in figure 9.

Data are integrated using XML, which allows interaction between both the underlying dataset and the three-dimensional model. The GIS is often viewed as a separate section to the model, yet by using XML, the data are an integrated part of the solution with the third dimension used to display information. Additional information such as Web sites from outside sources and other visual data, such as panoramas, can be displayed in a separate window. This window is optional, and the full browser environment can be given over to the three-dimensional model with data displayed via XML hotspots and icons. In developing such a system we do not propose to neglect the requirement for a two-dimensional plan, for as we have noted, this is an integral part of the data-query process. Instead, 3-D objects can be shown in plan form or with their height removed in the three-dimensional model but shown using nadir (vertical) views of the model. Data analysis takes place via a series of options and tools built into the browser, which sends queries to the attribute data.

These techniques represent a move towards networked 3-D GIS. Current GIS software packages concentrate on the ability to add roof morphology and basic textures to three-dimensional models. Our research takes this one step further towards fully textured mapped models linked to other data using open Internet standards. By using various optimisation paths we can distribute such models via a modem and thus move towards the possibility of creating an online 3-D GIS. These developments are allowing data-rich, three-dimensional virtual cities to develop. This is still an emerging field, but one that is central to extending GIS to a wide range of applications involving the built environment at ever finer spatial scales.

Chapter

Automata-based models of urban systems

Paul M. Torrens

Simulation models of urban-activity location are currently undergoing a transition from large scale, aggregate spatial representation in a static equilibrium to much finer scale disaggregate forms where dynamic processes are the prime focus of the simulation. This chapter reviews the developments where automata and agents feature centrally and suggests a framework for geographical automata systems that combines the key features of cellular automata approaches to urban development with multi-agent modelling. The framework is developed and applied to urban growth, pointing to new ways of integrating demand and supply through agent-based modelling.

1 History of automata

Automata were first conceived of in the 1930s by the British mathematician Alan Turing. Since then, the idea has been expanded and used for a variety of purposes: automata form the basic principle on which the digital computer is based, they are the mainstay of artificial intelligence (AI) and artificial life (ALife), and authors have suggested that the universe may even be regarded as an automaton itself (Wolfram 2002). Recently, automata have seen application in the realm of model development, where they are used as building blocks for the computer simulation of complex systems. Researchers in geography and urban studies have also begun to use automata to develop models of urban systems (Batty, Couclelis and Eichen 1997; Benenson and Torrens forthcoming; O'Sullivan and Torrens 2000; Torrens 2002a).

Simply stated, an automaton is a processing unit, which itself can be characterized using variables of any description. In addition, an automaton is endowed with the ability to process information input to it from external sources, generally understood to be the information contained in other neighbouring automata (figures 1 and 2). Various rules can be designed to determine how an automaton processes the information contained in its own characteristics, as well as that which it receives as input from neighbouring automata. These rules are time-dependent and can be considered as transition rules governing how automata should adapt and change over time in reaction to information in their surroundings. Herein lies the power of the automata concept: any mechanism, process or action that can be expressed in computable terms can be used to process information in an automaton. Practically speaking, this means that automata can be used to mimic just about any process. They are, therefore, very powerful tools for simulation.

Geographers became interested in automata in the 1980s, with earlier contributions from researchers in urban studies. One class of automata in particular, cellular automata (CA), have become especially popular for urban simulation in recent years. Also, another form of automata, multi-agent systems (MAS), are beginning to be adopted for use in urban modelling (Benenson and Torrens forthcoming; Benenson 1998; Benenson, Omer and Hatna 2002). However, automata tools do not simply 'port' across to geographical applications from their origins in mathematics and computer science, partially because investigation with these tools has not usually focused on the spatial properties, or the spatial applications, of the tools. Nevertheless, automata offer significant advantages for representing space and space-time dynamics, and research into the modification of automata for geographical use, application to spatial systems, and development of spatially explicit automata-based software is active (Torrens and O'Sullivan 2001).

This chapter presents an overview of research projects in the field of automata-based modelling of urban systems that the author is engaged in. Section 2 describes basic automata, cellular automata and multi-agent systems and section 3 generalizes these concepts to the application of urban systems. Section 4 describes ongoing research in automata-based modelling of urban systems, discussing the derivation of spatially-explicit automata tools and their application to urban systems. Section 5 concludes the chapter with a discussion of remaining limitations and opportunity for work in the field.

2 Automata, cellular automata and multi-agent systems

Basic automata, such as in Turing machines, are simple processing mechanisms, albeit with surprising power and functionality despite their simple specification. Basic automata are composed of a few components: states, an input stream, rules and a 'clock'. States describe internal attributes of an automaton: on, off, 1, 0, road, rail, etc. The input stream to a given automaton consists of information gleaned from outside the automaton, which the automaton will then process using its rule-set. Input can take any form, although it is generally formulated as information derived from the states of neighbouring automata (for example, the automaton to the left is 'on' and the automaton to the right is 'off'). Rules are conditional statements (which may also take the form of mathematical operators) that determine how an automaton should react to the information in its input stream. Generally, these rules are linked to an automaton's clock, governing how an automaton should alter its own internal states between time-steps based on information delivered via an input stream.

CA operate in much the same way as the basic automata described above. In the CA approach, however, individual automata are understood to be bounded by a cellular structure, for example, a grid, a hexagon or an irregular polygon. The cell represents the discrete confines of an automaton. Collections of CA can thus be understood to form a lattice structure. Also, the neighbourhoods from which an individual cellular automaton draws input can be defined as a structure of adjacent cells in some arrangement around an automaton (figure 1).

MAS operate on the same principles as basic automata. Individual automata in a MAS are understood to be autonomous agent-automata, and their state

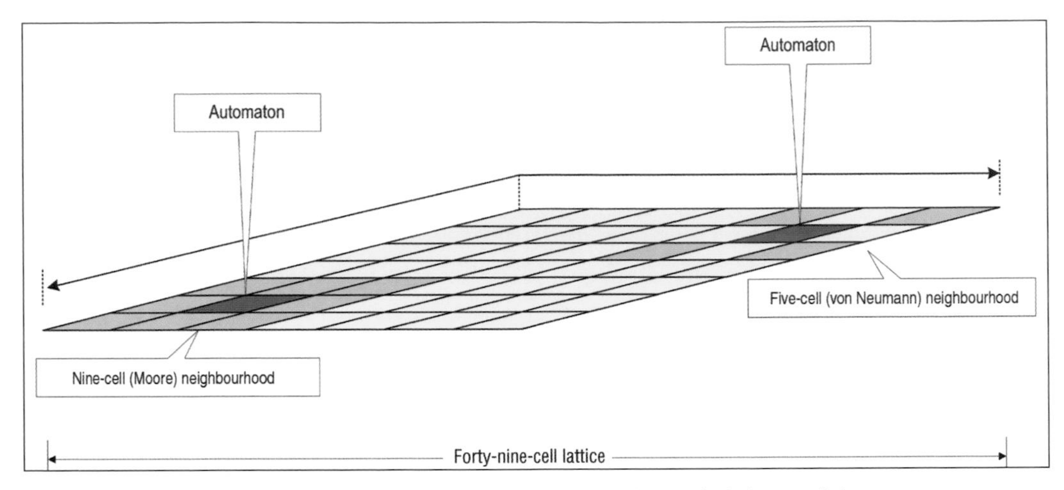

Figure 1 Two-dimensional cellular automata arranged in a regular lattice tessellation

descriptors generally reflect some agent-based characteristics. Likewise, transition rules for MAS are generally formulated in such a way as to represent behavioural characteristics or in some instances to simulate intelligence. Individual agent-automata may also be bounded by a discrete cellular space, although this is not a requirement.

MAS differ from CA in one important respect: individual automata are free to move within the spaces that they 'inhabit'. With CA, information moves between cells, propagating by diffusion through neighbourhoods as input streams between automata. However, individual cells themselves remain fixed in the CA lattice. In contrast, with MAS, automata are mobile. This has obvious consequences for the representation of spatial systems, and this is a topic that will be explored later in the chapter. The movement of agent-automata has implications for other components of the automata system, however. Neighbourhood relationships in agent-automata are dynamic: when individual agents themselves alter their location in the simulated space, the measures of adjacency between them also change (figure 2).

3 Automata as urban simulation tools

Assembling artificial cities from automata building-blocks

The components of automata listed above have close analogies with cities. Most urban entities, phenomena and systems can be specified as automata. State variables can be used to encode a wide variety of properties of urban systems into an

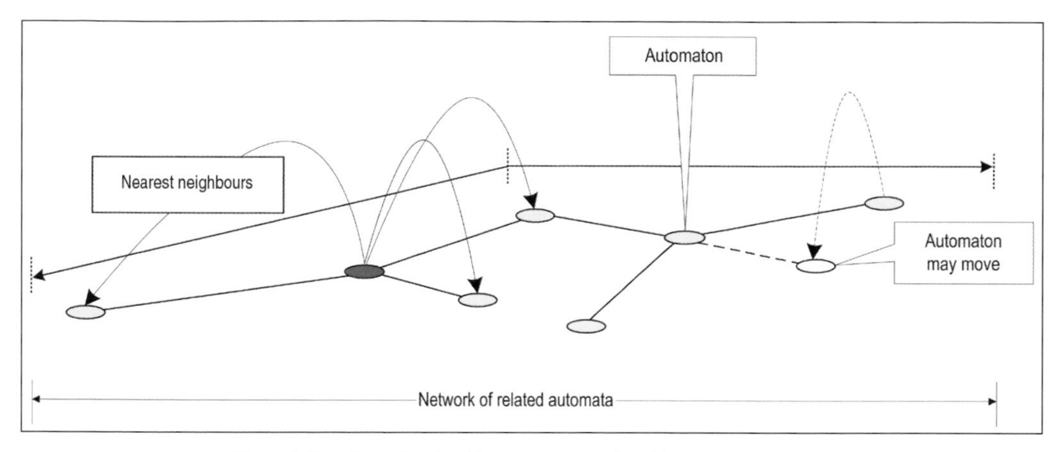

Figure 2 Two-dimensional multi-agent systems related by nearest neighbours

automaton model: land cover, land use, population density, etc. Similarly, a variety of urban objects can be represented as cells: land parcels, vehicles, road links, etc. Others can be represented without cellular boundaries: property centroids, trip waypoints, transport nodes, etc. Neighbourhood functions can be specified based on well-understood geographical relations such as market catchment areas, commuting watersheds, walking distance buffers, etc. Transition rules may be specified in such a way that they incorporate any geographic theory or methodology, for example, bid-rent theory, spatial cognition, space-time budgets, etc. Likewise, a variety of spatially-explicit movement rules can be designed for non-fixed automata, including collision avoidance, lane-changing and navigation. Also, an assortment of automata clocks can be designed to mimic the temporal attributes of real-world urban systems.

The advantages of automata as simulation tools

Automata models offer significant advantages for simulation construction in general and spatial simulation in particular. First, automata systems are decentralized in nature, with individual automata retaining autonomy within the system. In this sense, automata can be designed to work, collectively, 'from the bottom up' to accomplish tasks. Decentralisation is beneficial for a number of reasons. Centralisation in urban models is often an artefact of the methodology being used rather than an intuitive attribute. Most systems of interest in social science are understood to operate in a decentralised fashion, and automata offer the flexibility to represent them in a simulation environment. Also, because automata act as independent processors, automata models can be designed as massively parallel systems, with associated advantages for computational efficiency, especially when simulation processing is distributed across machines or processors.

Automata models can be designed to represent simulated objects at very high resolutions. They can be thought of as microscopic or atomic models, with objects represented at entity-level resolutions: pedestrians, households, vehicles, houses, etc. (Benenson and Torrens forthcoming). This has obvious advantages for representing human systems: it allows model developers to part with notions of average individuals and to avoid problems of ecological fallacy. In terms of representing spatial systems, it also offers opportunities for circumnavigating modifiable areal unit problems. If coarser resolutions are required, they can be aggregated on an intuitive basis from collections of objects at finer resolutions (Benenson and Torrens forthcoming).

The emphasis on interaction in automata models is another important property for simulation, particularly because the autonomous treatment of individual automata permits the simulation of entity-level interaction and any emergent behaviour that may result. Previous popular methodologies for urban simulation represented interaction in terms of aggregate flows; the spatial interaction model is an example.

The dynamic nature of automata is another attractive property for simulation-building. As mentioned, individual automata can be designed with internal clocks. Time in automata moves in discrete steps, and these steps can be specified at resolutions that approximate real time. State transition rules can be tied to simulated time clocks so that processes may be simulated along realistic time scales. For example, a model of traffic on a highway could be made to simulate real-time traffic movement, as well as diurnal periods of peak congestion, and even more long-term patterns such as holiday-influenced volume surges.

Simplicity is one of the often-recommended advantages of automata as a simulation tool. As with the Game of Life and Boids, automata simulations designed with simple specifications and few rules are often capable of yielding complex behaviours (figures 3 and 4). The decentralised and autonomous characteristics of automata allow for a variety of complex characteristics to be generated from simple conditions: chaos, emergence, self-organisation, non-linearity, phase transition, etc. This is appropriate for many urban systems where almost chaotic end-states are understood to result from simple initial conditions.

Of course, one of the motivations behind using automata as an urban simulation tool is the advantage that they offer for spatial modelling. In the context of CA, cellular boundaries can be designed to mimic various geographies: standard

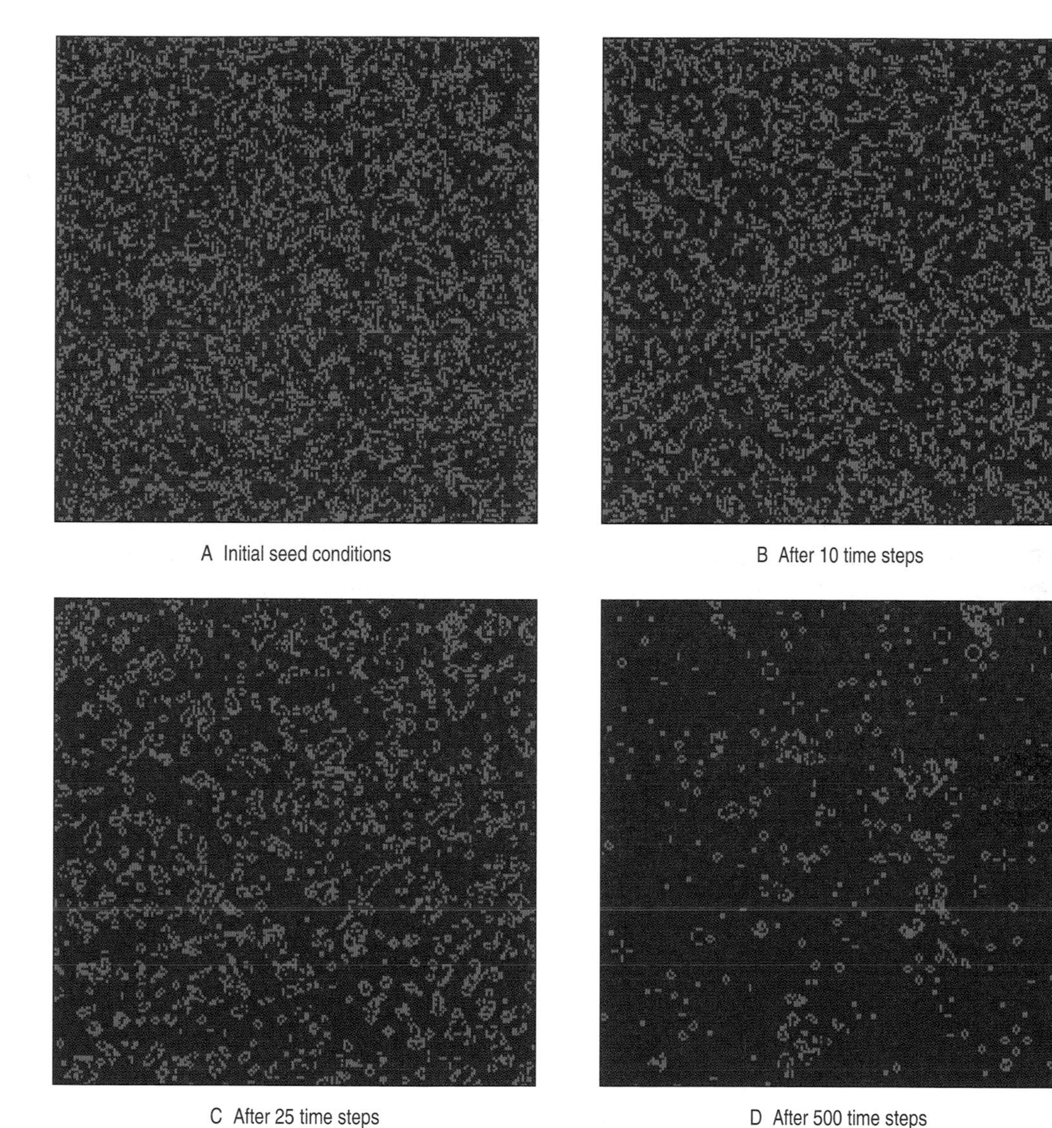

A Initial seed conditions

B After 10 time steps

C After 25 time steps

D After 500 time steps

Figure 3 Game of Life at several stages of evolution.

The model was developed using the RePast Java libraries (University of Chicago 2003). The model was defined as a 100 by 100 toroid space, with a cell size of two pixels. The initial seed conditions were based on the string 20010204593, with the space set to 50% fill capacity at the initial time step

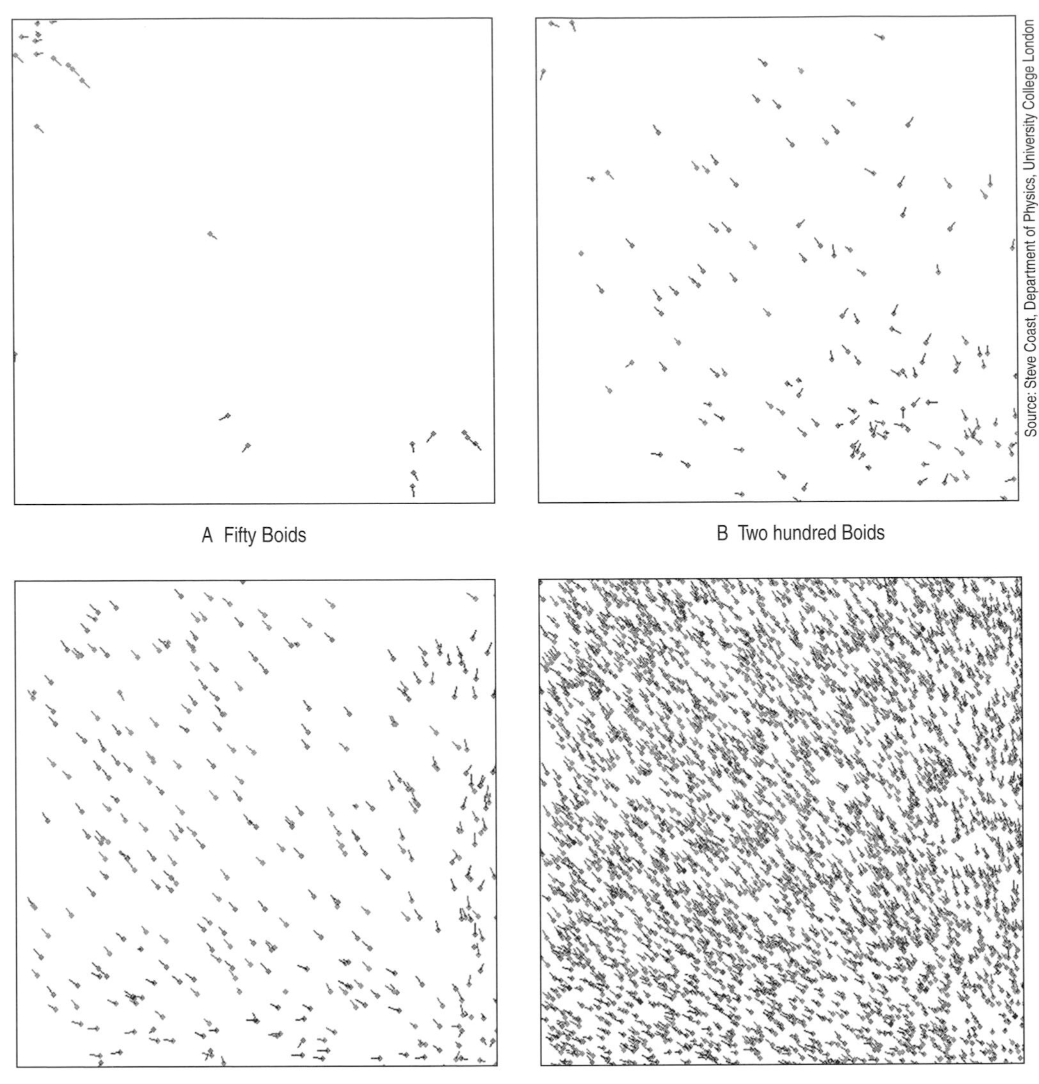

Figure 4 Multi-agent systems specified with a varying number of Boids.
The transition rules for the Boids model are detailed in Reynolds (1987)

polygons, Delauny triangulations, Voronoi polygons, etc. Various spatial analysis techniques from geographic information science (GISc) are available for generating such tessellations. Likewise, neighbourhood expressions can be used to encode spatial topologies and structures into automata models: networks, lattices, graphs, etc. Transition rules allow for the infusion of urban and geographic theory directly into model designs. In particular, transition rules permit the expression of form and function in a symbiotic relationship: the processes that drive systems can be expressed with the patterns that they generate. Again, there are formal geographic expressions that can be used to articulate these relationships: GIS operators, geo-algebra, etc.

Application of automata tools to urban systems

Automata tools have been used to build urban simulations for an array of applications, both as pedagogic models and planning support systems (Torrens 2002a). CA, in particular, have enjoyed widespread use as tools for urban development modelling, land-use simulation, and land-use/land-cover change modelling. MAS models have been less popular in urban simulation, although there seems no reason why this might be the case. Nevertheless, MAS models of urban systems have been constructed for a variety of purposes: simulating residential location dynamics (Torrens 2001a; Benenson, Omer and Hatna 2002), traffic systems (Barrett et al 2001) and urban population dynamics (Benenson and Torrens forthcoming). However, in most cases, these models have actually been formulated as CA and simply reinterpreted as MAS.

4 Using automata-based tools to advance urban simulation research

Despite the numerous advantages of automata for urban simulation and the volume of applications to urban systems, the field is very much in its infancy as an area of research. In papers published elsewhere the author has outlined potential research threads for automata-based modelling of urban systems (O'Sullivan and Torrens 2000; Torrens and O'Sullivan 2001). The following sections will detail some of the work that the author has been pursuing at CASA along these lines of inquiry. The next section deals with the design of spatial automata systems. This is followed by sections that outline ongoing work of applying this framework to urban systems.

Geographic automata systems as an expressly spatial simulation technology

There exists some justification for developing new and patently spatial automata-based simulation technologies for urban simulation. There is somewhat of a disjoint in the current literature between CA and MAS models in urban applications. CA are often used to model processes that would be better modelled using MAS and vice-versa. In many cases, MAS models are specified as CA models, with cells being paraphrased as agents. There is nothing particularly wrong with this approach, but it may be more useful to use the right tool for the right job. To some extent, the confusion between CA and MAS can be cleared by looking at their spatial attributes. CA are fixed and MAS can move. CA transmit information through their neighbourhoods, and the propagation of information is constrained by this function; in MAS, information can be exchanged through agent-neighbourhoods, but information can also travel with the agent in which it is housed and so may be less constrained than in CA (figure 2). This makes CA an appropriate tool for fixed entities that influence their environments by diffusive processes; MAS are more appropriate for representing non-fixed objects that affect their surroundings by exchange mechanisms such as communication, memes and stigmergy. However, both properties are complementary, and an explicitly spatial framework for automata model-building could reconcile the functionality of CA and MAS in a seamless fashion.

Often, automata-based tools are borrowed from origins in other disciplines outside geography: computer science, mathematics, physical sciences, etc. For the most part, the tools were originally developed for non-spatial uses and their application to geographical systems necessitates modification. Geographers have long been uncomfortable with modifying automata, uneasy about departing from strict formalisms (Torrens and O'Sullivan 2001). Nevertheless, by modifying automata-based tools along spatial lines, there is an opportunity to learn a great deal about the real-world systems that they are abstracting, as well as developing useful simulation environments for exploring geographic ideas and hypotheses. To date, most of the modification to automata-based tools has been rather ad hoc in nature, and there is a need for a formal framework to develop explicitly spatial automata models.

A number of authors have hinted at the rich potential for integrating automata and GIS. For the most part, current applications constitute a loose coupling of the two. The possibility of more tightly coupled models has received relatively little attention. In particular, there remains much opportunity for connecting automata with principles from GISc for manipulating data for use in automata models, tessellating automata spaces, formulating neighbourhood functions and expressing other spatial relationships between automata.

Finally, there is a strong need for geographically-rooted software libraries for developing automata-based models. A number of libraries enjoy widespread use, including Swarm (Swarm Development Group 2001), RePast (University of Chicago 2003), and Ascape (Center on Social and Economic Dynamics 2001), however, they were not designed explicitly for spatial applications and their representation of space is often weak. This is unfortunate, particularly considering the symmetry between object-oriented programming paradigms, urban models, and GIS.

With these ideas in mind, the author has been collaborating with the Environmental Simulation Laboratory at Tel Aviv University's Department of Geography and Human Environment to build a patently spatial automata-based simulation framework for urban modelling: geographic automata systems (GAS). Briefly put, GAS offer much of the functionality of basic automata, CA and MAS, but are specified in a spatially-specific fashion. GAS models are constructed from individual geographic automata (GA) building blocks and GAS are defined with geographic components:

- A typology of automata: GAS may comprise GA of different types, for example, spatially fixed and non-fixed.
- Automata states: GA can be characterized with state variables, as is the case with all automata. Variables of uniquely geographic significance such as heading, velocity, progress from an origin, etc., may be introduced.
- General state transition rules: GA state dynamics are driven by general state transition rules, akin to those of the other automata discussed previously. However, in GAS complex relationships between rule-sets for different automata types can be specified, for example, stigmergetic associations between fixed and non-fixed automata.
- Georeferencing rules: These rules determine the placement of automata in simulated spaces. Automata can be georeferenced, following GIS approaches, directly using coordinate arrays. GA can also be georeferenced using indirect rules that point to automata in a variety of ways, even when they operate dynamically in space and time.

- Movement rules: A dedicated rule-set for controlling the movement of automata is introduced in GAS, allowing for a plethora of fluid-like and migratory motions.
- Neighbourhood rules: Rather than relying on predefined neighbourhood topologies, GAS use a neighbourhood rule-set for determining adjacency between automata. This allows for the dynamic specification of neighbourhoods in space and time, as well as the linking of neighbourhoods to other properties of the model.

It is hoped that the development of GAS tools and software will advance geographic research in automata-based modelling. The following sections outline examples of GAS models applied to urban systems.

Hybrid automata for simulating urban development

The previous section offered a rationale for developing hybrid CA-MAS models in a GAS framework. However, there are also compelling reasons for constructing hybridisations between automata and non-automata models, such as traditional large-scale land-use and transport models. No single technique can capture the richness of an urban system, and it makes sense to use a diversity of tools to build virtual environments particularly when they are to be employed as planning support tools (Torrens 2002a). Despite the flexibility of the automata approach, limitations remain.

Automata are not particularly useful for simulating top-down processes, planning regimes being an obvious example. Also, automata-based models are closed systems. Closed systems are generally easier to model than open systems; there are simply less unknowns to worry about (Batty and Torrens 2001). However, system closure is an inappropriate assumption for most urban systems, which may be sensitive to a host of exogenous influences: national boom and bust cycles, regional inequalities, meteorological phenomena, etc.

The GAS approach goes some way toward resolving these problems, but in many instances it is necessary to interface automata models with exogenous simulations. Several authors have built models in this fashion, most commonly by introducing exogenously-specified growth rates, and in some instances tying automata models to a complicated chain of related simulation modules (Barrett et al 2001).

The author has been researching a general hybrid framework for modelling urban systems. The framework is hybrid in a number of ways. First, it is demarcated spatially with macro-scale elements of the simulated system relegated to exogenously-specified models. Second, this demarcation also serves to segment the simulation in terms of the direction of modelled processes: top-down events are handled largely exogenously, while bottom-up dynamics are simulated endogenously with automata-based tools. Third, the automata framework is itself a hybrid, along the lines of the GAS principles previously discussed. CA and MAS are fused in a unified and symbiotic manner. The next section discusses the specification of a pedagogic model designed to simulate general urban growth.

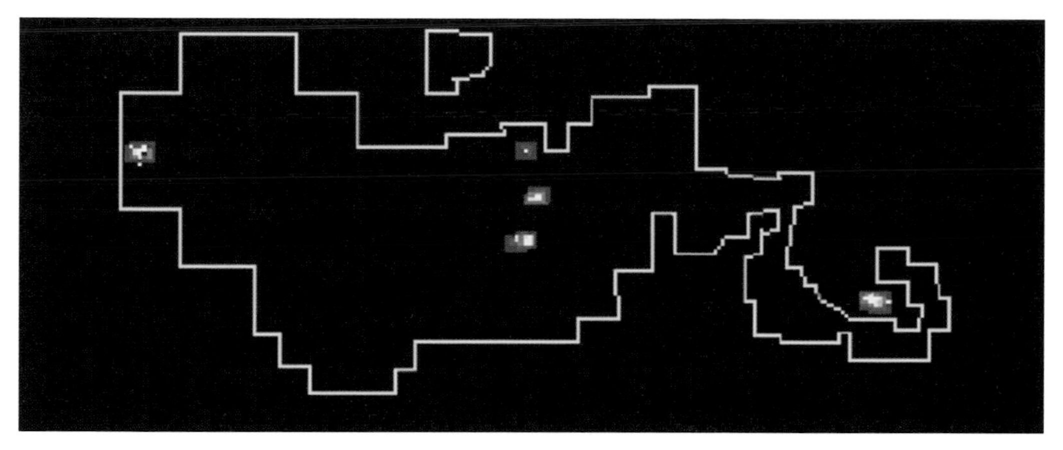

A Initial conditions

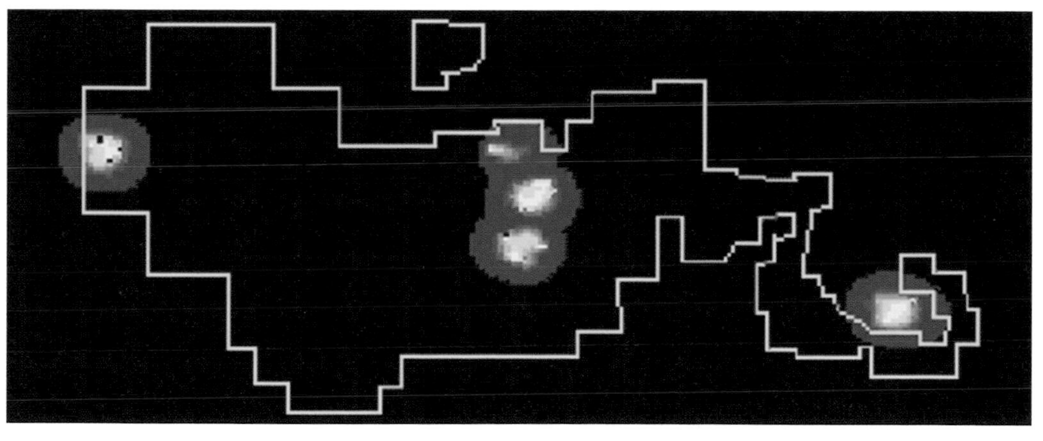

B After 50 iterations

Figure 5 The evolution of a hybrid model of urban growth from initial seed conditions

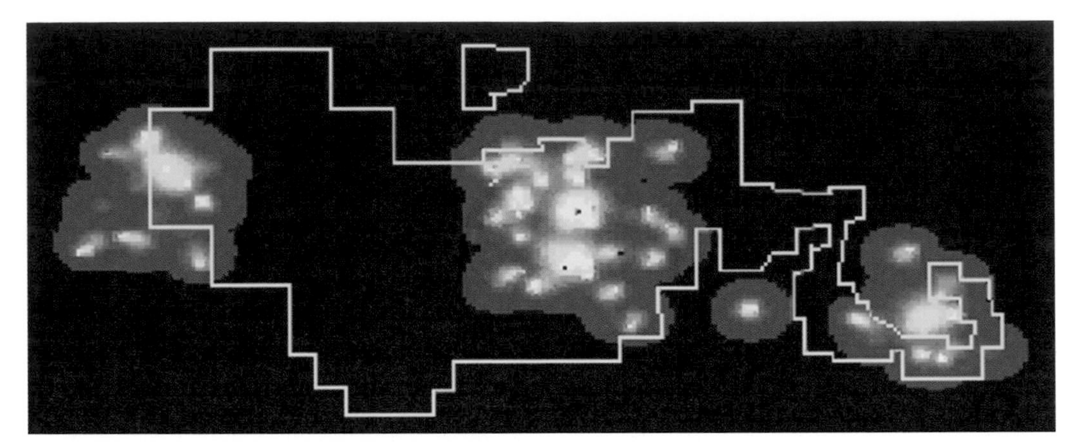

C After 100 iterations

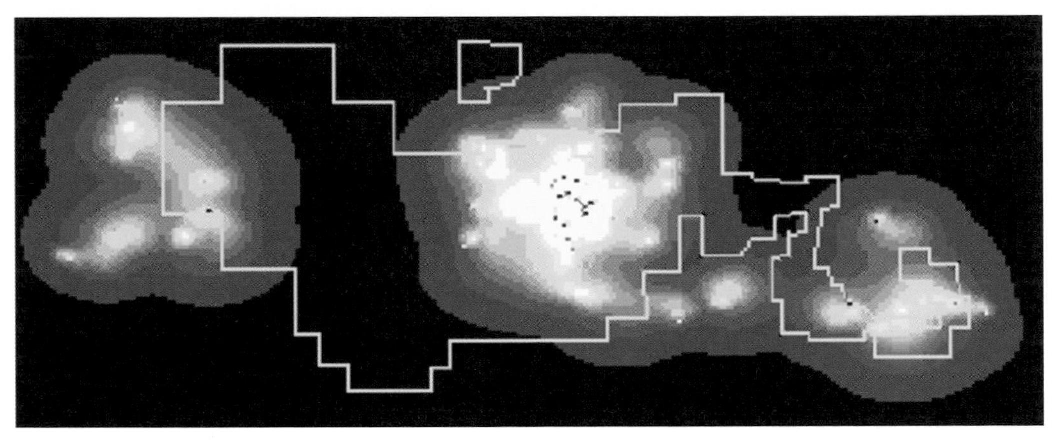

D After 250 iterations

Figure 5 cont. The evolution of a hybrid model of urban growth from initial seed conditions

An application to urban growth

This section discusses the implementation of a hybrid automata framework applied to the simulation of general urban growth in a pedagogic environment. The model simulates the spatial evolution of an urban system over time with emphasis on the patterns of urbanisation that it generates and the pace with which the landscape is urbanised. In the simulation, a city-system evolves from initial seed settlements, going through processes of compaction, poly-nucleation, in-fill, peripheral sprawl and densification of the central city (figure 5).

The structure of the model is outlined in figure 6. The model is divided into separate modules that simulate various components of urban growth at three levels of geography. At a macrolevel, the simulation is supplied from the top-down with exogenously-defined growth rates. These serve as the general 'metabolism' for the subsequent simulation. Generally, growth is defined in terms of population, with population being delivered through various gateways that are coded into the model as state variables. Because the example illustrated in figure 5 is pedagogic in nature, growth rates are defined with arbitrary values. However, in other implementations those rates have been linked to census data (Torrens 2002b).

At a meso-scale, constraint data is introduced to confine the simulation dynamics within reasonable bounds. This is particularly relevant to location in the model, as automata simulations tend to be quite sensitive to initial conditions. In the example illustrated in figure 5, the aforementioned gateways serve as constraints.

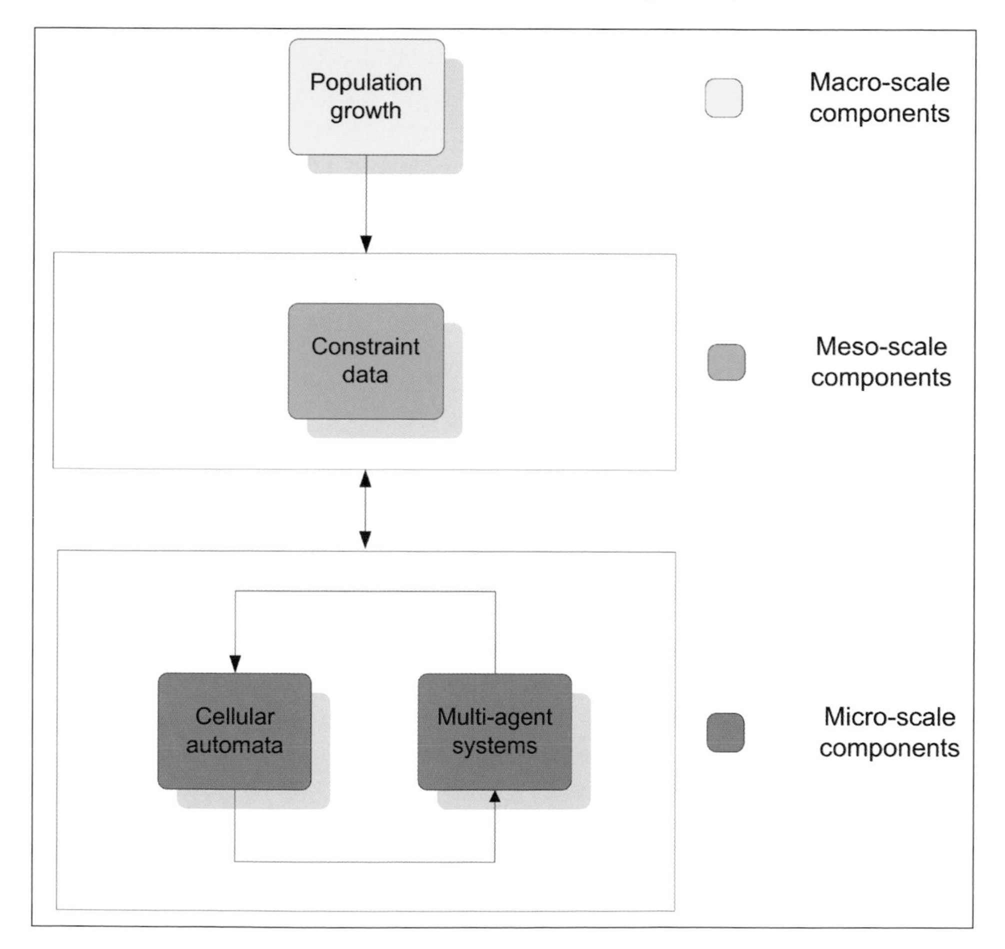

Figure 6 The structure of a hybrid urban growth model

The micro-level module then takes over, simulating the spatial distribution of this population locally around these gateways.

Micro-scale dynamics are simulated using hybrid CA-MAS. Agents, supplied to the micro-system from higher-level modules, are granted life-like functionality to mimic the location behaviour of developers and settlers in real urban systems. The microsimulation framework differs from standard CA and MAS implementations. It is specified as a GAS, but retains the individual properties of CA and MAS, as well as offering new functionality. Essentially, agents are laid on top of a CA layer but granted functionality that enables them to initiate state transitions in CA directly.

Two types of automata are considered: mobile automata (developer-settler agents) and infrastructure automata (representing the landscape and built environment). These automata have a set of states that describe their characteristics in relation to the simulation. Mobile automata have state descriptors that represent their movement (moving, static); infrastructure automata have states describing their development (occupied, density of occupation). A series of georeferencing conventions have been introduced to track automata in the simulated space. Automata have a set of coordinates that register their position in the lattice; mobile automata are also aware of their distance from the seed gateway from which they originated.

For fixed automata, neighborhoods are defined as a static nine-cell Moore neighborhood. Mobile automata are free to roam the system, constrained by

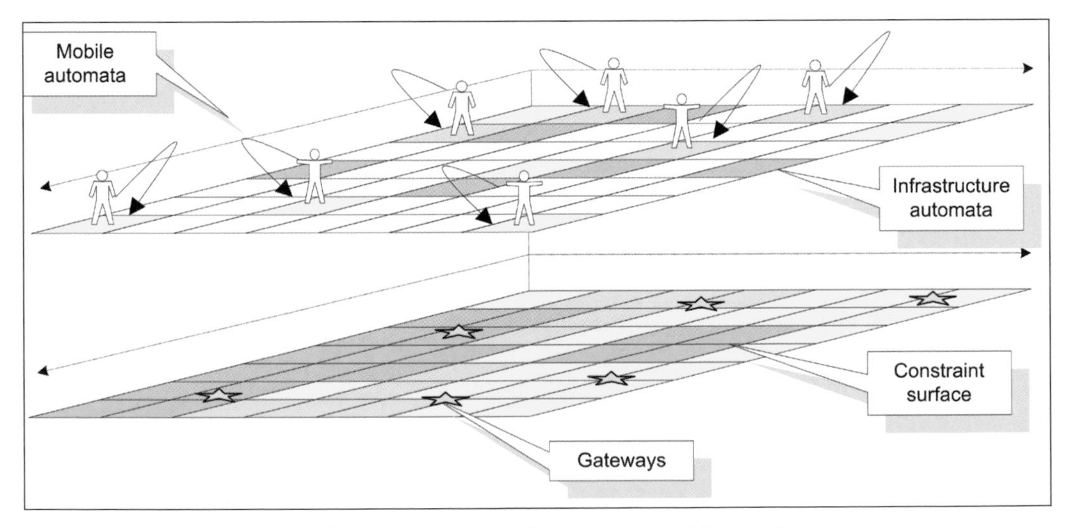

Figure 7 Cellular automata and multi-agent systems as information layers

user-defined weights applied to the distance with which the movement rules are exercised.

A number of transition rules are used to mimic the spatial patterns of development and settlement that govern the dynamics of growth in an urban system. The movement of mobile developer-settler agents in the simulated space is exercised subject to several rules of motion. Mobile agents exercise a given movement rule, based on predefined probabilities that can be defined elastically to weight certain patterns of behaviour over others. Once a movement has been exercised and an agent comes to a standstill, it develops and settles the infrastructure beneath it, that is, it exercises a single transition rule that allows it to access the state variables of the static infrastructure automata underneath, affixing a population value to the set of states that describe that cell. There are two local movement rules: one for development and settlement activity that occurs in the immediate vicinity of a given cell and another for that which takes place in a slightly extended area. Movement by leap-frogging is modelled, representing staggered, speculative and piecemeal settlement and development behaviours. Irregular movement is introduced to correspond to organic development and settlement in irregular linear patterns. A road-influenced movement rule assumes the gradual development of infrastructure to link clusters; accordingly, movement under this rule takes place in a linear fashion between nodes.

In addition to the rule that allows mobile automata to settle static automata cells, two other general state transition rules are included. A single state transition rule is available to static automata in the model. This simply diffuses population values between static cells in a nine-cell neighbourhood, thereby smoothing the distribution of settlement in the simulation. A random function that causes the population of given cells to decline in a given time step is also applied, related to the volume of development in the model overall. The larger a city grows, the greater the probability of population decline at a local scale. This function is designed to mimic decline due to overcrowding and associated problems of urban blight, etc.

Using these simple specifications, a realistic pattern of urban evolution can be simulated. The city-system starts from initial seed conditions, slowly extending and forming isolated clusters on the periphery which then coalesce over time and spawn additional clusters. As the simulation progresses, the original central seed sites, and their immediate periphery, grow denser and more compact while peripheral areas sprawl at lower densities and in a more fragmented pattern (figure 5).

There is at least one obvious limitation to this approach, however. The rule-sets driving model dynamics generate realistic patterns of urbanisation, but their representation of the processes operating within the system are somewhat unrealistic. The rules mimic the end results of system processes—the spatial manifestation of the phenomena—rather than the root behaviour of the system. Certainly developers and relocating households build and settle in a fragmented manner, agglomerate around existing settlements, and so on. It is appropriate to represent the system in this fashion when simulating the spatial dynamics of urban growth in a general sense. However, there are fundamental behaviours that motivate these actions: economic impulses leading to agglomeration, lifecycle motivations for moving to the urban fringe, etc.

In light of this consideration, the author is also working to build microscopic behaviour modules at an atomic scale that will simulate the sort of principal behavioural components characteristic of real-world systems. Currently, a small-scale residential location (demand) model has been built and a complementary development decision (supply) model is also planned. The specification and application of these models are beyond the scope of this chapter, but are reported elsewhere (Torrens 2001b, 2002a).

5 Caveats for the future

Thus far, the discussion about the use of automata tools in urban simulation has been reasonably sanguine, however, the optimism comes equipped with some caveats. A major issue of concern with models of such complexity and resolution is one of validation. Most urban simulations, whether developed as pedagogic tools or not, are built in applied contexts. A model must be applied in practice in order to test its validity. For previous generations of aggregate urban models, well-understood statistical tools are available for assigning predictability. Automata models cannot be thought of as predictive tools in the same sense simply because they are so non-linear in nature (Batty and Torrens 2001). They are better thought of as game-playing environments or 'tools to think with'. That is not to say that they are incapable of validation—a lot of research effort is being expended on developing validation techniques (Benenson and Torrens forthcoming). One approach is to validate the patterns that the simulations generate. However, more process-oriented mechanisms are needed.

The next obvious caveat concerns data. High-resolution models require fine-scale data. However, data on individual-level urban objects are not always available. Other authors have developed synthetic population generation routines for automata models (Barrett et al 2001), but data availability is still a concern. Privacy issues associated with the use of microlevel data should be considered. From a computational standpoint, processing is another concern. Ironically, advances in computer software and hardware catalysed the popularity of automata for simulation, but processing power and bandwidth remain problematic when compiling and running simulations. The experiments described in this chapter have a modest volume of simulated entities (a few thousand). To be useful as planning support tools, models with millions of interacting entities may need to be built. At these scales, most desktop machines begin to smoke around the exhaust fan. Elsewhere, authors are experimenting with distributed processing over networks of machines or Beowulf clusters, but the techniques have yet to trickle down to popular use.

Finally, the issue of practical application must be raised. If automata models are so good, why are they not widely used? The answer to that question is that they are, for the most part, still academic in nature. The field is in its infancy and the technology lags behind its practical application. Nevertheless, automata models of this nature have seen successful application in pilot studies (Barrett et al 2001) and their future as planning support tools looks promising.

Despite these caveats, automata modelling has the potential to contribute to spatial analysis, urban studies and urban planning in significant ways, both in terms of new tools for model development, new understanding of spatial systems and the virtual evaluation of hypotheses and public policy.[1]

Endnotes

1 The GAS idea was developed, collaboratively, with Prof. Itzhak Benenson at Tel Aviv University. He deserves at least 50 per cent of any complaints that its description prompts.

Acknowledgements

Thanks to Steve Coast, Department of Physics, University College London for the base Java code used to generate the images for figure 4.

Chapter

Agent-based pedestrian modelling

Michael Batty

When the focus of interest in geographical systems is at the very fine scale, at the level of streets and buildings for example, movement becomes central to simulations of spatial activities. Recent advances in computing power and the acquisition of fine-scale digital data now mean that we can attempt to understand and predict such phenomena, with the focus in spatial modelling changing to dynamic simulations of the individual and collective behaviour of individual decision making at these scales. In this chapter, we develop ideas about how such phenomena can be modelled showing first how randomness and geometry are all important to local movement and how ordered spatial structures emerge from such actions. We focus on these ideas with pedestrians, showing how random walks constrained by geometry but aided by what agents can see, determine how individuals respond to locational patterns. We illustrate these ideas with three examples: first for local scale street scenes where congestion and flocking is all important, second for coarser scale shopping centres such as malls where economic preference interferes much more with local geometry, and finally for semi-organised street festivals where management and control by police and related authorities is integral to the way crowds move.

1 Spatial dynamics at the fine scale: agents and infrastructure

As we approach scales where geometry in the form of streets, buildings and land parcels becomes significant, the focus of inquiry in urban modelling changes from one where location is dominant to situations where the local dynamics of movement are much more significant. This is easily illustrated in retailing. Firms fix their general location according to geodemographic patterns of demand and the availability and capacity of infrastructure, but their precise locations are determined according to the volumes of trade generated by their competitors. In turn, these depend not only on footfall in shopping streets but also on the ways in which goods are positioned relative to movement patterns of shoppers within the stores themselves. These ideas are manifest in the rules of thumb used by retail planners in organising the layout of shopping malls where stores position themselves according to the way their products relate to each other and to the position of the anchor stores that heavily influence the overall character of any given retail development.

Until quite recently these kinds of problem were seen as lying outside the remit of urban modelling and GIS. However, advances in computation—in terms of the acquisition of digital data using sensors, improved storage and processing of disaggregate data and advances in object-oriented programming that make it possible to represent and simulate large numbers of generic objects or agents—have enabled us to make substantial progress, some of which we discuss in this chapter. Conceptual developments dealing with dynamic systems at aggregate (see Batty and Shiode, this volume) and disaggregate levels (see Torrens, this volume) combined with ideas from complexity theory that enable us to link seemingly uncoordinated actions at the local level with the emergence of more global structures (see Longley et al, this volume) are providing the intellectual framework. Central to this is the idea of agents having behaviour that is simulated as a trade-off between selfish and unselfish actions, between individual and cooperative decision making. The classic case can be seen in local movement where the desire to reach some goal often leads to turbulence in crowd situations, while the desire to see what others are doing leads to flocking. Both can end either in congestion and panic or in smooth movement if local actions develop at speeds which allow synchronisation.

These models supplement rather than replace those used to model location and movement at coarser scales of granularity (see Longley et al, this volume) and add to the range of tools available to planners and engineers, notwithstanding the need for strategies to enable their integration. They also pose important challenges to GIS for they involve geometry in a way that has not been central to spatial analysis, and they invoke a concern for dynamics that has long been an Achilles' heel of geography and GIS. The models that we will sketch out here all deal with pedestrian movement in streets, buildings and related complexes and introduce ideas about variability and heterogeneity in urban systems which are largely defined away in the more aggregative applications that characterise mainstream GIS applications. The way in which randomness, geometry, and economic and social preferences combine and collide will in fact be our starting point and this introduces a very different mix of modelling styles than any developed in this field hitherto. The nexus of this emerging field arises from the synthesis between randomness and geometry in physics, related to ideas concerning economic rationality and social preference.

We will begin with ideas about hypothetical walks—random walks, adding geometry and economy as we go, thus constructing a generic model of local movement. We illustrate the approach in the rest of the chapter by applying it to three very different situations: very fine-scale street situations dominated entirely by geometry and the collective phenomena of crowding and panic; more economically structured movements characteristic of shopping behaviour in centres and malls; and managed situations such as festivals and street parades where control is essential to public safety.

2 Random walks, geometry and locational preference

Before we define agent behaviour, we should say a little more about what agents actually are—for the ideas we will introduce in this section apply with equal force to aggregate as well as disaggregate behaviour. Agents in our definition are mobile and their behaviour relates largely to this mobility. This behaviour in human systems is not simply determined by preferences, intentions or desires. Rather, it is also determined by the environment that reflects the spatial or geometric structure in which the agents function as well as variability between agents, in terms of their intrinsic differences and the uncertainty that they have to deal with in making any response. This variability we treat as randomness. These could all apply to aggregate behaviour but at the disaggregate level, agent-based approaches have become popular in several fields. In the virtual world for example, there are software agents or 'bots'—pieces of code that are transmitted across networks and respond to requests which they are programmed to act upon; in the physical world there are particles; in the natural world, plants and animals; and in our own domain—the human world—there are not only ourselves but also the agencies and institutions that we create to produce collective responses. The way in which randomness, geometry and intention combine is different in each of these domains. Modern physics, for example, is largely a product of randomness with the constraining effect of geometry, whereas modern economics is the product of randomness with distinct preferences that imply sentient behaviour. In human spatial systems at the level of the agent, behaviour is more complex in that it is clearly some product of randomness, geometry, economic intentions and social preference.

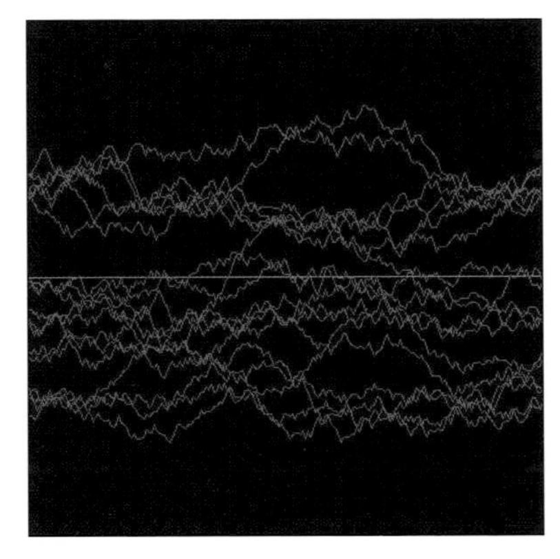

A The walk wraps around from left to right every 200 time periods

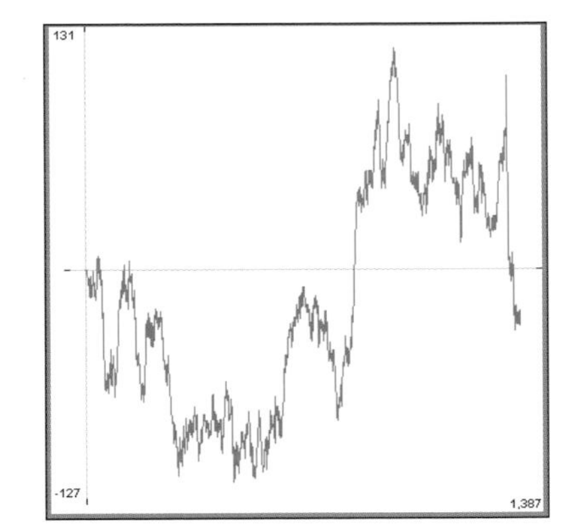

B The walk over its 1387 time periods

Figure 1 A one-dimensional random walk

We begin with the simplest one-dimensional random walk, in which an agent responds randomly to change in one dimension but moves consistently forward in the other. A typical example is a time series portraying the random variation of a stock or share price over time. Such a walk depends on the random variation from a preset position whose coordinates are calculated using the generic equation: new position = old (previous) position + random variation. In the case of a time series, the variation is in one direction through time where the coordinate of time is always known and increments inexorably while the change is random. Formally we can write $v(t+1) = v(t)+\varepsilon_v$ where $v(t)$ is the value of the stock or related market index at time t and ε_v is the random error in the stock $v(t)$. This is a first order process where the lag in value of the stock is one time period (one order) and is often called a Markov process. In figure 1 we show an example of the behaviour of this process where we have simply simulated variation from the starting point of $v(0) = 0$ for some 1400 time periods. Every 200 time periods the value wraps around in figure 1A, but we show the entire series in figure 1B where it is clear that if you examine a small part of the series and then scale this up—aggregate it over time—the series is self-similar or fractal, a well-known characteristic of this kind of constrained randomness.

To think of this kind of walk as a walk in human space, we only need to think of figure 1 as simulating how a human agent walks forward along the $y = 0$ line. Time thus becomes the forward direction and the randomness is a sideways deviation. If the apparent magnitude of the random variation in this figure suggests that our human agent is intoxicated, do remember that the amplitude of the random variation would appear much less if the y axis were marked in larger increments, and also that there is always some random deviation to help us steady ourselves when walking. Thus in figure 2A, we show four different walks all from the same starting position where the blue walk has the least variation, the yellow a little more with the green more still: the red line might be associated with someone who has had too much to drink! In figure 2B, we keep the variance the same but start a series of walkers from the same point and track their paths. Because change is random, some vary quite a lot, but there is a clear tendency for the walks to bunch together with the deviations from the straight line following a Normal distribution.

It is easy to write our generic model now for this kind of walk. Using x and y for the coordinates of each point on the walks in figures 2A and 2B, then

$$\left.\begin{aligned} x(t+1) &= x(t) + \varepsilon_x \\ y(t+1) &= y(t) + 1 \end{aligned}\right\}$$

Equation 1

where is it very clear that our walk is highly constrained in the y direction. It is now simple to turn this walk into a two-dimensional random walk. In a sense it already is, but if we fix change in the y direction randomly just as we have done in the x direction, that is, we set $y(t+1) = y(t) + \varepsilon_y$, then we can simulate how the walkers walk all over the space not just in one direction across it. In figure 2C we show the track of a walker who moves everywhere randomly but at least does not go outside the space—that is the walker avoids the edges—and in figure 2D, we show the course of four walkers doing the same. If we let these walkers continue in this way, then at the scale of resolution at which we are viewing the space, all the pixels in the space will soon be visited and the 1-D tracks will come to fill the

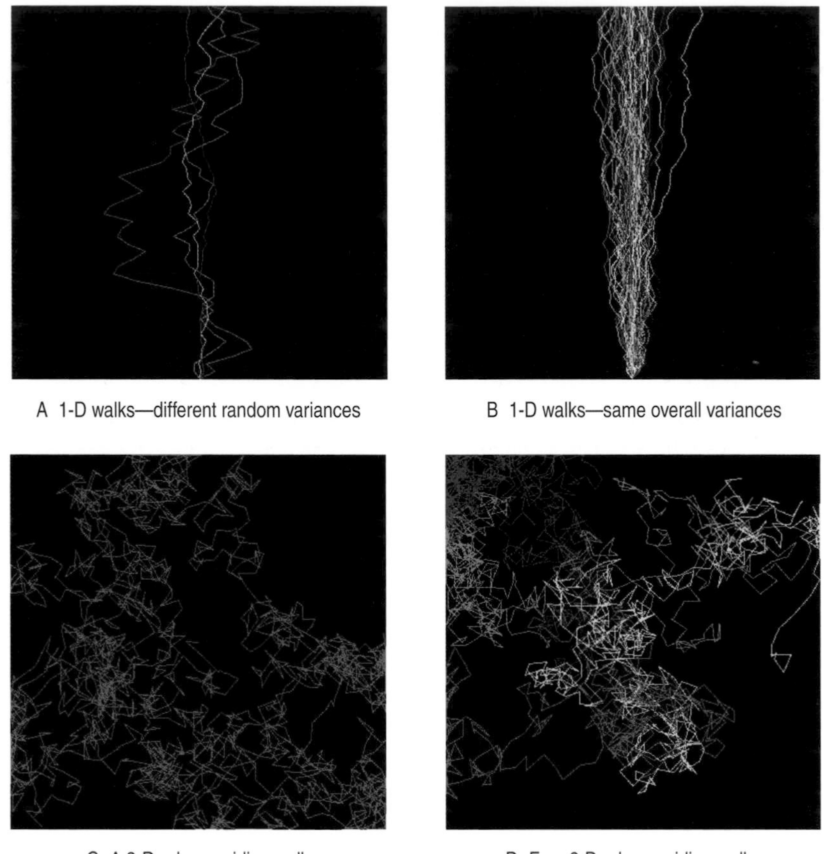

A 1-D walks—different random variances

B 1-D walks—same overall variances

C A 2-D edge-avoiding walk

D Four 2-D edge-avoiding walks

Figure 2 Multiple one- and two-dimensional random walks in the same 2-D space

2-D space. This is another example of randomness being fractal in that here is an example of a space-filling curve—a line of one dimension filling a space of dimension two. We say that the resulting object has a Euclidean dimension of one but a fractal dimension of two.

This is not a chapter on geography and fractals, however, (see Batty and Longley 1994) and much as this may be of interest, it is not our central quest. We now need to show how the geometry of the environment begins to constrain and give structure to such random walks. First though, let us note that as soon as we begin to make assumptions about how the walker moves, we, unwittingly perhaps, introduce behaviour. The walks in figure 2 twist too violently in terms of direction for these to be characteristic of the way human beings respond, and if we constrain such twists, producing the sort of gentler walks that we show in figure 3A, then we simulate greater realism. It is clear that there are two very different ways in which the walk can be constrained to avoid crossing the edge of the space. The rather blunt way is that whenever a walker bumps against the edge, the walker then shifts away from the edge, while the more intelligent way is that the walker sees the edge long before it is reached and takes evasive action. The first way, shown in figure 3B, simply prevents the walker from crossing the edge but is dominated by movement along the edge. The intelligent way requires a lot more computation. From any position the walker must look ahead and determine how far from the edge it is, then alter its heading in such a way that as it gets nearer to the edge, it veers away from it, thus retaining its smooth behaviour. We show such a strategy in figure 3C where the grey lines are lines of sight computed from every point where the walker is located. Note how the track, shown in red, can be altered according to how far the walker is from the edge. This is akin to introducing vision into our system, but this is expensive in terms of computer time for it means that before a walker moves, it must have information about how far every possible move is from every possible obstacle.

Our last example involves the interaction of agents with one another. In figure 3D we show how agents flock together from initial positions where they are entirely independent. Such flocking is based on the kind of curiosity involved in wondering what one's neighbour is doing, where s/he is going, and thence shadowing the neighbour's movement. Three factors determine how walkers flock. First, there is cohesion which is the direction pointing towards the centre of the neighbourhood containing the walkers who the walker in question is shadowing. Second, there is the separation in that a walker does not want to be too close to any of those in its neighbourhood and thus wishes to point away from those who are too close. And last, there is the alignment of walkers in their neighbourhood which is taken as the average direction in which all walkers are pointing. If these three criteria, which imply different headings, are weighted differentially different kinds of flocking take place and highly realistic simulations can be generated through their fine tuning. This kind of model was first proposed for 'boids', computer-simulated objects in flight, by Reynolds (1987) for simulating flocks of birds in computer movies, but has wide applicability as a model of the way groups of people coalesce and disperse in situations of crowding. We show the tracks of such behaviour for four agents who flock together and wrap from the top to the bottom of the screen in figure 3D.

So far, we have shown how a random walk can be given more structure by responding to the geometry of its environment, but it is quite possible to create

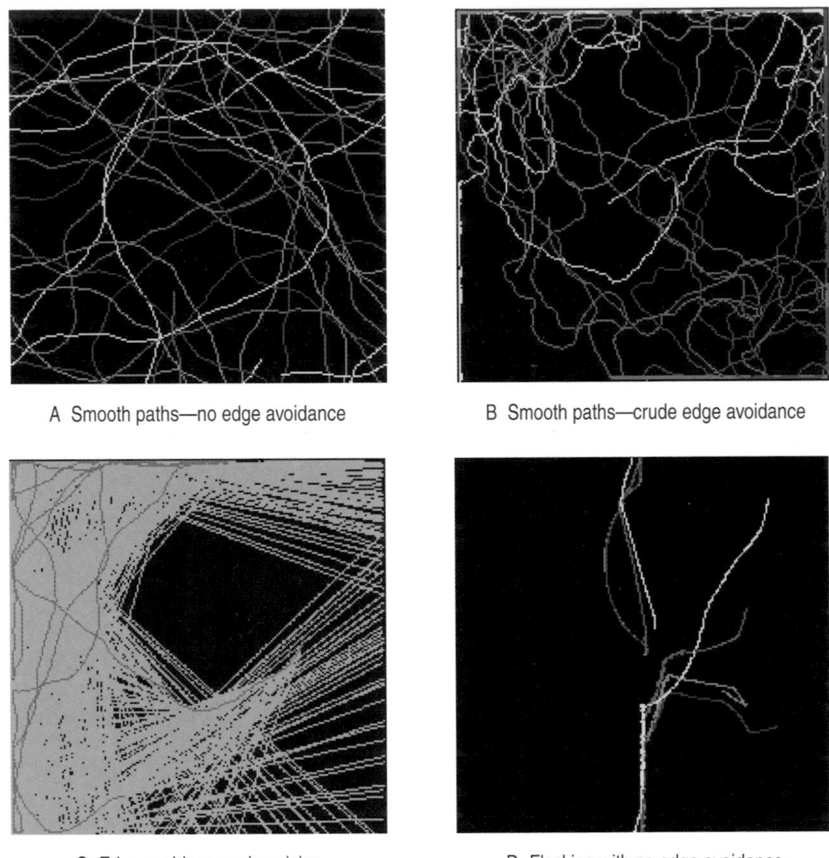

A Smooth paths—no edge avoidance
B Smooth paths—crude edge avoidance
C Edge avoidance using vision
D Flocking with no edge avoidance

Figure 3 Smooth walks A, edge avoiding B, with vision C and with clustering-flocking D

highly ordered geometric structures from the interaction of the random walkers themselves. In flocking, for example, the trail that is marked out becomes a dominant path from a set of unrelated paths in the first instance. There is, however, a dramatic example of this based on constraining each walk to terminate once the walkers reach any fixed point. For example, plant a fixed point at the centre of a space and then launch a series of walkers at a far distance from this point, letting them walk randomly in the space. As soon as one of the walkers 'touches' this point, it creates another adjacent point and then moves far away from the growing structure. If this process continues, what happens is that the structure grows in a treelike manner away from the initial point, creating a growing structure such as that shown in figure 4 for three time periods.

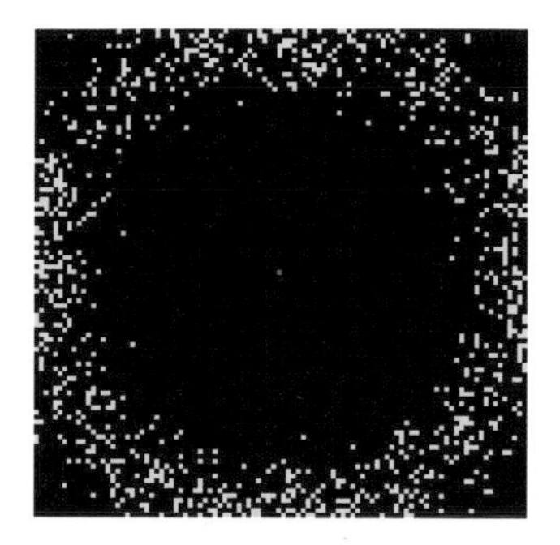

A The seed (red) at the centre; the agents (yellow) begin their random walk

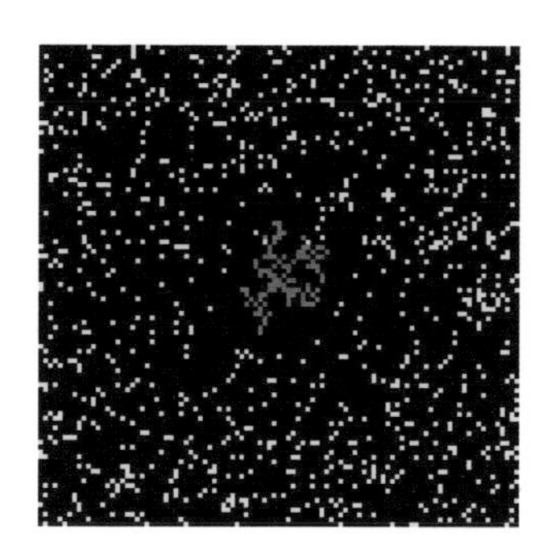

B Creation of structure: when 100 agents have touched the growing mass

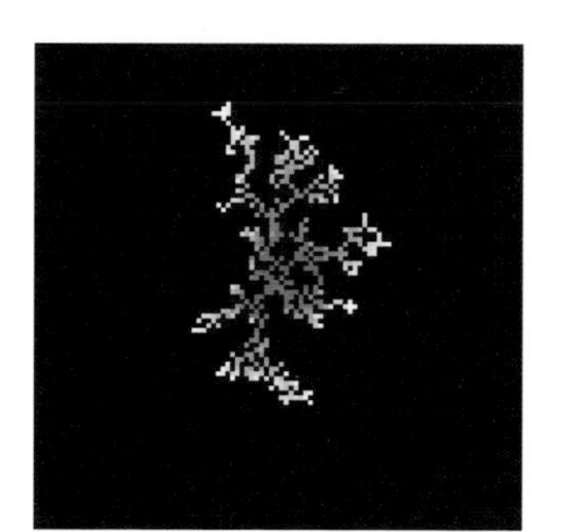

C The structure after 400 agents have come into contact with it

D The structure after 1000 agents have come into contact with it

Figure 4 The creation of structure: constrained diffusion-limited aggregation (DLA) from random walks

Basically, this is a good model of how a street system might grow in a town over a long time period. People migrate into the space and wander around until they come into contact with each other. Once they do so, they decide to settle and if the walkers are launched one at a time, then eventually the growing structure tends to fill the space available. The criterion for settlement is immediate adjacency to the existing structure and, as the walkers are more likely to come into contact with its edges (because the edges shield the centre of the structure), growth mainly takes place on the edge of the structure's growing tentacles. This is called diffusion-limited aggregation (DLA). At the most local level in terms of movement within streets, there is some sense in which such patterns reflect the location of shops on radial roads emanating from a town centre, thus reflecting location which is the product of movement (migration) over much longer time periods than those used in the models produced here. Once again such structures are fractals, this DLA model being at the core of much recent work on thinking of city morphologies as fractals (Batty and Longley 1997).

3 Movement at the very fine scale: a generic model

We now need to stand back and derive a generic model for local movement from the ideas we have already introduced. All the elements for this have been stated, at least implicitly, and there are at least four features that direct movement. First, geometric obstacles need to be negotiated. Second, agents repel each other when congestion and crowding builds up, while third, agents are attracted to each—these second and third features being the elements used in flocking. The fourth element, though in one sense the most important, relates to the desired direction in which the walker wishes to travel. In our random walks, we assumed that this was either straight ahead or completely random, but in real situations, we must generate such directions as the product of preferences and intentions. A useful formulation of these ideas has been developed over several years by Helbing and his group (Helbing et al 2001). Helbing (1991) refers to his generic model of pedestrian movement as a social force model in which each of these four features is associated with a force that pushes the walker in a particular direction. In general, we might think of movement to a new location as being formed from

$$\begin{bmatrix}\text{new}\\ \text{position}\end{bmatrix}=\begin{bmatrix}\text{old}\\ \text{position}\end{bmatrix}+\begin{bmatrix}\text{desired}\\ \text{position}\end{bmatrix}+\begin{bmatrix}\text{geometric}\\ \text{repulsion}\end{bmatrix}+\begin{bmatrix}\text{social}\\ \text{repulsion}\end{bmatrix}+\begin{bmatrix}\text{social}\\ \text{attraction}\end{bmatrix}+\varepsilon$$

Equation 2

where each of these components is some function that is 'added' into the algorithm that is used to compute movement. Desired position is a behavioural variable that relates to locational preferences while geometric repulsion is the summation of forces which stop a walker from bumping into some obstacle. Vision is important in its computation. Social repulsion and attraction are summations of all the interaction effects with other agents that are within the neighbourhoods or fields where such effects are relevant to movement.

There are several ways in which this formulation can be made operational. At very fine scales where panic and crowding are the main behaviours as in evacuation and emergency situations, the model can be formulated traditionally in terms of the physics of motion where velocity and acceleration play a central role

(Helbing , Farkas and Vicsek 2000). Where speed is not important then the coordinates of position can form the essence of computation (or headings if this be the preferred form: see Batty, Jiang and Thurstain-Goodwin 1998) while in situations where location is relative, then pixel location relative to local neighbourhoods can be used as in traditional spatial-interaction modelling (Batty, Desyllas and Duxbury forthcoming). In fact, in computation associated with such models, there is no simple equation structure, because motion must be computed through a sequence of decision rules that incrementally update position.

To show some simple examples of what this means, in figure 5 we illustrate a narrow street where walkers are launched to the left (west) and stream towards a desired location at the right (east). Figure 5B shows that there is an attraction surface around the most desirable point, and one way of walkers being attracted is to compute their position according to the steepest gradients. The geometry of the street constrains their movement too and in figures 5C and 5D, we show what happens when we reduce the variability—random movement—which is also an essential component of the way we handle crowding and dispersion. In figure 6, we take this simple street and add obstacles by narrowing it. In figure 6A, the street decreases in width immediately to a narrow outlet, while in figure 6B and figure 6C we show the effect of funnelling on the speed at which walkers are able to negotiate such a barrier. Finally, in figure 6D, we show what happens if the street becomes in effect a series of rooms linked by narrow entrances and exits. This is the kind of analysis that these types of models are able to generate; Helbing, Farkas and Vicsek (2000) have produced some extremely impressive results on the geometry of streets and corridors and the effects these have on different kinds of panic and its control.

In figure 7, we show a situation where there are two types of pedestrian: those forming a parade which moves around the corner of a street intersection (agents in white) with pedestrians watching (agents in red). The model enables us to assess the build up in pressure between the two types of walker and to set a threshold which, if breached, leads to one crowd mixing with the other. This is achieved in figure 7B where the red agents build up pressure in trying to see the parade (social attraction) but then try to diffuse (social repulsion) with serious consequences in that the only way this can happen is for the crowd to panic and break through the parade into freer space. This is the kind of disaster scenario that such models can be used to predict.

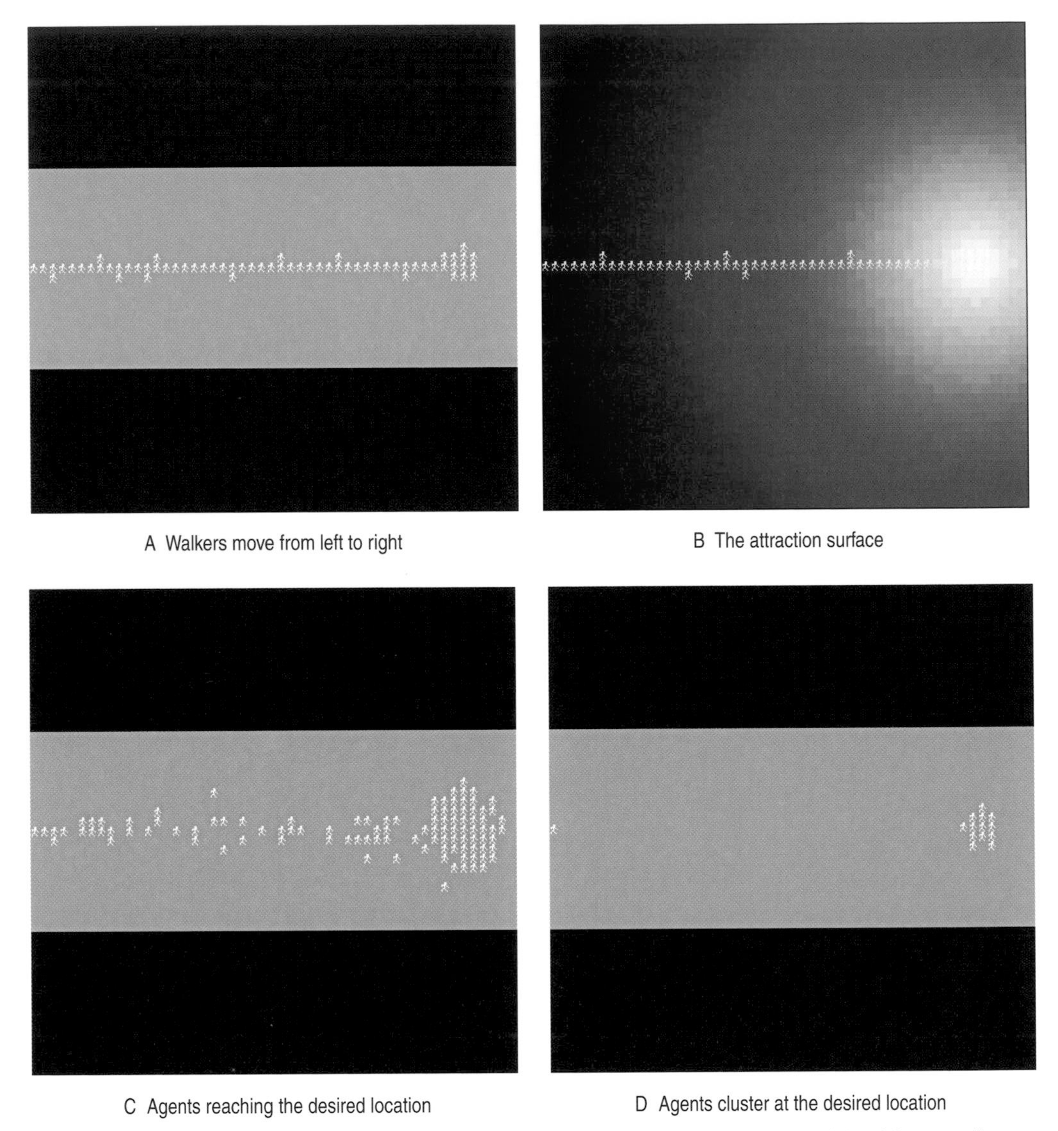

A Walkers move from left to right

B The attraction surface

C Agents reaching the desired location

D Agents cluster at the desired location

Figure 5 Walkers head towards the most desired location along a narrow street or corridor within limits of the centre line

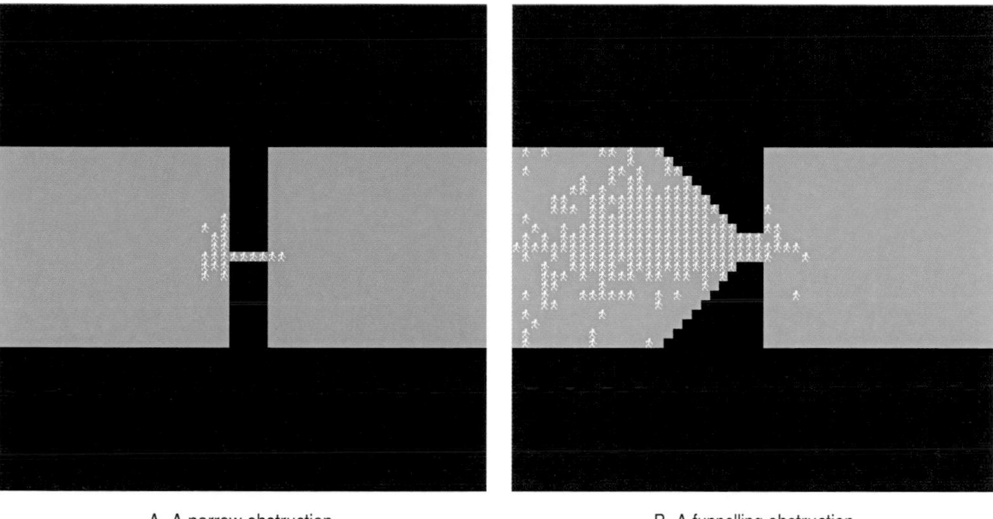

A A narrow obstruction

B A funnelling obstruction

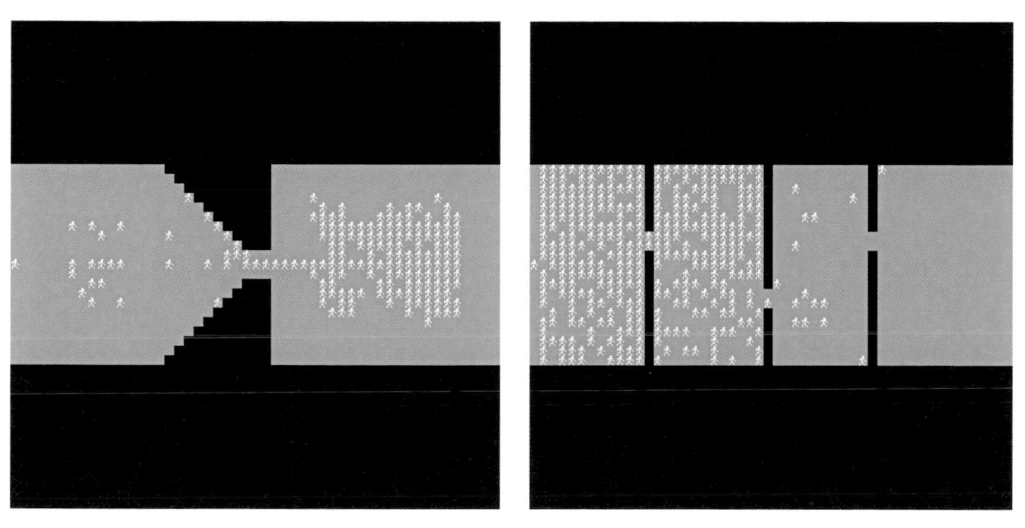

C Moving through the funnel

D A series of narrow obstructions

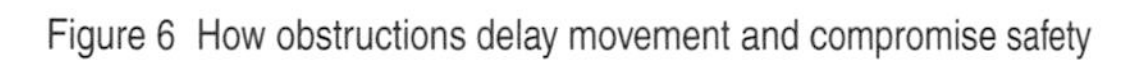

Figure 6 How obstructions delay movement and compromise safety

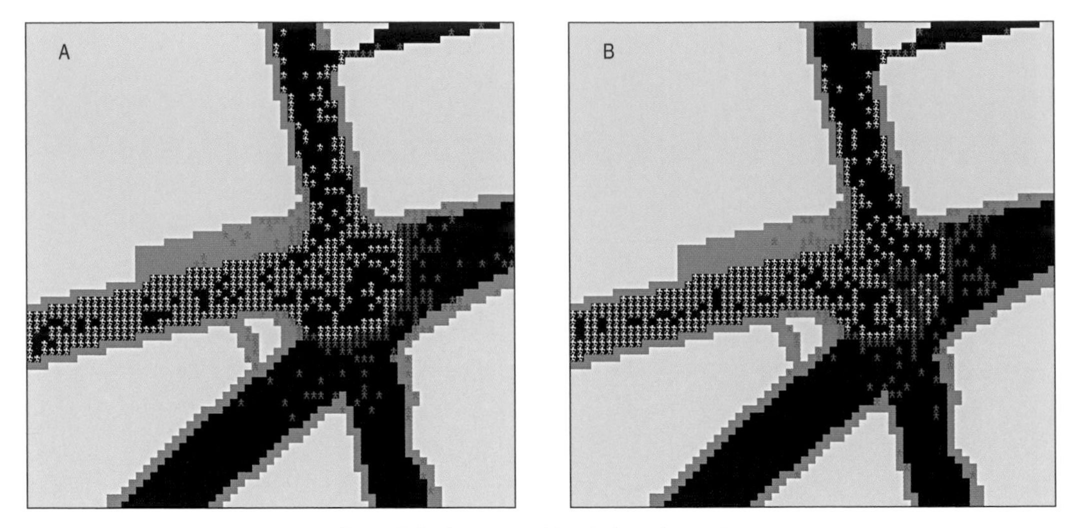

Figure 7 Panic as a crowd breaks through a cordon

The agents colored white are part of the parade while the red agents are those watching the parade. In A, these fringe the parade but in B, the pressure of the crowd on the paraders leads to them breaking into the parade.

4 Individual and collective behaviour: pedestrians in buildings, malls and centres

Our first application is to a complex building—the Tate Britain art gallery in London—that currently houses the gallery's classical collection, although when our study was undertaken, the gallery also housed the modern collection now in the Tate Modern. The Space Syntax group at UCL (www.bartlett.ucl.ac.uk/research/space/overview.htm) specialises in recording pedestrian movement in buildings, and they carried out a survey of movement through the rooms of the gallery over a one-week period in August 1995. The building provides an excellent test bed for simulating local movement as there is only one entrance but many rooms and corridors which contain very different paintings and sculptures—thus providing a clearly differentiated set of destinations for art lovers and the more casual visitor alike. In many respects, the kind of movements associated with the gallery are not too dissimilar from those associated with browsing when shopping and thus the applications provided a rather good problem on which to fine tune the model.

Figure 8A shows the pattern of movements recorded by the Space Syntax group over one hour in August 1995 using a technique of person following which the group had perfected in many other studies. It is clear from these patterns that the gallery has a greater number of visitors on the left side of its central axis where the classical and British collection was housed, in contrast to the rather sparser number of person movements associated with the right side where the modern collection was kept. It is unclear whether this reflects the preferences of the visitors or the more convoluted room structure associated with the right side—or indeed whether it reflects various perceptual clues associated with light in the gallery once they enter. The book shop—in red in figure 8B—is also a major focus for movement having a somewhat different purpose from the rest of the gallery.

The model that we applied was built around Helbing's generic structure indicated in equation 2 on page 90 but with several variations to match the problem. To begin, the attraction surface was configured in two related parts: first, each room had a certain attraction while we also used a more global attractor to spread visitors throughout the gallery on the assumption that if a visitor found a room attractive, they would not spend all their time there but would wish to visit other areas as well. In terms of geometric repulsions, we developed two variants—one in which walkers moved right up to an obstacle and negotiated it in the rather blunt way by 'bumping' into it; and a second, much more elaborate version in which we incorporated vision into the structure by preprocessing. The preprocessing stage involved computing the visual (isovist) fields associated with each pixel point describing the gallery (Batty 2001a). Social repulsion related directly to density limits on crowding and, if breached, resulted in dispersion within a local neighbourhood; social attraction was based on a flocking-like algorithm that enabled visitors to follow others if it appeared that flows to particular rooms were increasing. Fluctuations based on the random walk behaviour that we used in the 1-D walks in figures 2A and 2B were also incorporated to reflect variations in walking behaviour within a large and heterogeneous population. We also experimented with some versions of the model where the number attracted to rooms changed the attraction surface associated with the rooms, starting with a version where all the rooms were equally attractive and seeing how a hierarchy of attraction evolved through movement. A typical output of the model is shown in figure 8B where the yellow dots relate to the 550 or so walkers observed in the gallery during the one-hour period when the observations were made.

The model was operationalised using a series of decision rules, which was operated in strict sequence but within a parallel processing structure in which each agent made its own computations simultaneously with all the other agents. The essence of this model, however, is not to mimic the actual paths that were observed by the Space Syntax Group but to examine the statistical physics of steady state behaviour associated with this kind of model in this kind of application. In figure

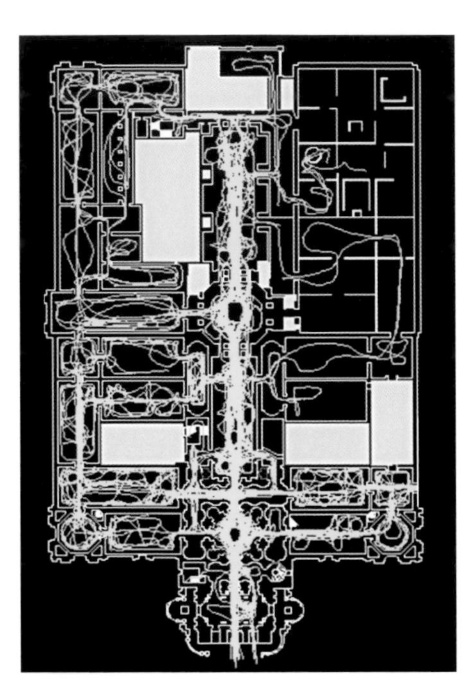

A Paths observed over one hour, August 1995

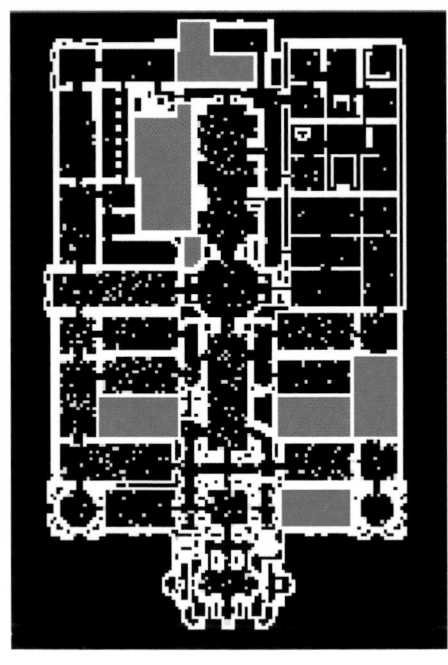

B Simulating movement within and between rooms. (Red denotes bookstore)

Figure 8 Observation and simulation of visitors within the Tate Britain Gallery.

In C, the density scale ranges from dark red at 1 persons per 20 square meters to white at 1 persons per 2 square meters

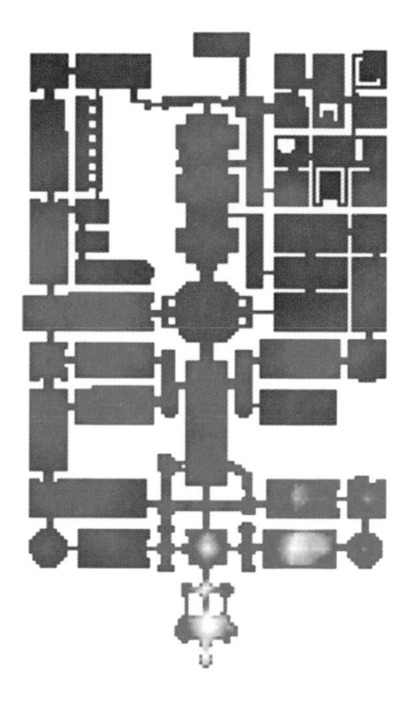

C Density of visits to rooms in the gallery

8C we show the pattern of room densities which was computed over many runs of the model where the walkers moved around the gallery for many time periods. Here it is clear that from a starting position where all the rooms had equal attraction, the asymmetry of movement actually observed in the gallery was simulated by the model in its steady state. We can say little more than this, although the implication is that it is the configuration of the gallery rather than what is on its walls and in its rooms that conditions how people move within it. We can also use the model rather effectively to close and open certain rooms. The bookshop, for example, is a case in point (see the red room in figure 8B), and if this is closed, then the pattern on the right-hand side of the gallery changes substantially.

The major problem that we face with any kind of pedestrian modelling is that the data that we are able to get are usually inadequate in many ways. Density counts are now easy from closed circuit television (CCTV), related sensors or gate counts, but the observation of actual paths is fraught with difficulty. Along with colleagues at the Shibasaki Laboratory University of Tokyo, we are experimenting with laser scanning technologies, but although these are able to pick up actual paths travelled, there are limits on the density of the scene in terms of numbers of walkers. To illustrate that our models are much further ahead than the data that we have, we have adapted the Tate model to a small town centre in the British

Midlands—Wolverhampton—for which we have only crude footfall data. Nevertheless, this centre is rather simple as most of the movement within the centre is on foot with pedestrians coming from a ring of car parks around the edge of the centre as well as the rail and bus stations. There is little on-street parking and this makes the calibration of the model feasible. We show an example of the output from the model in figure 9 where the streets within the town centre are clearly shown and where the movement originates from the car parks and stations around the edge. Wolverhampton is one of the few towns in England bounded by a complete ring road with no residents living in the centre, and thus all movement is associated with shopping and office work. The structure of the model differs from the Tate in that the global attraction surface is more complex albeit centred on the prime retail pitch. Local attractions are also built into the model based on the idea that once a shopper visits a particular shop, the probability of shopping in an adjacent shop is increased or decreased dependent upon that shop and the type of shopper. In short, movement, although conditioned by a general pattern of moving from edge to the centre, is complicated by the actual visits made. In this way the physical and locational structure of activities in the centre becomes associated with the preference of the shopper.

This framework has been elaborated in CASA with the well-known Swarm modelling system that was designed to simulate artificial life forms. Haklay et al (2001) developed such a model where preferences and geometry were combined in a much more sophisticated way than in the model shown here. In fact, in the model whose typical output is shown in figure 9, shoppers only visit the centre for a fixed period of time with an increasing number returning to their rail and bus stations and car parks while others enter the centre. In this way, we are able to build the routine dynamics—the ebb and flow of shoppers during the shopping day—into the model and thus begin to assess critical densities at particular points in time. Thus we can begin to get a handle on locational attraction not only in space but also in time, and this leads us full circle to ideas about the 24-hour city and the way retailers are beginning to compete in time.

5 Highly managed spatial events: street parades and carnivals

To apply a full version of the model, we will now adapt it to a problem with major implications for crowd safety. Unlike shopping trips which have quite distinct purposes reflecting various preferences for places and goods, movements associated with street festivals and parades are in a sense simpler, more focused locationally

Figure 9 Simulating pedestrian shopping trips from car parks in Wolverhampton Town Centre (within the ring road—mapped extent 1 km square)

but in another sense, more complex in that walkers and visitors respond less predictably to the stimuli that together comprise the event. We have built a model of an annual street festival in west central London known as the Notting Hill Carnival. The Carnival has grown from a small West Indian street celebration first held in 1964 to a two-day international event attracting 710 000 visitors in 2001. It consists of a continuous parade along a circular route of nearly 5 kms in which 90 floats and 60 support vehicles move from noon until dusk each day. Within the 3 km^2 parade area, there are 40 static sound systems, and 250 street stalls selling food. The peak crowds occur on the second day between 4 P.M. and 5 P.M. when in 2001, there were some 260 000 carnival visitors in the area. In 2001 over the two days there were 500 accidents, 100 requiring hospital treatment with 30 per cent related to wounding, and 430 crimes committed with 130 arrests. Some 3500 police and stewards were required each day to manage the event.

The safety problems posed by the event are considerable. There are many routing conflicts because of cross movements between the parade and sound systems, while access to the carnival area from public transport is uneven with four roads into the area taking over 50 per cent of the traffic. A vehicle exclusion zone rings the area and thus all visitors walk to the carnival. Crowd densities are high, overall at about 0.25 persons per m^2; a density of 0.47 ppm^2 line the carnival route, while there are 0.83 ppm^2 inside the route where the sounds systems are located. We have good data for crowd densities from our own cordon survey, data on entry/exit volumes at London Underground (subway) stations and 1022 images of the parade taken by police helicopters in the early afternoon of the second day. We show some of these images in figure 10 from which we have extracted detailed crowd density-statistics throughout the area. What we do not have are good data on the paths taken by the visitors from their points of entry into the carnival area to the various attractions that comprise the event. However, most visitors enter using one of 38 entry points with half the volume associated with five entry points related to subway stations, and this simplifies the application.

The variant we propose is based largely on Helbing's (1991) social force model, but it also incorporates ideas from fluid flow, queuing theory, event simulation, scheduling and trip accessibility (Still 2001). It is built within a cellular automata structure in which agents embody self-organising behaviour in the form of flocking and swarming (Burstedde et al 2001; Dijkstra, Jessurun and Timmermans 2002). In essence, the model generates the relative accessibility of different attractions which make up the entire event with respect to the points where visitors enter and then simulates how these agents walk to the event from these entry locations. In events such as these, we are never in a position to observe the flow of pedestrians in an unobstructed manner because the events are always highly controlled. We have thus designed our model in three stages. First, we build accessibility surfaces from information inferred about how walkers reach their entry

Figure 10A An ambulance attempting to move injured visitors through the crowd surrounding the parade

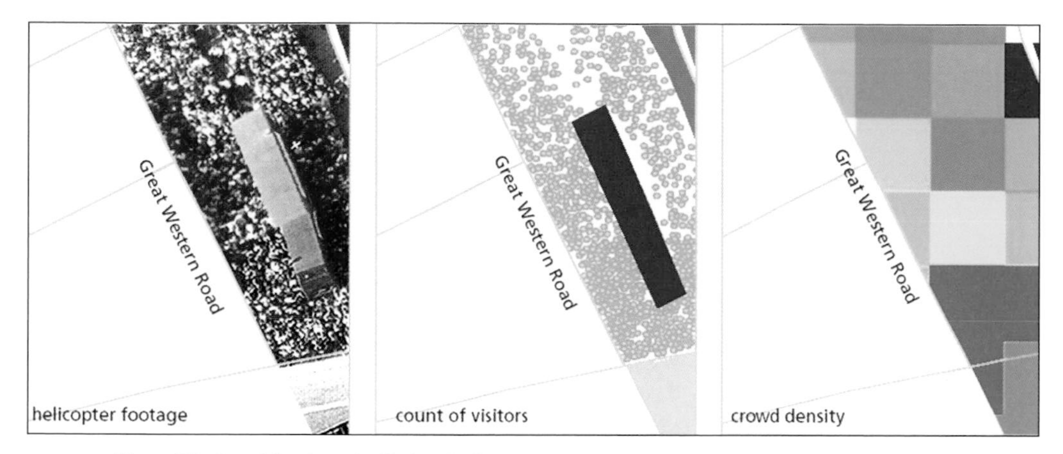

Figure 10B Crowd density at the Notting Hill Carnival—calculating crowd densities from aerial photographs

points (origins) relative to their ultimate destinations at the carnival. Second, we use these surfaces to direct how walkers reach the event from their entry points and then assess the crowding that occurs. Finally we introduce controls to reduce crowding, changing the street geometry and volume of walkers entering the event, and fine tune this process iteratively until an acceptable solution is reached. These three stages loosely correspond to exploration, simulation and optimisation.

In the exploratory stage, we begin with walkers located at their ultimate destinations in the carnival area. Walkers move randomly from location to location (which we refer to as cells), avoiding inaccessible cells that are obstacles such as buildings and barriers. The probability of moving from one cell to another is computed according to the accessibility of that cell, which in turn is based on the number of walkers having already visited the cell. At the beginning of this process, all cells have the same probability but eventually a walker will discover an entry point to the carnival—an origin—and when it does, the walker switches from exploratory to discovery mode and returns to the destination with knowledge of the discovery. As the destination is known, in that the walker has come from this, it lays a trail back to its source akin to the way ants drop pheromone once they have discovered a food source and head back to their nest (Camazine et al 2001). When the walker enters the neighbourhood of its destination, it switches back to exploration mode and the search begins over again.

It takes some time before agents discover an origin. Before this, the search is a random walk with the route accessibility surface set as a uniform distribution. If a walker crosses the edge of the event space, the walker is absorbed, regenerates at its destination and begins its search again. In its early stages, this is a random walk with absorbing barriers, much like that shown earlier in figures 2C and 2D, with the variance of its lengths roughly proportional to the time taken so far to traverse the terrain. As the process continues, more and more origins are discovered, while during exploration, walkers direct their search at routes to origins already discovered. Those origins closest to destinations are discovered first and a hierarchy of 'shortest routes' is built up, continually reinforced by this positive feedback. This is a variant of a generic algorithm predicting trail formation and collective foraging behaviour amongst animal populations. It is extremely efficient for predicting shortest routes in geometrically constrained systems (Bonabeau, Dorigo and Theraulaz 1999). In figure 11 we illustrate this for the accessibility surface to various destinations and shortest routes to the tube (subway) stations in Notting Hill. To impress its efficiency, we first show the simulation without obstacles to movement—with the street pattern—in figures 11C and 11D which shows the true nature of the swarming that takes place, and then with the real street pattern imposed in figures 11E and 11F.

The exploratory stage finishes when the walkers converge on an efficient and unchanging set of paths between the entry points (origins) and the attractions (destinations) which define the carnival events. In the second stage, we launch walkers from their entry points, and these walkers move towards these events using the accessibility surface based on the shortest paths from stage one to condition the probabilities of movement. There are two effects that complicate this and the first is flocking (social attraction), which directs movement as an average of all the movement in the immediate neighbourhood of a walker (Reynolds 1987, Vicsek et al 1995). But the second effect indicates that a move only takes place if the density of walkers in any cell is less than some threshold based on the accepted standard of two persons/m^2. If this is exceeded, the walker evaluates the next best direction and, if no movement is possible, remains stationary until the algorithm frees up space on subsequent iterations. These rules are ordered to ensure reasonable walking behaviour. This second stage is terminated when the change in the density of walkers in each cell converges to within some threshold where it is assumed a steady state has emerged.

We can assess how good the model is at predicting the observed distribution of crowds using a battery of statistical tests based on densities and paths and if the model survives, we move to the third stage that is more informal. Note that so far we have not introduced controls on where people are able to move for we begin with no street closures or barriers used in crowd control. In a sense, the carnival is never without any control so our third stage is to introduce such controls, one

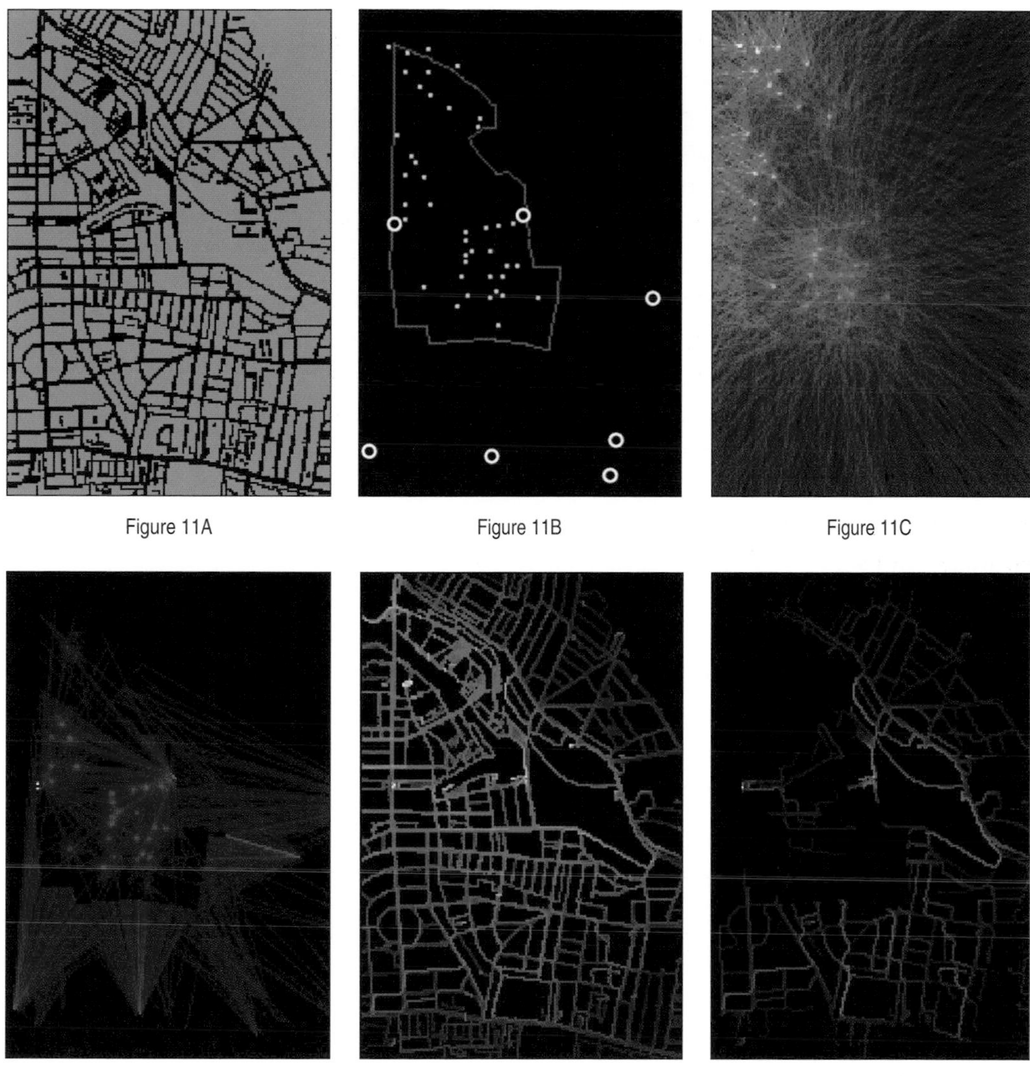

Figure 11 Exploration of the street system and discovery of entry points (tube stations) in Notting Hill: A, the street geometry; B, the parade route (red), sound systems (yellow) and subway stations (white); C, accessibility from parade and sound systems without streets; D, shortest routes to subway stations without streets; E, accessibility with streets; and F, shortest routes with streets. Relative intensities (of accessibility) are shown on a red scale (light = high; dark = low)

by one, to ensure that we produce a safe simulation as well as showing what might be achieved under certain strategies. We do this by examining statistics from the second stage and gradually making changes to reduce the population at risk by introducing barriers, capacitating entry points and closing streets. As the repercussions of this are not immediately obvious, we make these changes one by one, rerunning the model until an acceptable solution emerges. In estimation, this stage may also be used to assess the efficacy of existing controls. It is not possible to develop a formal optimisation procedure as so many additional factors such as resources for policing, etc., cannot be embodied in the model. Nevertheless, we consider this interactive method of introducing control the best approach so far for assessing alternative routes.

In implementing the actual model, we start by finding the shortest routes from the parade and static sound systems to the 38 entry points located on the edge of the traffic exclusion zone shown in figures 12A and 12B. The swarm algorithm predicts the numbers of walkers that 'find' each entry point and we compare this uncontrolled prediction to the cordon survey, explaining 64 per cent of the variance. We then use the observed volumes of visitors at entry points to launch these as agents who 'climb' the accessibility surface produced in this first stage. We show the crowd density distribution in its steady state in figure 12C and this identifies significant points of crowding. We predict around 72 per cent of the variance of the observed densities for 120 locations where good data are available. At the third stage, we rerun the model with the official street closures and barriers imposed as illustrated in figure 12D. At this stage, we have increased the variance explained to 78 per cent, but not all the points of extreme crowding have been removed. This suggests that even in estimating the model, it can be used in a diagnostic manner to identify vulnerable locations as we show in figures 12E and 12F.

We can now use the model to test alternative routes as we did in the full project (ISP 2002). After considerable political debate, an interim change in route was agreed on for Carnival 2002 where the northernmost section of the parade (shown in figure 12A in green) was removed. Running the model leads to slightly reduced average crowding, but other problems associated with starting and finishing the parade that are not included in this model emerge. The process of changing the carnival route is still under review with better data being collected each year. Many of the problems of using this model interactively with those who manage the event are being improved. As we gain more experience with this approach, we are better able to adapt this kind of model to different policy-making situations.

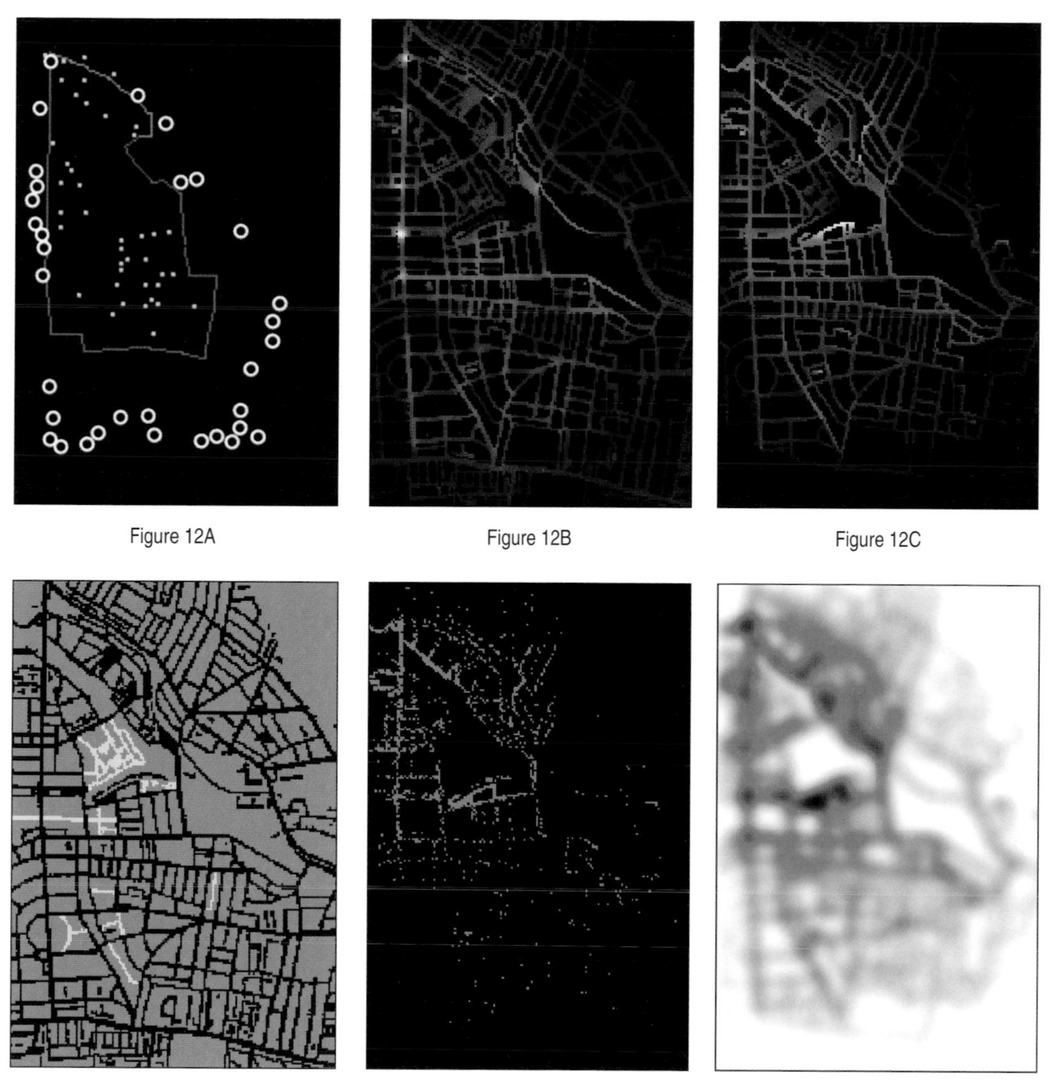

Figure 12A Figure 12B Figure 12C

Figure 12D

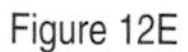

Figure 12E

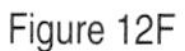

Figure 12F

Figure 12 The full modelling sequence and identification of vulnerable locations: A, the 2001 parade route (red and green) with proposed 2002 route in red, sound systems (yellow), and entry points (white); B, composite accessibility surface from stage one; C, traffic density from stage two; D, areas closed by the police used in stage three; E, location of walkers in the stage three steady state; and F, vulnerability of locations predicted from stage three, on a red scale (lightest red low to darkest red high: the white background are areas not relevant to such vulnerability).

6 The future: fine-scale dynamics and GIS

In this chapter we have shown how a focus on finer scales of granularity than GIS is typically concerned with creates problems of representation and dynamics which are hard to embody within traditional GIS. Clearly, the models we have been dealing with are part of an enhanced geographic information science but a focus on the fine, localised scales with human action and motion at the fore, reveals that traditional GIS is largely concerned with systems represented in terms of their inanimate characteristics. Insofar as human activities are represented in GIS, these are in static, aggregative terms. Moreover, the kinds of events that this chapter has been concerned with are much shorter lived than those that are traditionally a part of our science. This is partly because we are only just beginning to get a handle on fine-scale events of short duration but it is also because such events now seem more important than they ever were, as is witnessed in our concern for developing an appropriate science to deal with safety, crime and leisure.

This chapter also reveals a wider concern with GIScience. Spatial analysis and modelling form very different traditions and, although it is not surprising that much of GIS and GIScience has been developed by those who are involved with these other traditions, it has always been difficult to generate an appropriate fusion of spatial analysis and spatial representation. This has been managed in ad hoc and satisfactory ways for static aggregative phenomena, but once the concern shifts to dynamics and to fine spatial scales, then fusing traditional GIS with these kinds of models and analysis becomes an even greater challenge. The kind of fusion that has been attempted quite successfully by Evans and Steadman (this volume) for land use-transport models is not possible with the models of this chapter because the processes and scales involved are too different from the way traditional software in information systems has evolved. Nevertheless, what we have shown is that the fine scale of granularity and its dynamics are rich in detail and that the visualisation involved in making sense of these events is consistent with the traditions of GIS. We have raised many new issues in this chapter which we see as some of those to which GIS needs to respond to as it becomes increasingly adopted and adapted in urban policy making where physical design and social action are integral to relevant human action.

Section *II*

Data systems, GIS and the new urban geography

The new urban geography is fundamentally about powerful generalisation founded upon new rich spatial data infrastructures. It is also about clear thinking and informed judgment. Elena Besussi and Nancy Chin illustrate these ideas in relation to the global phenomenon of urban sprawl—a term which has both similar negative connotations and originating processes across the urban world, but which needs to be understood with reference to regional and local priorities as well. Sensitivity of urban simulation models to regional context is the theme of Joana Barros and Sinesio Alves Junior's analysis of rapid urbanisation in Latin America. The best policies require the best data, both of human activities and of built form. Mark Thurstain-Goodwin illustrates how GIS representations can inform the policy aspirations of decision makers while allaying the confidentiality and ethical concerns of data suppliers. With regard to built form, Sarah Smith describes how developments in urban remote sensing make it possible to strip away the built forms of cities, as a precursor to reintroducing artificial and vegetative structures in detailed representations of urban morphology.

Chapter

Identifying and measuring urban sprawl

Elena Besussi and Nancy Chin

While urban sprawl is recognised as rapid and uncoordinated growth at the urban fringe, there is little broader consensus as to its causes and consequences. Sprawl can be recognised in terms of urban forms, land-use patterns, and the related movement patterns that are necessary for urban areas to function. Several types of sprawl can be identified, namely, continuous suburban growth, linear/ribbon development, and scattered/leapfrog development. Yet sprawl cannot be defined through physical form alone. This chapter describes the qualitative and quantitative methods that we are developing for the comparative analysis of sprawl in Europe within the SCATTER Project. Qualitative methods consist of in-depth examination of the opinions of local authority practitioners gathered through interviews and synthesised by content analysis and concept mapping. Quantitative methods based on density and land-use patterns are also being developed from space-time series data that reflect changes in population, employment and mobility.

1 How to understand urban sprawl

Urban sprawl is generally agreed to be rapid and uncoordinated growth at the urban fringe. However, the exact meaning is always debatable for it is often used as a synonym for contemporary urban growth and as Ewing (1994) implies, it is often easier to define sprawl by what it is not. It is sometimes implicitly defined by comparison to the ideal of the compact city, and for the most part, emerges as its poor cousin. Urban sprawl has become a hot topic at the present time. This is largely because of the perception of it as a force eroding the countryside, thus marking the final passing from an urban-rural world to an entirely urbanised one with all the negative connotations that this implies for the visual environment, as well as a growing concern for the impacts posed to long-term urban sustainability.

These perceived negative effects are being tackled with growth management policies that attempt to restore a more compact urban form by channelling development to historical downtowns and attempting to set physical limits on growth through growth boundaries and land preservation. Confusion over the characteristics and impacts of sprawl stems largely from inadequate definition. One more reason for the lack of a clear definition is the theoretical pluralism surrounding the basis of the topic. Different perspectives overlap and provide unique insights and analytical approaches that range from qualitative methods to more formal quantitative research models. It is in this context that, in this chapter, we appraise the contribution of several qualitative and quantitative methods for identifying sprawl.

2 General definitions

As we have already implied, sprawl has become an umbrella term, encompassing a wide variety of urban forms. Given that there is no agreed comprehensive definition, it is not surprising that there is also little agreement on the characteristics, causes and impacts of sprawl. The various elements that feed into a definition of sprawl are discussed here as urban forms, land-use patterns and their impacts.

Definitions of form

A variety of urban forms have been covered by the term 'urban sprawl' ranging from contiguous suburban growth, linear patterns of strip development, leapfrog and scattered development, all being set within a pattern of clustered, non-traditional centres based on out-of-town malls, edge cities, and new towns and communities (Ewing 1994; Pendall 1999; Razin and Rosentraub 2000; Peiser 2001). In terms of urban form, discussion of sprawl is juxtaposed against the ideal of the compact city that has high-density, centralised development, and a spatial mixture of functions, but what is considered to be sprawl ranges along a continuum of more compact to completely dispersed development.

At the more compact end of the scale, the traditional pattern of suburban growth has been identified as sprawl. Suburban growth is defined as a contiguous expansion of existing development from a central core (Hall 1997). Scattered or

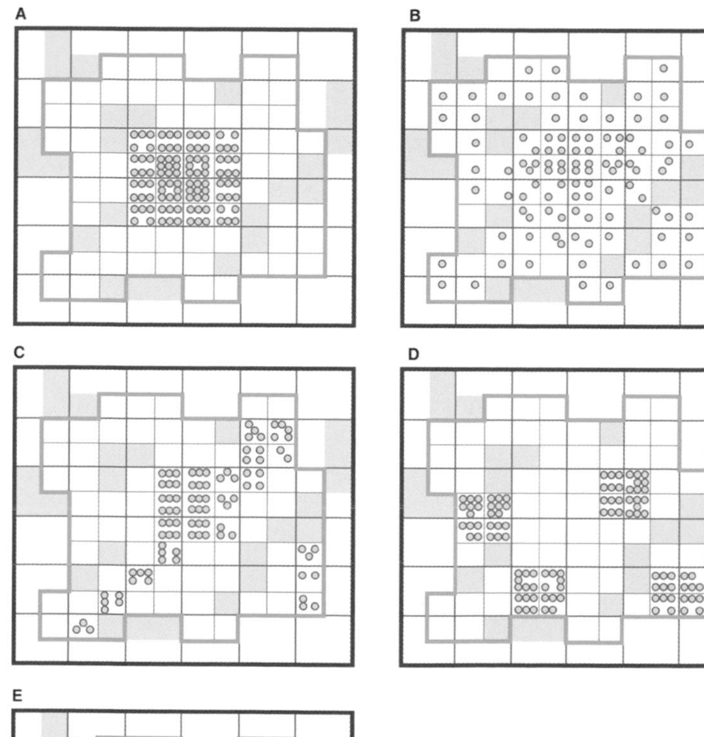

Figure 1 Typologies of urban development (after Galster et al 2001)

leapfrog development lies at the other end of the scale (Harvey and Clark 1965). The leapfrog form exhibits discontinuous development away from an older central core, with the areas of development interspersed with vacant land. This is generally described as sprawl in the current literature, although less extreme forms are also included under the term. Other forms that are classified as sprawl are compact growth around a number of smaller centres, located at a distance from the main urban core and linear urban forms, such as strip development along major transport routes. Figure 1 provides an illustration of these typologies of sprawl.

One problem with these definitions is that developments as diverse as contiguous suburban growth and scattered development are both classified as sprawl, however, the forms and resulting impacts are vastly different. It may, therefore, be more useful to define sprawl, not in any absolute way, but as a continuum of development, from compact to completely scattered. This idea is acknowledged by Harvey and Clark (1965) who identify three forms of sprawl: low-density continuous development, ribbon development and leapfrog development.

Definitions based on land use

Land-use patterns are the second element used to define sprawl. The Transportation Research Board (1998) lists the characteristics of sprawl that apply to the United States as low-density residential development; unlimited and non-contiguous development; homogenous single-family residential development with scattered units; non-residential uses of shopping centres, strip retail, freestanding industry, office buildings, schools and other community uses; and land uses which are spatially segregated. Sprawl can be further characterised by heavy consumption of exurban agricultural and environmentally sensitive land, reliance on the automobile for transport, construction by small developers and lack of integrated land-use planning.

The characteristics provided by the Transportation Research Board (1998) are broad and cover almost all post-World War II development in the United States. The authors themselves claim that 'sprawl is almost impossible to separate from all conventional development' (Transportation Research Board 1998: 7). Unfortunately, while this ensures that no aspect of sprawl is omitted, it does little to differentiate sprawl from other urban forms. Sprawl is most commonly identified as low-density development with a segregation of uses, however, it is not clear which other land-use characteristics must be present for an area to be classified as sprawl.

Definitions based on impacts

The third method of definition focuses on the impacts of sprawl. This idea was first put forward by Ewing (1994) and later by Razin and Rosentraub (2000). It provides an alternative to definitions based on urban form and is based on the idea that the distinction between sprawl and other forms is a matter of degree. Sprawl is acknowledged as difficult to distinguish from other forms, but the focus is upon the undesirable impacts of sprawl rather than sprawling form per se. Ewing (1994) has identified poor accessibility of related land uses and lack of functional open space as a way to identify and define sprawl. More generally, Ewing has suggested that sprawl can be defined as any development pattern with poor accessibility between related land uses and that it results from dispersed development of homogenous land uses.

The problem with a definition based on function is that it assumes there are negative consequences to sprawl. It is important to note that causal links between the form of the sprawl and these negative impacts is not always identified in research case studies, and there is a temptation to describe any development with negative impacts as urban sprawl. There is clearly a lot of confusion surrounding the definition and meaning of urban sprawl. Despite this diversity of forms and definitions, the usual assumption is that the typical urban form is monocentric (despite widespread evidence to the contrary) and that a single principal centre of economic and service functions is surrounded by residential development. Most definitions also focus on the density of development and distance from the city centre, rather than on land-use patterns.

3 Urban sprawl and sustainability

Urban sustainability has now been adopted as an explicit policy objective in many parts of Europe, including the European Commission (EC). In this policy field, the sustainable indicators movement is a rapidly strengthening force. One of the key problems facing policy makers is that although there are hundreds of different indicators to assess sustainability, 'there is no single, correct set of indicators which should be collected, no "one size fits all" solution' (Thurstain-Goodwin and Batty 2001; see also Lloyd et al, this volume; Thurstain-Goodwin, this volume).

A major strand of the sustainability debate focuses on whether cities can become more sustainable. In particular, much attention has focused upon the question of whether the arrangement of a city's physical elements affects its capacity to function in a sustainable way. Answers to this question are inherently ambiguous.

However, it is now generally accepted that there is a relationship between urban form and sustainability. 'It is clear that a major strategic factor determining sustainability is urban form, that is, the shape of settlement patterns in cities, towns and villages' (Breheny and Rookwood 1993: 151). What this precise relationship might be is less certain, and the debate on the sustainability of different city forms, grouped into compact and diffused models, is nevertheless still an open question. The complexity embedded in the concept of sustainability does not allow for a straightforward assessment of the different impacts of urban sprawl and hampers the definition of policy measures. Following the sustainability goals promoted by the European Union (EU), indicators assessing the impacts of urban sprawl have been grouped into three categories: environmental, economic and social. These indicators provide a useful framework around which to structure further research on the definition of urban sprawl.

Environmental Impacts

Land consumption

Land consumption depends on the relative compactness and density of human settlements. Orfeuil (1996) measured land consumption as the amount of land used by each inhabitant for residential purposes and showed how this has increased in the last 20 years by at least 200 per cent. However, there is no consensus over the use of this measurement method. Camagni, Gibelli and Rigamonti (2002) have calculated land consumption in urban development as the ratio of land area developed for residential and service uses between 1981 and 1991 in each commune of the Milan metropolitan area to the number of dwellings developed during the same period. 'This indicator was preferred to the per capita consumption of land because the latter may increase in cases where the population of a commune declines, giving a false indication' (Camagni, Gibelli and Rigamonti 2002: 205). The results of their analysis (Camagni, Gibelli and Rigamonti 2002: 206) suggested that by this measure, land consumption is actually declining rather than increasing.

Another factor to be considered is the consumption of land for road infrastructure, which accounts for 25 per cent of the total urban area in Europe and 30 per cent of that in the United States. Research carried out in the Paris region showed that the private car, which accounts for 33 per cent of total trips, consumes 94 per cent of road space, while the bus, with 19 per cent of total trips, consumes only 2.3 per cent. In other words, a moving bus consumes 24 times less space per passenger than a single car (Servant 1996).

Energy consumption

Energy consumption depends on the characteristics of mobility patterns such as trip length and modal choice between private and public means. Per capita levels of gasoline consumption have increased constantly since the late 1970s. Opinions about the risks of depleting this non-renewable energy source through urban sprawl tend to differ. However, both the United Nations and the European Union have moved in favour of the compact city model embracing the position, supported by research (Newman and Kenworthy 1999), that the densest cities consume the lowest amount of energy for public and private transport.

Atmospheric pollution

Researchers have proved that atmospheric pollutants have dramatically increased the level of risk for human health. However, it is hard to establish a direct connection between urban density and form and increases in the amount of atmospheric pollution. Studies have sometimes suggested conflicting results. Some support the hypothesis that more compact city forms limit the number of journeys and the length of trips and suggest that dense areas produce far lower per capita emissions than less dense areas. Yet according to Burton (2000), it is possible that the compact city may present a health risk due to localized air pollution, particularly from traffic, but also from the closer proximity of residential and industrial uses. Irrespective of the precise implications of such results for well-being, it is clear that measures of density alone are insufficient to explain observed levels of pollution. There is, thus, a need to investigate further the relationship between density level and pollution in order to understand which urban activities should be more concentrated.

Economic sustainability

The economic sustainability of the dispersed city model must be addressed at two different scales. Urban sprawl tends to impose several different and often hidden costs (notably transport costs) on individuals and households. A study of the area of Ile-de-France has shown strong positive correlation between the distance from the city centre and the percentages of household budget devoted respectively to housing and transport (Guérois 2002). (The farther a family lives from the city centre, the more they spend on housing, including utilities, and transportation.) Urban sprawl is often associated with high costs of urbanisation, such as the realisation and supply of network utilities and public services. The low density of housing and population within a scattered pattern of urbanisation undermines the economic feasibility of public services, especially transport services.

At the macro-economic level, this has implications for issues of economic efficiency and economic performance of cities even though the debate remains largely theoretical. It is difficult to establish a causal link between the size and density of cities and their economic efficiency. Recent studies (Prud'homme and Lee 1999; Cervero 2001) indicate that places with sprawling, car-based landscapes are poor economic performers, while the economic advantages of agglomeration and higher employment densities still persist for large cities where various innovations (notably new information and communication technologies) can help to overcome the restrictions on growth related to congestion.

Social impacts

In European cities mostly affected by dynamics of suburbanisation and sprawl, urban structure has developed according to clear patterns of social ecology. Population distributes spatially according to age, family size, social class and professional class. The degree of spatial segregation between households varies sharply as one moves from the central city towards the suburbs, and there is evidence that this has greatly accentuated during the 1980s, particularly in France and Britain (Gober 1990). The European city, the very crucible of social interaction, innovation and exchange risks losing its fundamental role as a result of the cumulative effect of these tendencies to decentralisation, increasing specialisation of land uses, and greater social segregation.

However, European cities do not completely conform to the North American model, according to which the lowest and highest income/social classes are located at the urban core, while middle class households characterise the suburban areas (Hall 1993). Moreover, there are additional differences depending on the size of cities. Large European cities display a different population distribution pattern from medium-size cities and seem to conform more to the North American model, while smaller cities show lower levels of spatial segregation and significant trends of social mixing and inclusion in their suburban areas. Studies for the Paris area and other minor French cities (Berger 1999) have demonstrated that professional qualifications, household size, and income are amongst the variables, which can also best explain location choice.

4 The SCATTER project

Urban sprawl cannot be diagnosed through an analysis of urban form alone, because as we have argued, it is comprised of a combination of form, land-use patterns, and their linked impacts. One project that addresses the meaning of urban sprawl in Europe and ways to deal with it is the SCATTER Project (Sprawling Cities and Transport: from Evaluation to Recommendations) which is part of the EC's Fifth Framework Programme: Energy Environment and Sustainable Development (Key Action 4 'City of Tomorrow and Sustainable Development'). The SCATTER project addresses sprawl through both qualitative and quantitative analysis and is being demonstrated for a sample of key cities in the European Union based on a consortium of six European partners in Belgium, Germany, the United Kingdom, Finland, France and Italy.

The qualitative analysis focuses on an in-depth study of the local authorities in the six case cities of Brussels, Stuttgart, Bristol, Helsinki, Rennes and Milan. The study entailed qualitative interviewing with local authority representatives and experts in land-use and transport planning and collecting associated descriptive information. The purpose of the interviews was to detect and understand the

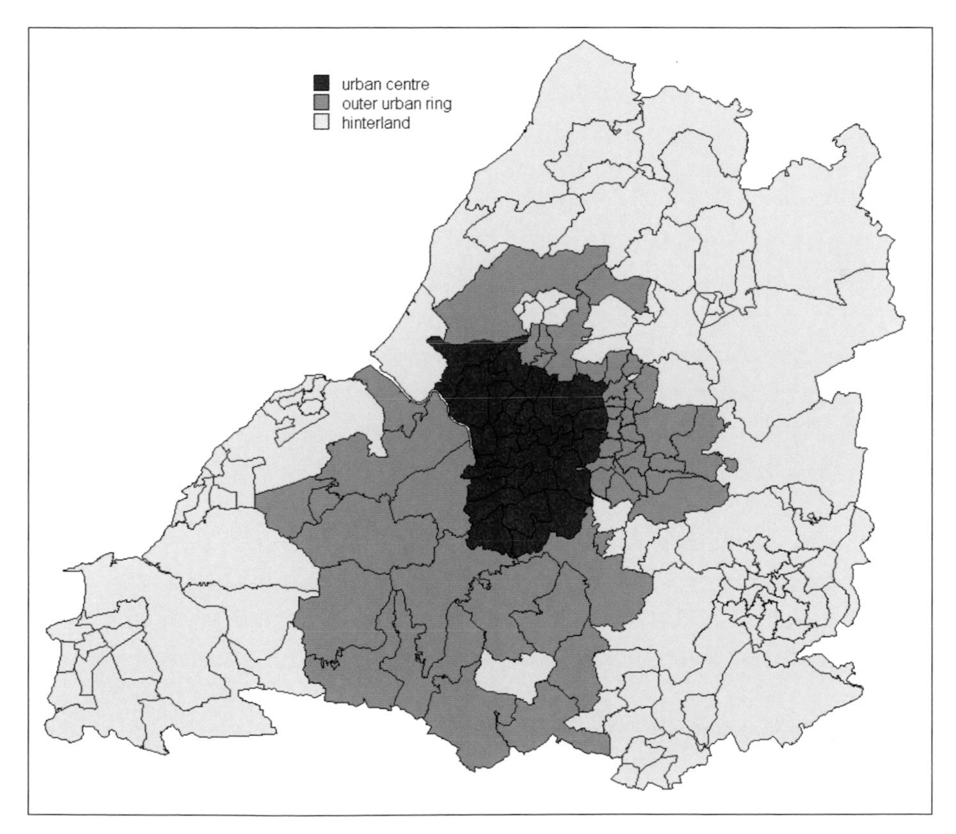

Figure 2 Urban zones in the Bristol region

local events and rationale involved in the emergence of urban sprawl, its relevance in the decision agenda of local authorities and experts, and the overall level of awareness of this particular urban phenomenon. The contribution of experts was valuable with regard to the urban system analysis, the description of the geographical, political, socio-economic, institutional context and the history of the urban sprawl, and especially with regard to the interactions between all these aspects in a global and systemic way.

Although it is difficult to assimilate these issues in terms of all their interactions through an exclusively statistical or quantitative analysis, a second key strand to the project is such statistical analysis of spatial change in small areas for each of the six cities. This is based on a statistical analysis of a before and after database that measures sprawl primarily through a shift-share analysis, in order to highlight the attractiveness or otherwise of particular zones for residential location or employment. The method divides each region into zones based on commuting patterns, which we illustrate for the Bristol region in figure 2. The indicators derived are temporal mean-growth rates, growth-concentration ratios, and overall indicators for economic development effects. In the following sections, we will discuss the preliminary findings from the SCATTER interviews followed by a discussion of other potential methods for quantitative analysis.

5 Typologies of sprawl

The different configurations of urban sprawl in European countries are often described as a local response to demographic, economic, social and policy trends. The emphasis is upon dynamic interactions between the pre-existing socio-economic and urban structures, which often display extraordinary levels of persistence (Batty 2001c) and impacts of the waves of urban growth and decline on the areas surrounding the city. Observed spatial and functional urban patterns are the results of these interactions.

However, only some of the spatial manifestations of urban growth in Europe may be described as sprawl. Rather than being an all-inclusive concept, as in most of the United States-based literature, sprawl is used more selectively as a label to describe some of the planned and existing types of urban growth. Disagreement, discordance and uncertainty remain in the literature on European case studies. Different classifications of development patterns have been suggested. Camagni (2002: 205) uses levels of land consumption as a quantitative measure to classify different types (T1 to T5) of urban development patterns:

	T1	T2	T3	T4	T5
T1	pure in-filling				
T2	in-filling /extension	pure extension			
T3	N/A	extension/ linear development	pure linear development		
T4	in-filling/ sprawl	extension/ sprawl	linear development/ sprawl	pure sprawl	
T5	N/A	N/A	N/A	N/A	large-scale projects

Source: Camagni, Gibelli and Rigamonti 2002

Table 1 Types of urban development patterns

- (T1)—in-filling, characterized by situations where the building growth occurs through the in-filling of free space remaining within the existing urban area;
- (T2)—extension which occurs in the immediately adjacent urban fringe;
- (T3)—linear development that follows the main axes of the metropolitan transport infrastructure;
- (T4)—sprawl that characterizes the new scattered development lots; and
- (T5)—large-scale projects, concerning the development of new lots of considerable size that are independent of the existing built-up urban area.

Through the combination of these simple and theoretical typologies of land uses, Camagni has identified ten prevalent urban forms (table 1).

The expansion of individual housing in the suburbs of many European cities throughout the 1970s and 1980s has created vast zones of low elevation and low-density urbanisation in the outer urban areas. These are sometimes planned as development areas and built as private housing estates, sometimes as individual houses built in a more scattered fashion. These developments can comprise not just residential activities but also retail and service buildings or industrial zones. Levels and spatial distributions of density differ depending on regional and local variations. In search of these variations, a second approach has widened the analytical framework to include the distribution and organisation of land uses and urban functions. As a result, different patterns of urban development have been identified: mixed- or single-land-use patterns, patterns of different rural-urban relationships, or compact, clustered or dispersed patterns as is shown in table 2.

Figure 3 Garden suburbs in France

Preliminary results from the qualitative investigations carried out for the SCATTER Project reveal quite different descriptions of urban sprawl by policy makers and local experts. One objective of the project is to understand the local events involved in the emergence of urban sprawl and its mechanisms. In order to gain this qualitative knowledge of sprawl, interviews carried out with local authorities and experts from the six case cities highlighted a wide range of factors responsible for the location and relocation choices of population and productive activities. For the purpose of synthesis, all identified factors have been grouped into four main categories: planning regulations and interventions, the changing structure of population and households, the structure of employment and of the economic sectors, and the negative externalities associated with urban agglomerations. This is shown in table 2.

The interviewees also share a perception of urban sprawl as the one type of urban and living environment characterised by a lack of factors that can contribute to a good quality of life. These lacking factors include the aesthetic quality and the level of maintenance of the built environment, the supply and standard of the public spaces, and the quality and typology of settled functions. The following sections describe the three main typologies of sprawl that have been identified through the analysis of the interviews.

Public policies and plans	The structure of population	The economic sector	Negative and positive externalities
• Regulations restricting or promoting new building development in selected areas • Location of business, industrial or commercial centres • Investment and realization of public transport infrastructures and services • Regeneration plans in the central areas • Regulations and fiscal measures on housing typologies	• Diversification and fragmentation of household's typologies • Increase of incomes • Diversification of lifestyles and work-styles (more flexibility) • Emergence of new ideological and cultural trends with regards to living and housing environments	• Restructuring and relocation of industrial activities • Growth of the service and business activities and decline or restructuring of agriculture	• Reduced access to services in urban agglomerations • Increasing pollution and criminality in urban centres • Increasing land values in central areas • Lack of open and green spaces in urban centres • Reduced travel times and travel costs from urban centres to peripheral areas • Increased accessibility of peripheral areas

Table 2 Causal factors of location dynamics

The sprawling residential suburb

This category of urban sprawl is characterised by an in-fill process where scattered and low-density housing developments are located between centres or between transport infrastructures that usually are marked by wedges between transport corridors. The quality of housing and of the residential environment is high, but there is a limited supply of open spaces and public services, the latter mainly composed of commercial centres that make this type of sprawl similar to the American suburb (figure 4). The impacts of the sprawling residential suburb are mainly on mobility patterns and can be attributed to the scattered and low-density nature of these developments.

This type of sprawl is seen principally as the result of a transition from altruistic, collective attitudes to more individualistic, self-gratifying behaviours connected to lifestyles and work-styles. This is demonstrated by the fact that, despite the lack of a sense of community and of urban identity, these residential neighbourhoods are still the preferred location of the majority of young families with children and

Courtesy of Corbis

Figure 4 Sprawling residential suburbs in the United States

a medium-to-high income (see Webber and Longley, this volume, for a discussion of geodemographic classification of household types).

The sprawl of deprived peripheries

The peripheral areas of main and secondary centres, left aside by regeneration policies and new planning interventions and built by public housing policies during the early stages of deconcentration, have been the main attractors to groups of population looking for residential areas at low prices and within affordable travelling distances from the main centre. In some cases, these population groups are comprised of illegal immigrants, students and elderly households who do not have access to the residential suburbs described above. In other cases, the process of peripheral growth that occurs in the first ring of available areas around the urban centres involves a wider range of types of population and households. This happens particularly in those situations where geographical infrastructure or planning constraints limit accessibility to suburban areas far from the historic urban centre. This type of sprawl, characterised mainly by high densities and deprived social and physical environments, takes the form of the public housing estates of the 1950s and 1960s, as we show in figure 5, as well as of the new

Source: Dal Pozzolo 2002:106

Figure 5 Public housing sprawl

Source: Dal Pozzolo 2002:106

Figure 6 Contemporary peripheries

Figure 7 Commercial centres in Northern Italy

peripheries shown in figure 6, often poorly designed by a private sector seeking speculative interventions.

Commercial strips and business centres

According to the interviewees, urban sprawl manifests itself in the form of commercial, service and business centres outside the boundaries of the compact city. These developments are mainly the result of a planning approach that sees land uses as mutually competitive. This approach can produce a system of market forces, which plays a critical role in determining land-price mechanisms. The location of these land-use functions follows a rationale based on accessibility, low cost of land, and agglomeration economies. As a result, these activities locate close to transport infrastructure like airports, ports and motorway junctions. Also, this sometimes happens in consequence of public or private/public planning decisions. Figure 7 shows the location of commercial centres for Northern Italy.

6 Quantitative measures of sprawl

From the preceding analysis it is clear that the term 'urban sprawl' is a multifaceted notion, and as such it is often difficult to identify such sprawl on the ground, and even more difficult to link this particular development form to outcomes affecting the quality of life of urban inhabitants. The use of quantitative measures of sprawl enables empirical investigation of sprawl's characteristics and consequences, and assists in the assessment of planning policies tackling sprawl. There is little work specifically targeted to measuring sprawl, and the majority of current work characterises sprawl in terms of density and urban form. It should be reiterated that low density is a necessary but not sufficient condition for identifying sprawl, that density is itself an incompletely specified concept, and that land-use patterns are an equally important component in identifying urban sprawl.

Average density

Urban sprawl or urban decentralisation is almost always defined as consisting of low-density development. However, density is often neither quantified nor adequately specified and would be better served by a more rigorous definition. The simplest measure is average density, which provides some indication of the intensity of land use and represents the relationship between the number of people living in or using an area and a given land area. Average density, however the numerator and denominator terms are defined, is not the most useful indicator of sprawl as it tends to smooth out patterns of distribution within the land area being measured and may cover up patterns of decentralisation.

When measuring urban sprawl, it is more appropriate to use residential units as the numerator rather than population since this is a better unit for measuring physical land use. The most useful term for the denominator would include all developable land, including residential, industrial, institutional, service, commercial, vacant land in leapfrogged tracts and agricultural land that has been withdrawn from active use for land speculation. Agricultural land that is in farm use, parks and land unsuitable for building, such as marshy land, would not be included as there is no immediate potential for development.

Density gradients

Average density measures often do not provide a direct discriminator of changes in density over space at scales that are appropriate to intra-urban analysis (Longley and Mesev 2002). Attempts to estimate these changes in density have been developed based on mathematical models. One classic method is the negative exponential density function popularized by Clark (1951). The main assumptions of the model are that the urban area is monocentric in form with a peak in density near the centre just outside the central business zone (the central zone itself has few residents and a very low density), with a progressive decrease in density to the outer areas of the city. This may have provided an appropriate fit in 1951 when

Clark was writing, but it is less clear today whether this provides an appropriate representation of change in density. Recent work questioning the assumption of a monocentric form has looked at the rise of the polycentric city. Anas, Arnott and Small (1998) have generalized the monocentric negative exponential density function to a polycentric city.

One difficulty with polycentric density functions is that regional change in the patterns of dispersion over time may not be clear if the subcentres change in different directions: for instance, some subcentres may show a steeper gradient with less dispersion, and others a more shallow gradient. Additional attempts to model changes in density using mathematical models have been conducted by Batty and Kwang (1992), who have investigated the applicability of the inverse power function. This tends to overpredict central densities, however, it is a better predictor of density for areas at the periphery of cities, providing a higher density than the negative exponential model. Longley and Mesev (2002) discuss the change in

	Galster et al 2001	Malpezzi and Guo Wen-Kai 2001	Spatial Statistics	Landscape Ecology
Continuity (measure of vacant space), for example, leapfrog development	X	X		X contagion
Concentration (spread of housing over total urban area), for example, ribbon development	X			X isolation of patches; dispersion of patches
Clustering (spread of housing over ¼ mile sq area)	X			
Centrality (location to the CBD)	X	X		
Nuclearity (mononuclear vs polynuclear development)	X		X G Statistic	

Table 3 Summary of measures of urban sprawl

density gradients over time, noting that the gradient may not decrease over time in the areas of the city where growth is complete; in this case, the gradient remains constant.

7 Configuration of development

In addition to identifying sprawl based on density, measures have been developed based on form of development put forward by Galster et al (2001) and Malpezzi and Guo Wen-Kai (2001). The measures put forward by Galster examine patterns of development based on density of housing units or population. Table 3 provides a summary of possible measures for these dimensions.

The measures developed by Galster et al (2001) decompose patterns of urban sprawl into eight components: density, continuity, concentration, clustering, centrality, nuclearity, mixed use and proximity. Low scores indicate more sprawling areas. These measures are very useful and target the major dimensions of sprawl, however, further work remains to create useful definitions and indicators of sprawl. We suggest combining the measures using factor analysis to provide indicators of specific types of sprawl. This would provide the first step to a more accurate linkage of the impacts of urban sprawl to particular development forms. Additionally, more cities both in the United States and Europe are recognised as being polycentric in nature, notwithstanding the fact that certain aspects of the measures are still related to the idea of the monocentric city. It would be useful to expand the range of measures to apply to polycentric cities.

The above measures focus on density and urban form rather than the functional uses of the city. However, crucial components in defining urban sprawl are the concepts of mixed uses and the accessibility of uses. The concept of a mixed land-use city entails: a varied and plentiful supply of facilities and services, with residential uses balanced by compatible non-residential uses; horizontal mix of uses through the urban area, focusing either on existing or new housing developments; the development of mixed subcentres and an improved mix of uses near public transport nodes; and the vertical mix of uses within buildings (Burton 2002). Work by Burton (2002) has measured mix of uses based on seven key facilities: newsagents, restaurants and cafes, takeaways, food stores, banks and building societies, drugstores and doctors surgeries. They investigate the distribution of those facilities through the city and their location relative to residential developments. The focus here, however, is on services rather than the integration of employment with residential uses, and while these indicators attempt to measure the spread of services through the city, there is no measure of their accessibility, another key characteristic of urban sprawl.

9 Conclusion

Urban sprawl is clearly a major barrier to achieving cities that are sustainable in terms of their transport and energy uses. While it is evident that sprawl is not sustainable in these terms, it is not clear what kind of urban form could develop to produce sustainable urban structures. In a sense, sprawl represents the essential conundrum of contemporary urbanisation and urban growth in that the link between higher living standards and the way these are translated into physical patterns of living in cities is almost impossible to gauge. There is even the notion that if the market for urban land that leads to sprawl is too distorted by planning policy, this may have severe and negative impacts on living standards and economic growth. Work in the SCATTER project is contributing to making more transparent links between city form and sustainability with the first step of identifying the typologies of sprawl found in Europe. What has emerged from interviews with local experts in six case cities is the suggestion that there are three types of sprawl: the sprawling residential suburb, the sprawl of deprived peripheries and the sprawl associated with commercial strips and business centres.

Having achieved the first project objective of devising a robust conceptual classification, the second goal is to use these concepts to classify urban forms on the ground. One set of tools that can assist with this is the use of quantitative measures of the kind we are developing at CASA based on growth and change at the small area scale. A more robust identification of urban form also holds the key to more accurate assessment of associated impacts and evaluation of policy measures. The development of quantitative measures of sprawl is still in its infancy, but useful measures have been developed, based on density, configuration of residential uses, and mixture of land uses which identify various elements of sprawl. Further work could expand this by creating composite indices that combine the measures of individual elements and provide measures based both on form and land-use patterns. Current measures are still largely targeted on the monocentric city, and so it is important to adapt these to deal with the kind of urban sprawl characteristic of polycentric urban regions, which are more typical in Europe than in North America.

Chapter

7

Simulating rapid urbanisation in Latin-American cities

Joana Barros and Sinesio P. Alves-Junior

The high rates of urban growth in Latin America during the 1960s and 1970s produced rapid urbanisation and severe housing problems. Planning policies have approached urban growth as a static problem rather than as a spatial form that emerges from the urban development process, part of an ongoing dynamic process. This chapter focuses on a specific kind of urban growth that occurs in Latin-American cities, called 'peripherisation', which is demonstrated by an agent-based simulation model. Preliminary results suggest that the actual development process is a manifestation of socio-economic inequality that is reproduced in space by the locational process. The simulation exercises allow us to articulate the problem of urban growth in Latin-American cities through their dynamics, thus presenting a new and important perspective on the phenomenon.

1 Introduction

Rapid urbanisation has been the main theme of urban studies in Latin America since the explosion of rates of growth in the 1960s and 1970s. While studies predicted an unprecedented rate of growth in these cities by the year 2000, the speed of development was blamed as the cause of spatial inequalities and housing problems. Yet, the actual rates of growth have slowed since the 1980s, and current studies forecast that growth rates will remain as such. The urban problems, however, have not disappeared in the last two decades, and, despite the currently lower rates of population growth, cities keep expanding and developing in the same way.

The present study looks at issues related to the rates of growth of Latin-American cities by investigating their actual mode of urban growth. The aim of this chapter is to explore the idea that not only has urban growth in Latin America been more rapid, but these cities grow in a fashion different than cities in the United States or Europe. It is this dynamic process that must be studied further to better support planning decisions. We use urban modelling techniques to study growth dynamics in order to unpack the problem of urban growth in Latin-American cities. Using very simple simulation exercises, we will show that some of the most important issues regarding urban growth in Latin-American cities can be investigated in new and innovative ways.

We present the Peripherisation Model, a simulation model that explores a specific mode of urban growth characteristic of Latin-American cities. We then relate these experiments to real examples using maps of two Latin-American cities, São Paulo and Belo Horizonte, Brazil. Finally, we revisit some of the assumptions about rapid urban growth based on the experiments presented and show how studying these themes can assume different perspectives when viewed dynamically.

2 Rapid urbanisation in Latin-American cities: an overview

While the problem of urban growth in Europe and North America has been formulated in terms of sprawl (as discussed by Besussi and Chin, this volume), in the Third World and more specifically in Latin America, the main focus has been on the rapidity of growth of cities. Indeed, Latin America has seen the fastest urban growth in history. During the period between 1950 and 1980 growth rates were dramatically high (see Hall 1983; Valladares and Coelho 1995) and, based on these data, studies anticipated continuing high rates of growth. It was believed that many areas would double in population and a few would triple, creating urban areas that by the year 2000 would be without parallel in history (Hall 1983). Latin-American countries went from being predominantly rural to predominantly urban within a couple of decades, with high concentrations of urban population in cities with more than one million inhabitants (UNCHS 1996). This rapid urbanisation produced various kinds of social problems, especially in terms of housing, since the governments of such countries did not manage to provide enough housing and urban infrastructure to accommodate all the migrants who fuelled the growth of these cities.

However, this population trend has changed since 1980. After decades of explosive urbanisation, urban growth rates have slowed, the rate of metropolitan growth has fallen and fertility rates have declined (Valladares and Coelho 1995). Moreover, rural to urban migration has come to assume a much smaller role in urban population growth and, most recently, the pace of urban expansion has been largely maintained by births in the cities. These new trends have been detected in the period between 1980 to 1990, and have been confirmed by recent censuses.

It is important to stress that urbanisation in Latin America as a whole has not been a homogeneous process and to recognise the differences that exist between Latin-American countries. Virtually all the cities in Latin America with a million or more inhabitants had much slower population growth rates during the 1980s than the average for the period 1950 to 1990. However, the major cities in the southern cone accounted for the slowest rates of population increase during these four decades (UNCHS 1996). The southern cone countries of Argentina, Chile and Uruguay reached high levels of urbanisation in the 1950s and this process has slowed down in the last decades (UNCHS 1995).

The changed demographic trends have had impacts on urbanisation patterns. Such patterns can be defined mainly by a process of population deconcentration including a fall in the overall rate of population growth, a reduced concentration of population in the core of metropolitan areas coupled with significant growth of small and medium-size municipalities, and a declining rate of demographic growth in regional capitals and major urban centres. The principal problem of urban growth is no longer the high rates of population growth and rural-urban migration. Rather, it is the spatial pattern of growth—the peripherisation process—which enlarges the peripheral rings of cities and metropolises despite reductions in overall urban growth rates. The peripherisation phenomenon is becoming an increasingly significant issue, particularly in the larger cities of Latin America. In those cities, the demographic growth rate has slowed down, migration has taken second place to natural increase, and the bulk of the housing stock now consists of upgraded (or in the process of upgrading) low-income residential areas, with a large number of spontaneous settlements.

The phenomenon of peripheral growth, which has been recognised by Latin-American researchers and planners and termed 'peripherisation', can now be considered the established process of growth of most Latin-American cities. Peripherisation can be defined as a growth process characterised by the expansion of the borders of the city through the massive formation of peripheral settlements which are, in most cases, low-income residential areas. These areas are incorporated to the city by a long-term process of expansion where some of the low-income areas are recontextualised within the urban system and occupied by more wealthy (higher status/income) economic groups, while new low-income settlements keep emerging on the periphery.

Peripherisation is an urban spatial problem with strong social and economic effects and that is unlikely to be alleviated or reduced without informed planning action. The peripheral ring in Latin-American cities consists mostly of low-income housing including large spontaneous settlements, which usually lack urban services of any kind. As such, peripherisation clearly constitutes a social problem. However, it is not simply a problem of extreme social inequalities appearing in the city in a very concrete spatial form. Rather, the problem is the perpetuation of such a form in space and time and, in this sense, peripherisation is a social problem of spatial order as shown in figure 1.

To better understand the process of peripheral growth, it is necessary to study the physical/spatial elements that comprise it and the dynamics of its formation, growth and consolidation. The study of spontaneous settlement in the global context of urban growth is important not only for the purposes of planning but also for understanding the nature and scope of urban problems in developing countries. The need to discuss the dimensions of urban growth in developing countries along with the implications of the corresponding mushrooming of spontaneous settlements has been addressed by UNCHS (1982).

As a static phenomenon, spontaneous settlements are seen only as a housing problem. From a dynamic perspective, however, they present a problem of urban development. So far, static analysis by both the research community and planners has been the major approach to understanding spontaneous settlements, and the connection between spontaneous settlements and urban growth has attracted

Photograph courtesy of Flavio Laurenza Fatigati

Figure 1 Aerial view of São Paulo, showing spontaneous settlement 'Favela Paraisopolis' and high-income neighbourhood 'Morumbi' in the background. August, 2000

little, if any, attention. The high rates of growth have been seen as the main cause of the formation of spontaneous settlements in Third-World cities, together with the inability of the state to deal with rapid urbanisation.

Although the formation of spontaneous settlements is considered a consequence of high rates of growth in recent years, its persistence, despite the recent slow down of these rates in Latin-American cities, suggests a consolidation of this process as the normal mode of urban growth in those cities. It must be noted that what is considered here is the consolidation of the process rather than of the spatial structure. This means that the topological structure of location (core-periphery) remains the same, while the spatial location of the periphery is modified in a constant movement towards the city's borders. Spontaneous settlements keep moving and expanding the urban frontiers as a consequence of the core's development. The spatial structure can be considered as 'a pattern in time' (Holland 1995), since it is a dynamic phenomenon in which the spatial pattern is constantly being reproduced.

In terms of urban planning policies, the phenomenon is also seen from a static point of view. The focus of government interventions is still on the local/housing scale, upgrading settlements and providing housing tracks for the low-income groups. There has been no focus either on the dynamics of the process or on the linkage between the local and the global scales, that is, the overall growth of the cities, which has been seen as a mere result of a demographic phenomenon.

Peripherisation, like urban sprawl, is a suburbanisation phenomenon. Whilst urban sprawl has been studied in detail and its main features seem to be broadly understood, in Latin America's case the need to understand the peripherisation process remains a central issue. In contrast to sprawl, which is an inherently spatial problem, urban peripherisation is essentially a social problem with spatial character. From a social point of view, peripherisation is not a simple problem to solve, neither is it in the hands of planners to attempt it. As a spatial problem, and more specifically as an urban development problem, the phenomenon requires considerably more investigation.

Like urban sprawl, peripherisation is fragmented and discontinuous development. It also presents problems related to urban sustainability, transportation, and the cost of infrastructure and urban services. Studies from the 1970s suggest that the lowest densities in Latin-American cities are found in high-income residential areas, the highest densities in middle-class areas, and the densities of spontaneous settlements somewhere between these two (Amato 1970). Finally, an interesting difference between urban sprawl and peripherisation is that, while urban sprawl is directly related to people's preference for suburban settings, peripherisation is not a direct consequence of locational preference. On the contrary, people who move to the city's border do not wish to live there but are impelled to do so.

3 Simulating urban phenomena

Urban systems have been traditionally studied using modelling techniques. Since the 1960s a number of models have been developed and, more recently with advances and the popularisation of computer tools, the possibilities of exploring urban systems in terms of their dynamics have increased considerably.

The computer has become an important research environment in urban geography. In urban applications, the use of automata-based models, more specifically cellular automata (CA), has replaced traditional transport and land-use interaction models, shifting the paradigm of urban models towards a complexity approach. The idea of a structure emerging from a bottom-up process where local actions and interactions produce the global pattern has been widely developed ever since automata-based models proved to be useful for a number of different urban applications. Cellular automata models have been successfully built as operational models, that is, for real-world applications (see Almeida et al in press; White and Engelen 1997) as have heuristic descriptive models, with their more theoretical objectives (see Batty 1998; Portugali 2000). Agent-based models have been used to study a number of different aspects of urban phenomena, from pedestrian movement (Batty, Desyllas and Duxbury 2002; Schelhorn, O'Sullivan and Thurstain-Goodwin 1999) to spatial change (Torrens 2001a; Barros and Sobreira 2002), as discussed elsewhere in this book (Batty, this volume; Torrens, this volume).

In the next section we will present some experiments with an agent-based model, which has been used to investigate different aspects of urban growth in Latin America. As discussed by Torrens (this volume), an agent-based model is an automata-based model based on cells in which the transition rules of the cellular automata (CA) are replaced by actual decision-making rules. For the purposes of the present chapter, we assume that an agent-based model is an automata-based model that allows us to explore three kinds of interaction: agent-agent, environment-environment and agent-environment. Cellular automata models explore only the spatial layer of the city (environment-environment) and, although transition rules are often based on representations of human decision making, this representation is not explicit. Agent-based models understand that human decision making plays a major role in urban processes and change, and we argue that the fact that agent-environment interactions are explicit in the model opens up an avenue to analyse the dynamic processes that link spatial development with social issues. This kind of analysis is of fundamental importance in cases of strong social differentiation, as in urban development in the Third World.

4 The simulation experiments

The aim of our simulation model is to develop heuristic-descriptive models of the decentralised process underlying the process of growth in Latin-American cities. Urban growth can be explored through a number of different perspectives. In particular, the intention is to develop and investigate a peripherisation model that is sensitive to the particular mode of urban growth that is characteristic of Latin-American cities.

To simplify the behaviour rules as much as possible, we based the model on the agent-environment relationship and we do not explore environment-environment or agent-agent relationships. The model was built with essentially speculative objectives; that is, it focuses on how simple local rules generate emergent and complex structures—in other words, how individual actions of people searching for a place to settle their own houses create complex spatial structures. The model was built on the StarLogo platform, a user-friendly parallel programming tool developed by the Epistemology and Learning Group at the Massachusetts Institute of Technology (education.mit.edu/starlogo).

Peripherisation can be defined as a kind of growth process characterised by the expansion of borders of the city through the formation of peripheral settlements, which are, in most cases, low-income residential areas. These areas are incorporated into the city through a long-term process of expansion in which some of the low-income areas are recontextualised within the urban system and are occupied by a higher economic group while new low-income settlements keep emerging on the periphery.

The model represents the process of expulsion and expansion by simulating the locational process of different economic groups in an attempt to reproduce the residential patterns of these cities. In the model, the population is divided in distinct economic groups according to the pyramidal model of distribution of income in Latin-American countries. This distribution suggests that society is divided into three economic groups (high, medium and low income) where the high-income group are a minority on the top of the triangle, the middle-income group are the middle part of the triangle and the low-income group is on the bottom of the triangle.

Our simulation model assumes that, despite these economic differences, all agents have the same locational preferences, that is, they all want to locate close to the areas that are served by infrastructure, with nearby commerce, job opportunities and so on. In Third-World cities these facilities are found mostly close to the high-income residential areas, reflected in the model's rules where the agents look for a place close to a high-income group residential area. What differentiates the behaviour of the three income groups are the restrictions imposed by their economic power. Thus, the high-income group (represented in the model in red) is able to locate in any place of its preference. The medium-income group (in yellow) can locate everywhere except where the high-income group is already located; and, in turn the low-income group (in blue) can locate only in otherwise vacant space.

In the model, agents are divided into three breeds (colours) in proportions based on the division of Latin-American society by income. All the agents have the same objective, that is, to be as close as possible to the red patches, but each breed (economic group) presents different restrictions on the places they can locate. Since some agents can occupy another agent's patch, it means that the latter is 'evicted' and must find another place to settle.

In what follows we present some of these experiments in which we have tested the behaviour of the model with different parameters. Two main parameters define the behaviour of the peripherisation model: 'steps' and 'proportion of agents per breed'. Step is the number of actual steps, that is, the number of pixels that the agent walks before trying to settle in a place (patch). Experiments were carried out testing different parameters in order to analyse the behaviour of the model.

In figure 2, we show the sequence of snapshots of the peripherisation model with different values for the parameter 'step'. The sequence of snapshots were taken with the same initial condition, that is, only one seed on the coordinates (0,0) and the same proportion of agents per breed, that is, 10 per cent red, 40 per cent yellow and 50 per cent blue. It is interesting to notice that the different steps determine a different spatial development with the same number of iterations (variable time—t). Experiment 2 presents a more dispersed spatial development than experiment 1, with less homogeneity within the patches of colours. Also, it tends to develop more rapidly in space than experiment 1. This arises because the larger the step, the more empty spaces are left between patches, making the search virtually easier; that is, the agents find an appropriate place to settle faster.

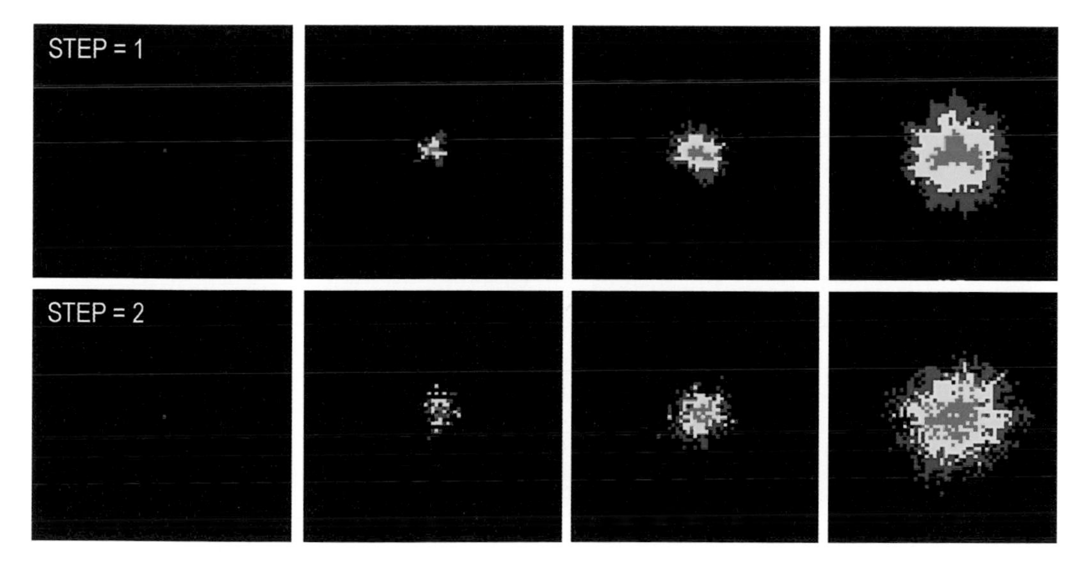

Figure 2 Sequences of snapshots from Peripherisation Model using different values for the parameter 'step'

We also tested the behaviour of the model with different proportions of agents per breed. Each scenario respects the pyramidal model, which is acknowledged as the distribution of population per economic group in Latin America. In the following experiments (figure 3), we tested three proportions 5-30-65 (5 per cent of red agents, 30 per cent of yellow agents and 65 per cent of blue agents, respectively), 10-40-50 and 10-30-60. The snapshots of these tests at the end of 2000 time periods are presented in figure 3. The number of steps was fixed and equal to two.

We can see the proportion of spatial development occurring at different speeds according to the proportion of agents per breed. This means that different breeds develop spatially at different speeds, as can be observed in figure 3, where we can see that the first sequence (proportion 5-30-65) presents much slower spatial development than the other three experiments. This is because there are fewer red agents in this experiment and the red agents settle faster than the other two because they can settle anywhere (any colour of patch). Also, according to the model's rules, the more red patches there are, the faster the other two agents will settle.

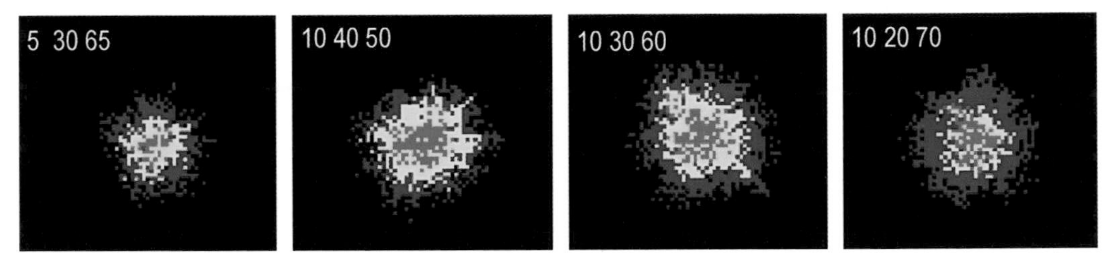

Figure 3 Snapshots from Peripherisation Model after 2000 time periods using different proportions of agents per breed

We also carried out some tests with different initial conditions to explore the idea of path dependence in the model's behaviour. All these experiments present fixed parameters (step = 2 and proportion of agents per breed = 10-40-50). Figure 4 shows some of these; experiments A and B explore the idea of multiple seeds. This is the case for metropolitan areas, which are the result of the combination of several cities or villages that ended up as a single spatial area because of their spatial proximity. The initial condition of the sequence C represents a path or a road, which is a very common situation in the real world. Sequence D presents as an initial condition an attempt to resemble a grid of colonial Portuguese and Spanish cities, typical in Latin America. It is interesting to note that the spatial development starts with a mixed structure, and as time passes, the core-periphery

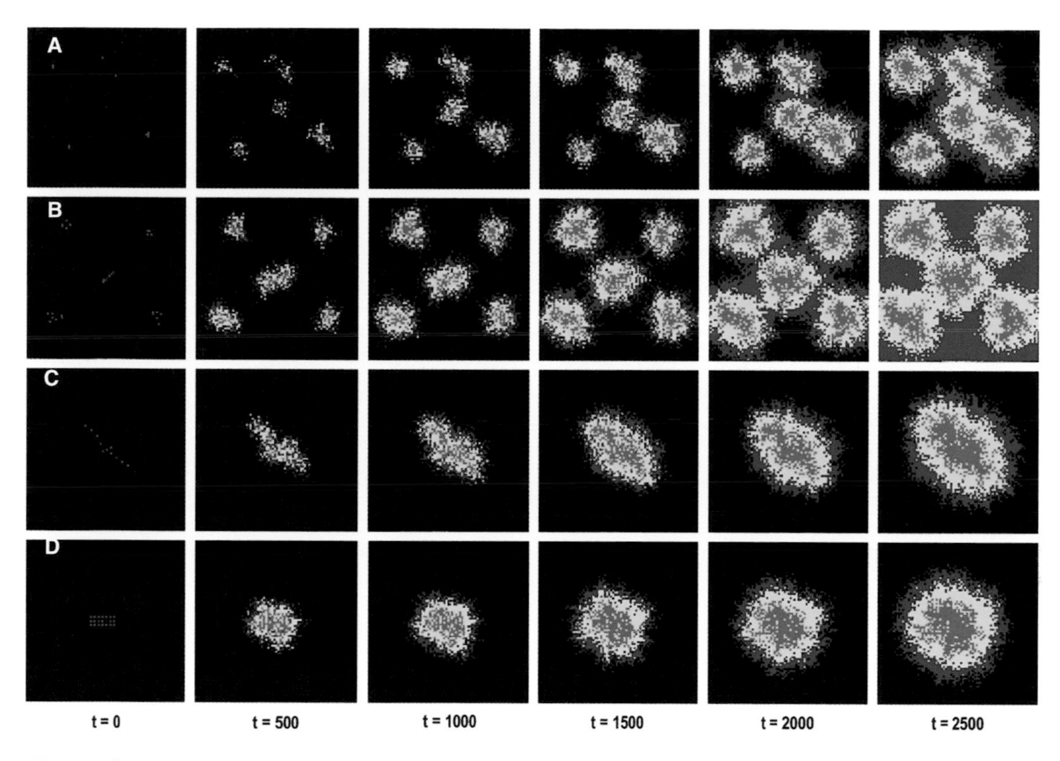

Figure 4 Sequences of snapshots from Peripherisation Model testing different initial conditions: multiple seeds (A and B), path or road (C), and grid of colonial city (D).

structure emerges. As in reality, this spatially segregated pattern is consolidated in the model and as the simulation runs, spatial development expands maintaining the core-periphery structure.

5 Comparison with reality

In what follows, we show some simple maps built from the Census 2000 dataset for two Latin American cities: São Paulo and Belo Horizonte. These maps represent patterns of income concentration that, together with the simulation model above, illustrate the process of peripherisation in Latin-American countries. Figure 5 shows maps of income distribution in these two Brazilian cities. The city of São Paulo has a population of over 10.4 million inhabitants and occupies an area of 1509 km^2, of which 900 km^2 is urbanised. Its metropolitan region is comprised of 39 autonomous cities with a resident population of more than 17.8 million inhabitants occupying an area of 8501 km^2. Belo Horizonte is today the

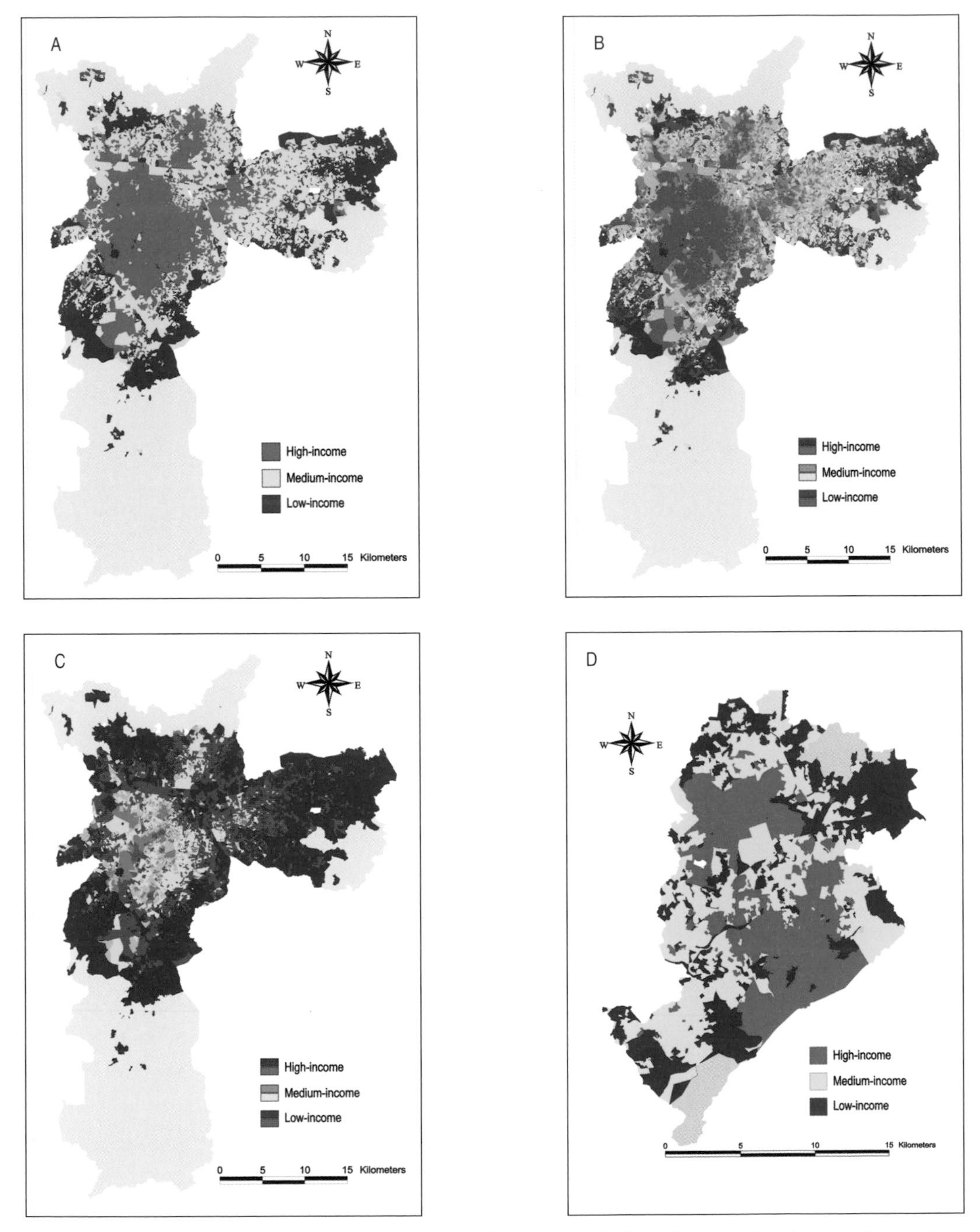

Figure 5 Maps of São Paulo (A to C) and Belo Horizonte (D to F), showing distributions of income in the urban area. Maps A, B, D and E use quantile breaks and maps C and F use natural breaks

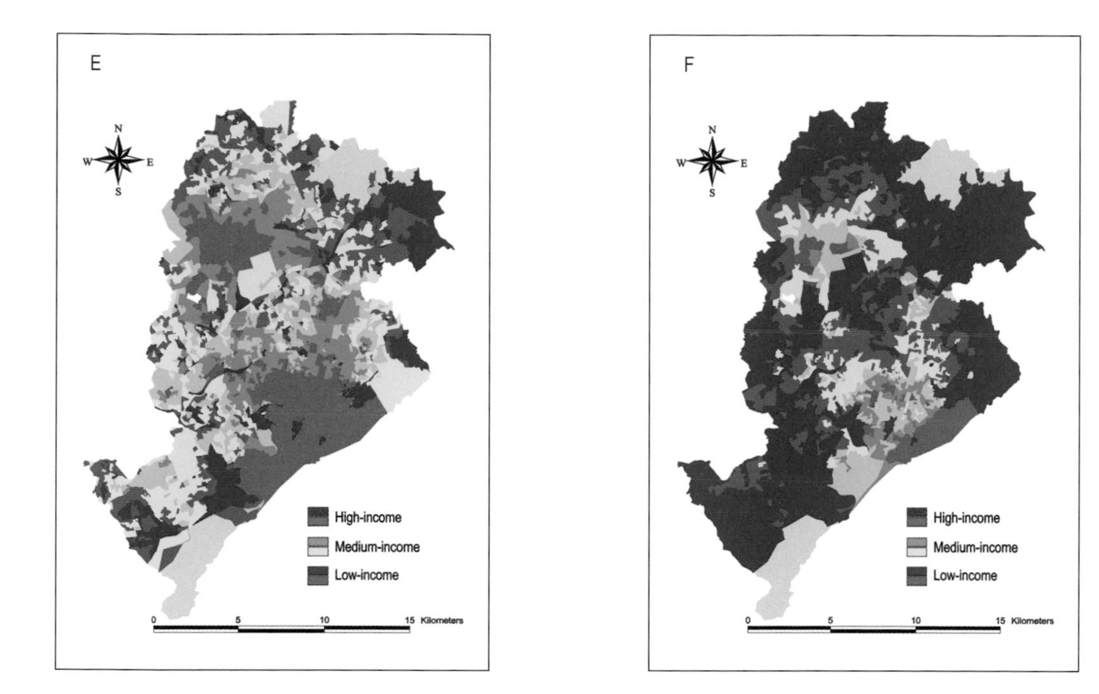

third largest metropolitan area of Brazil and is comprised of 33 municipalities that have a population of 4.8 million inhabitants over an area of 9179 km^2. The city of Belo Horizonte itself is far more compact than its metropolitan region with a population of 2.2 million inhabitants in an area of 330 km^2.

The maps in figure 5 show the distribution of income per census sector in the urbanised area of São Paulo (A to C) and Belo Horizonte (D to F). The limits of the maps are the administrative boundaries of the cities and, therefore, the metropolitan areas (Greater São Paulo and Greater Belo Horizonte) are not included in the maps.

The data used here are monthly householder incomes for census sectors (the finest granularity of published census data, equivalent to census blocks in the United States and enumeration districts in the United Kingdom) and are part of the Census 2000 dataset provided by the Brazilian Institute of Geography and Statistics (IBGE). This variable was chosen because of its similarities to the rules of the Peripherisation Model, which is based on the division of agents into economic groups.

For the maps of the city of São Paulo we used a total of 12 428 census sectors out of the 13 278 sectors available and for Belo Horizonte's maps we used a total of 2549 sectors out of 2564. Each census sector contains an average of 250 households or 750 inhabitants. The remaining sectors, coloured grey in the maps, represent either non-urbanised areas or sectors without available information for the population variable.

The aggregated data for each urban census sector were normalised by the number of householders in each sector and then classified into three ranges (figures 5A and 5D) or six ranges (figures 5B, 5C, 5E and 5F). The maps use red for the higher-income groups, yellow for middle-income groups, and blue for the lower-income groups, as in the simulation model to aid comparison. As in the images produced by the model, we can easily identify a concentric pattern in figures 5A and 5D, in which the high-income groups are concentrated towards the centre of the urban area and the concentration of high-income groups decreases towards the urban periphery. The graduation is most easily observed in figures 5B and 5E where the same data were graduated into six classes, showing a decrease in the incidence of higher-income groups towards the edge of the city.

Figures 5C and 5F show a different classification of income, where the number of red areas is fewer in comparison to the two other maps. In these maps, the classification was performed according to the natural groupings of the data (income) values, and the map actually shows that there are very few people who belong to a high-income group and a lot of people who belong to the low-income group. It should be noted, however, that we have not used established definitions of income groups either in the simulation model or in the maps in figure 5. Our focus is on the relative locational pattern of these groups within the city only and, as such, the actual number in each income group is not directly relevant.

Of course, the spatial patterning of the maps is not as concentric as the patterns produced by the simulation model. This has various causes, such as initial conditions, topography, the presence of water bodies, etc. In particular, the topography of these areas has strong influences on the spatial development of these cities. Belo Horizonte, for example, was built in a valley which constrained development to the east and forced development to the north and south.

It is also important to mention that the maps shown here do not encompass the metropolitan areas of these two cities, but are restricted to the cities' administrative boundaries. This means that São Paulo and Belo Horizonte are actually part of polycentric urban areas like the ones shown in figure 5 and, therefore, the analysis of their urban form is restricted.

6 Conclusions

The preliminary results obtained from experiments with different parameters of the model resulted in important insights related to the speed of spatial development of Latin-American cities. However, it is necessary to investigate whether the behaviour of the model can be related to trends in the urban reality. For example, can the fragmented spatial pattern of Latin-American cities be related to the speed of spatial development, or is there a difference in the speed that people from different income groups settle?

It seems clear that the actual development process of Latin-American cities consists of socio-economic inequality that is reproduced in space by the locational process. The peripherisation process was initially caused by the high rates of urban growth in these countries, but it is now consolidated as the normal process of development of the individual cities. The result is an emergent pattern of spatial segregation characterised by stark differences between core and periphery, which is consolidated as the residential spatial pattern of Latin-American cities. The perpetuations of both process and spatial pattern reinforce the social inequality, which was their cause in the first place, working as a vicious cycle.

The simulation exercises allowed us to develop the understanding of the rapid urbanisation process and to investigate its dynamics, changing our perspective of the problem from a demographic viewpoint to a dynamic and morphological one. This research has taken a step in the direction of bringing a new perspective to an old problem. However, the peripherisation model is at a very preliminary stage of development and is still a crude simplification of the phenomenon. It is possible to refine the model and encompass more complex behaviour as well as introduce spatial constraints on development. Thus, our next steps are to further develop the model and enhance the spatial analysis using real data to further comparison with real-world structures.

The need for a better understanding of urban spatial phenomenon in Third-World cities is essential in order to provide a basis for future planning actions and policies. The approach outlined in this study is part of an ongoing research project which is still in its early stages and is much broader than any particular individual effort. Nevertheless, we believe it provides good evidence that urban modelling tools together with GIS and fine-scale data can provide an appropriate basis for research on Latin-American urban processes. We also believe it is necessary to approach the problem by relating morphology to dynamics and that automata-based models and complexity theory provide the appropriate means.

Acknowledgements

Joana Barros and Sinesio P. Alves-Junior are supported by the Brazilian government agency CAPES (processes 1711-989 and 1675-001 respectively). The authors would like to thank Flavio Fatigati (São Paulo city environment department) for kindly providing the aerial photo of São Paulo.

Chapter

8

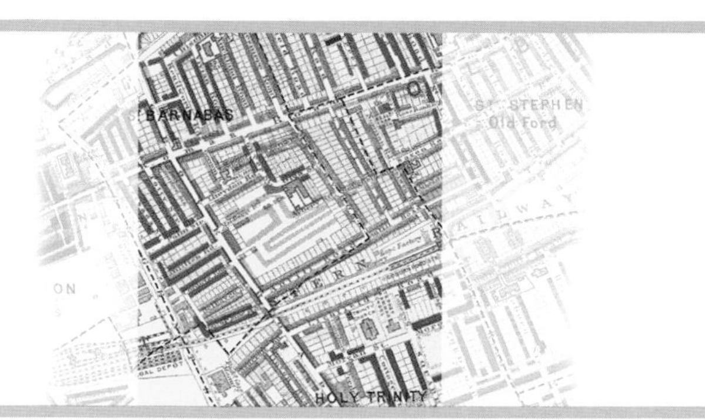

Data surfaces for a new policy geography

Mark Thurstain-Goodwin

Maps are the main ways in which spatial policy is communicated at many scales from the global to the local. However, to understand maps that are based on large numbers of individual features or attributes we need to use different styles of communication at different levels. In principle, analysts and policy makers require detailed access to individual data so that they can discern patterns across many scales but, in practice, representing the data at the finest scale can be confusing. We explore this problem in the context of a new urban geography of town centres that we have been working with for government. We suggest that the surface metaphor is one of the most fruitful ways to visualise individual point data where the density and intensity of those data varies dramatically over short distances.

1 Introduction

Maps have influenced policy, directly or indirectly, through the centuries. They have been used for reasons commercial and political—to aid navigation and depict trade routes, to reinforce ideologies, and to control populations. In this chapter, we will look at how maps have been used to inform policy and how the presentation of maps influences the way in which these policies are formulated and implemented. While all the examples in this chapter come from Britain, the lesson learned can be applied anywhere maps are used by policy makers.

2 Policy maps through history

An early example of how policy makers have used maps is the Mappa Mundi. Completed around 1251 and currently residing in Hereford Cathedral, it depicts a flat, circular Earth at whose centre rests Jerusalem. It ties the Judeo-Christian message directly to the land and elevates the church, which commissioned the map, even further. Its sponsor's message is implicit—this is God's Earth, and we are its custodians.

Maps have been used to exert less subtle means of social control. Elizabethan Britain's anti-Catholic policy was coordinated via the map base. Burghley's 1590 map of Lancashire (figure 1) identified the location of the homes of the Catholic gentry within the county, facilitating swift retribution should there be an uprising. The locations of potential landing sites for the Spanish army are marked, as are hilltops on which warning beacons would be lit in the event of a Papist invasion. We can see this as an early example of a state database, a catalogue of Britain's Catholic aristocracy, using the map as a way of storing and communicating policy information.

Such examples could persuade us that maps are only used to design and enforce coercive policy. Of course, they have also been used to effect policy change for the benefit of society. Concerned about the long-term impacts of urban poverty on Victorian Britain, a shipping magnate Charles Booth undertook to establish the full extent of the problem in London: he made a map.

Booth's 1889 map of poverty in London (shown in figures 2A and 2B at different scales) classified neighbourhoods into one of eight classifications, from home to the lowest class of criminals (shown in black and dark blue) whose only pleasure was cheap gin to the wealthy upper-middle and upper classes who enjoyed the finest champagne (shown in red and yellow). This provided a means of

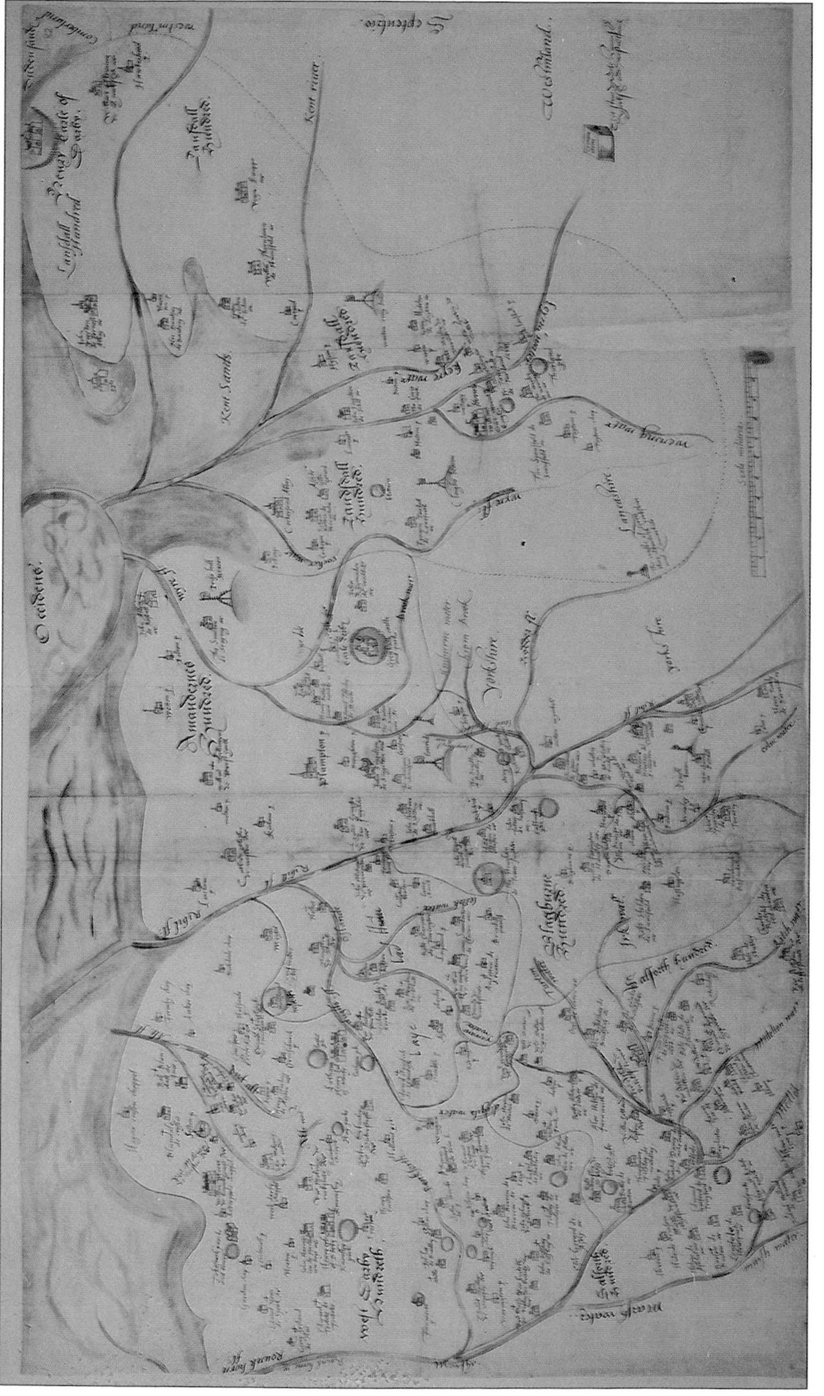

Figure 1 Burghley's Lancashire map

Figure 2A Booth's poverty map of London

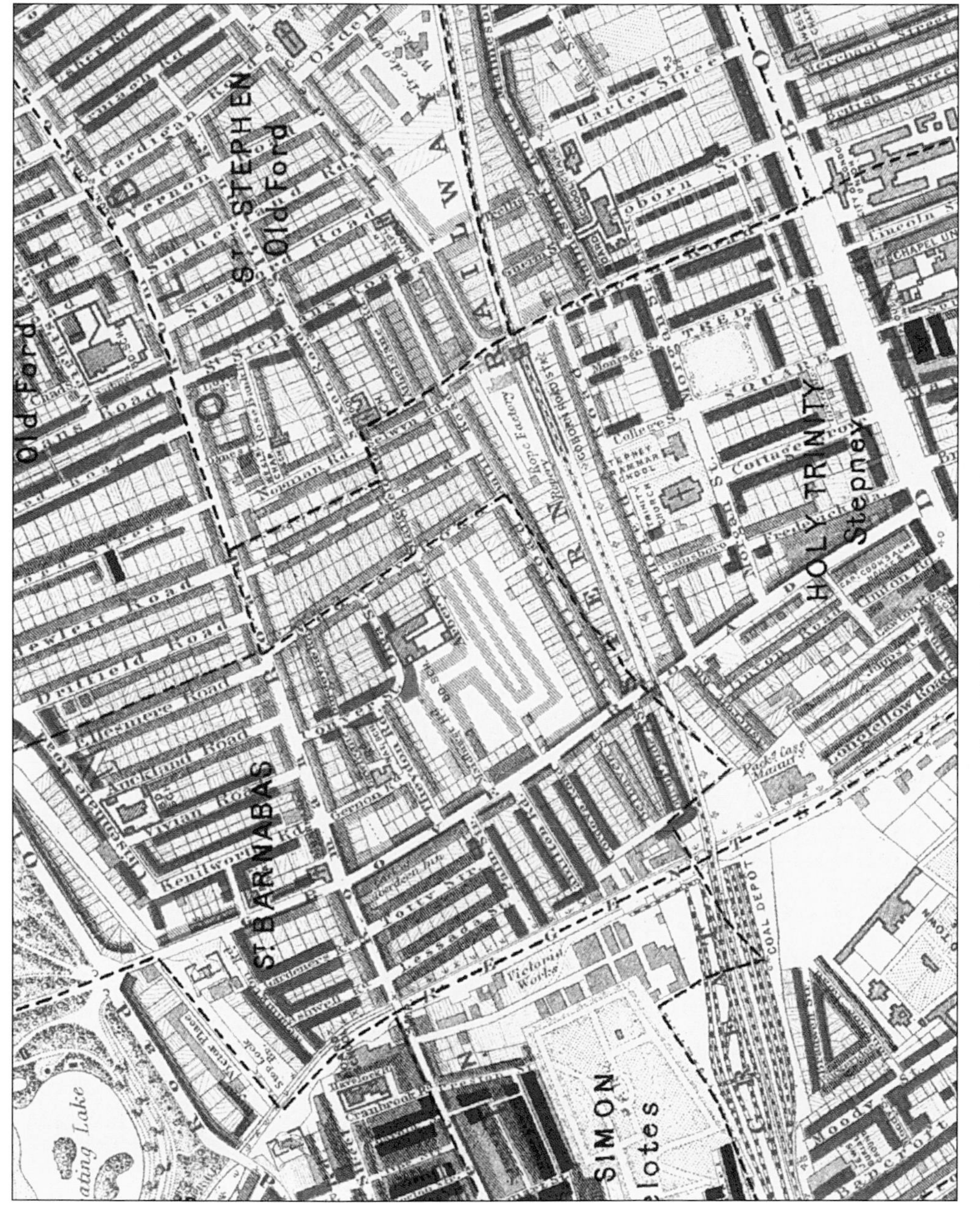

Figure 2B Poverty in Bow, east London

graphically assessing the distribution of poverty in the capital city of the British Empire. (These maps can be found in the Charles Booth archive at booth.lse.ac.uk). Discovering that over 30 per cent of London's population was living in poverty at the start of the 1890s, Booth was to undertake a life-long campaign for a state pension to alleviate poverty amongst the elderly. His maps would lie at the centre of his campaign and it is in this more positive vein that this chapter will continue—although always with one eye on the less charitable ways in which maps can be used.

3 Representing the world

Our conventions for representing the world have not changed fundamentally since the early days of mapping. The basic representational forms based on the points, lines and polygons that sit at the heart of most GIS were here long before the technology was invented. Point features have traditionally been used to map discrete entities: at coarse levels they can map large objects such as cities while at finer scales they are equally suited to showing the location of individuals. Lines have been used to navigate the tangle of streets in our towns and cities while polygons have been used to indicate the extent of land, its ownership and control. All three feature types have been brought together in composite maps, allowing the data to be modified and new patterns to be identified. Whether represented in a computer or by the pins and strings in Churchill's Cabinet War Room maps, these feature types remain basically the same.

Of the three representations, it is perhaps the polygonal representation that has been predominant in the field of policy making. One of the key reasons for this is that zonal geographies are the main ways of graphically defining states and administrative areas. The hierarchy of administrative zones that makes up the state—its regions, counties, districts and wards—are used to map, describe and perhaps even define it. They conform to the policy maker's perception of how land and society is structured and should be managed. They have the added benefit of providing a framework for the collection of statistics, allowing the administration to monitor the performance of these zones and implement policies.

Booth's maps, for example, are structured around parish boundaries, the edges of the paper maps following the historic spatial organisation of the Church in London. It is possible that Booth arranged the maps in this way to not only reflect an obvious zonal geography, but also because he realised that the key users of his maps would be the clergy on whose shoulders the care of the poor and destitute often rested. His maps were therefore targeted at the level where restorative policy would ultimately be implemented.

More generally, many of the roots of the framework of Britain's policy geography can be traced back to an Anglo-Saxon heritage. The shire county become the main policy unit of England in the tenth century onwards and delimited the extent and control of the ruling classes. It was further divided into hundreds, divisions by which the Saxon overlords were able to manage the population. These divisions would later prove invaluable in the collation of statistics.

William the Conqueror's Domesday Book was the first successful attempt at a collection of statistics in England. It used the administrative infrastructure of the defeated Anglo-Saxons, organised around the spatial hierarchy of shires and hundreds, to frame England's first census. The Domesday Book recorded the wealth of England, before and after the Conquest, and offered the means by which William was able to repay the supporters who had joined his invasion force on the promise of land and fiefdoms. The Domesday Book was used to divide the spoils, an administrative device successfully used by the Normans to subjugate the Anglo-Saxon population. This made William the 'first database king' (Schama 2000).

No contemporary map representations of the Domesday Book remain, if indeed any were made. When the statistics are mapped at the scale of the shire, it becomes clear that the population of England at that time was concentrated in the shires of East Anglia, the country's most productive agricultural region (figure 3A). As the boundaries of the shires remained largely unchanged in the following centuries, it is possible to track the changing population geography of the country. When population data from the Poll Tax returns of 1377 are mapped (figure 3B), it is possible to see that the eastern part of England is still the most heavily populated, although other important wheat-growing areas such as Leicestershire and Northamptonshire were becoming more heavily populated (Darby 1963).

4 Problems with zones

While zonal systems offer effective means of collating data and statistics, as well as providing a handy method for mapping these data, these types of representation can become frustrating to both the geographer and the policy maker. One of the most limiting factors is that zonal geographies have a scale threshold below which data are not presented. In the maps shown in figures 3A and 3B, for example, it is not possible to see the more subtle variations in medieval England's population geography. The burgeoning population of the cities and boroughs are

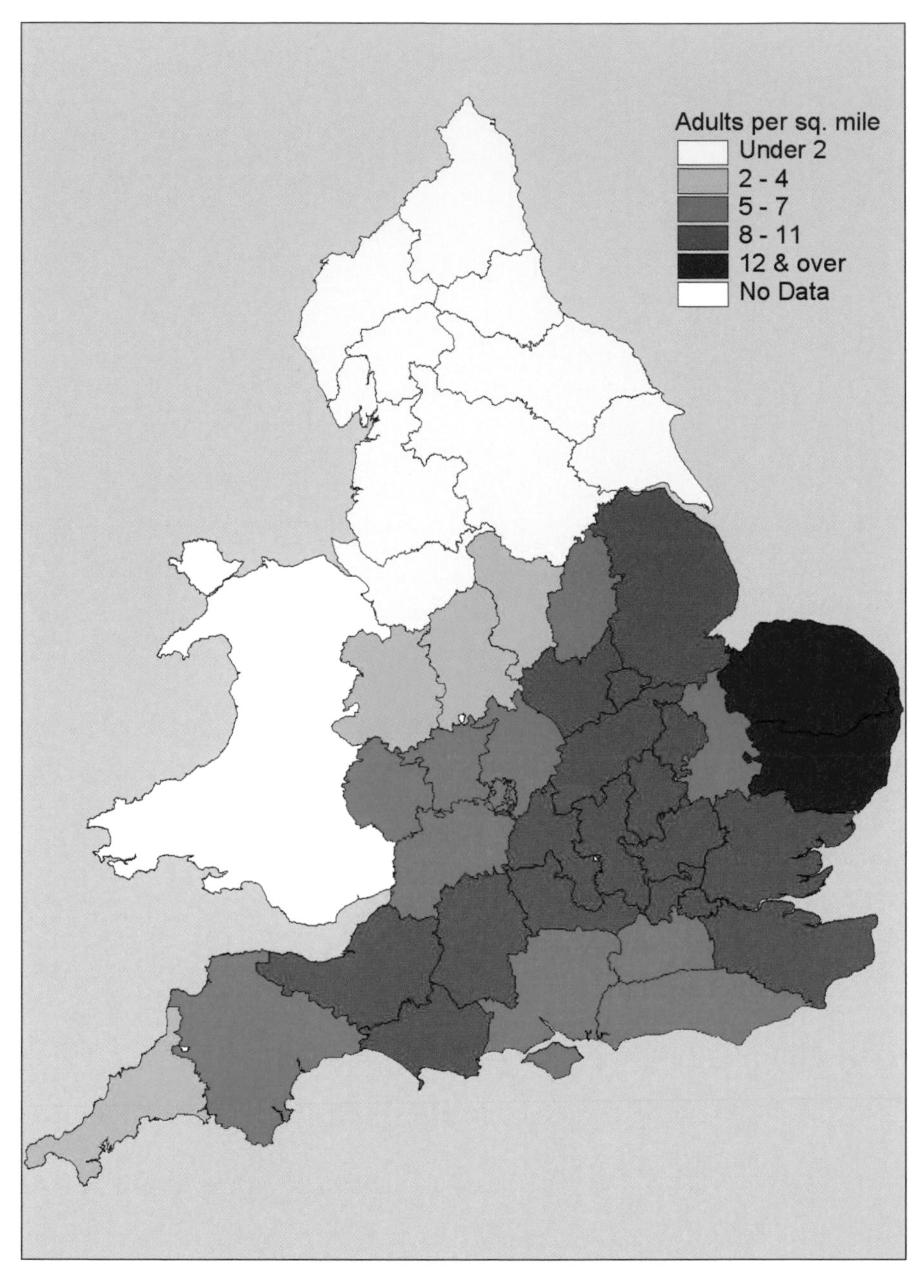

Figure 3A England's Domesday population (after Darby 1963)

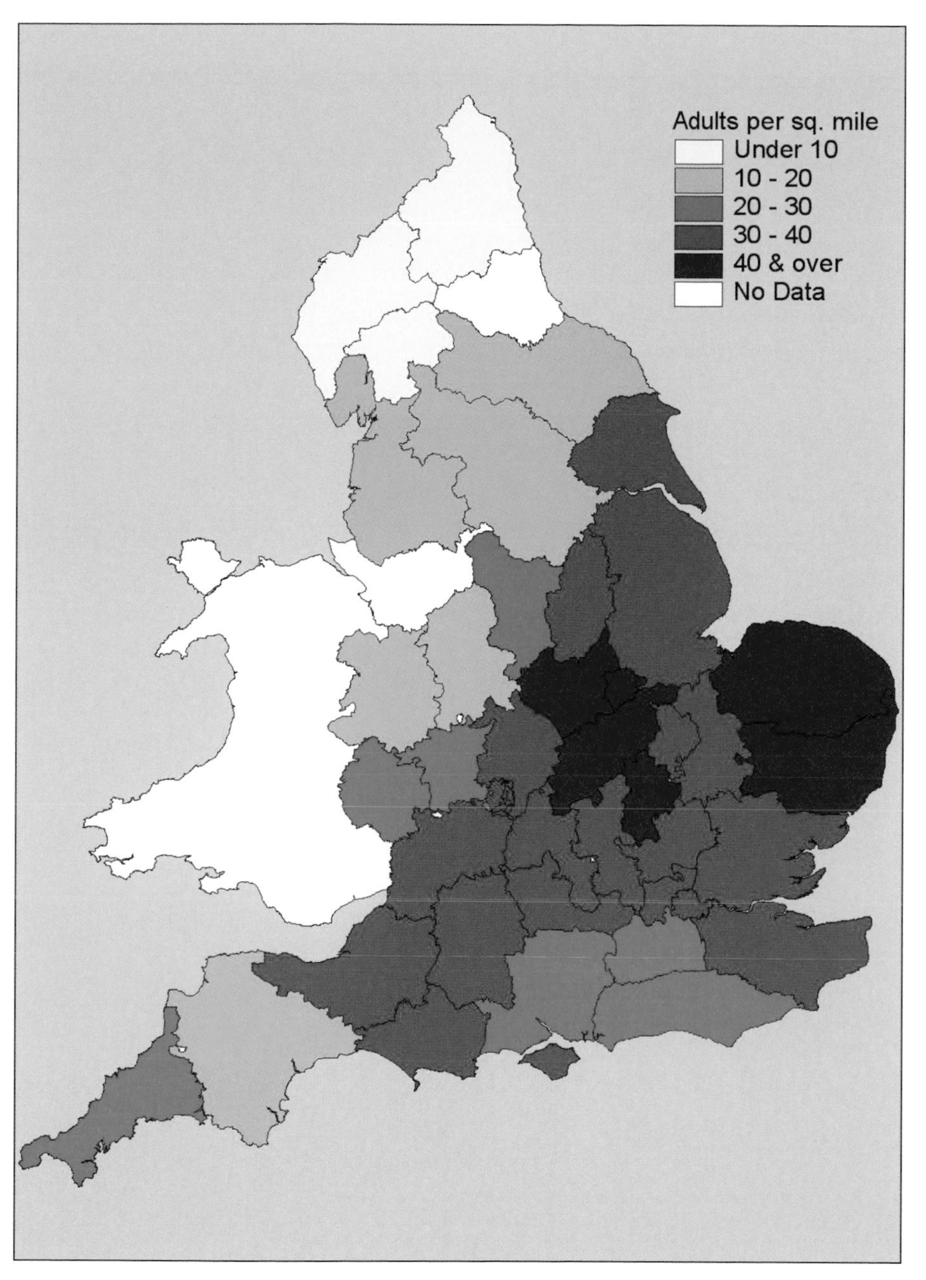

Figure 3B England's population in 1377 (after Darby 1963)

hidden from our view, as are the many important concentrations of population along the coastal fringes of the southwest peninsula. We are instead presented with an implied homogeneity of space across the zones, within each of which the people, architecture and nature are assumed to be the same. We are drawn into the temptation to ascribe broad aggregate characteristics of these zones to all of the people that fall within them, when in fact we cannot be certain they hold at scales below the mapped units.

More commonly known as the ecological fallacy, this problem can have major consequences on the formulation of policy. Perhaps one of the best illustrations of this fallacy was provided by W. S. Robinson, who identified a correlation between the number of African Americans in American states and literacy levels at the state level. The 'obvious' assumption derived from the analysis of these data would be that African Americans are racially inclined to illiteracy; an assumption that is both simplistic and untrue and one which he goes on to demolish in the rest of the paper (Robinson 1950).

Other problems associated with the analysis of areal data are well documented in the GIS literature. Related to the notion of ecological fallacy is the oft-cited Modifiable Areal Unit Problem (MAUP). This is the phenomenon by which the choice of zonal boundaries within which statistics are mapped and aggregated can have more effect on the output analysis than the real distribution of the phenomenon being observed. The size and shape of areal units is often dictated by the need to simplify the collection of census data and, thus, it is unlikely that these areas '. . . have any intrinsic geographical meaning'. Furthermore, '. . . the value of any work based upon them must be in some doubt, and may not possess any validity independent of the units which are being studied' (Openshaw 1984: 4).

If the choice of a zonal boundary system affects policy makers' abilities to understand these phenomena, even bigger problems occur when the boundaries themselves are changed. This can occur for many reasons: to improve the efficiency of data collection, perhaps, or to attribute changes in the real, underlying geography. Of course, the objectives may be more nefarious: elections can be fixed by gerrymandering—the process by which electoral boundaries are manipulated in order to optimise the voting distribution of a particular party, while undermining that of the others (see Schietzelt and Densham, this volume).

These issues do not prevent areal geographies being commonly used to support, define and implement policy. We have seen the reasons why this is so: because defined spatial areas often neatly correspond with the ways in which society is controlled and administered and because it is relatively easy to collate statistics within them. Perhaps tradition is the most powerful influence—policy makers are just used to seeing data presented in this way.

5 Mapping the individual

If this is the case, then it is a relatively modern tradition. Early proponents of spatial analysis did not invariably take a zonal approach. The long tradition of point-pattern analysis, pioneered by John Snow's analysis of typhoid cases in the middle of the nineteenth century, illustrates that we also tend to understand the world in terms of its discrete units. For this reason, and for the quality of analysis that is possible where data are available at this scale, it makes sense wherever possible to base our policy on data mapped at the individual level, or as close to it as possible. Here 'real' patterns in the data become evident, safer conclusions can be drawn, and more effective policy be developed.

Data availability at as fine a scale as possible is crucial to achieve this. For example, John Snow was able to locate cholera cases relative to the various water pumps because there were so few pumps. Furthermore, it has been argued he knew what he was looking for and was able to target his data collection exercise accordingly. The frequently cited story of Snow having his eureka moment that cholera was a water-borne disease after analysing the data on the map, an early form of geospatial data mining, has been disputed by Brody et al who contend that he had developed his hypothesis prior to the mapping exercise (Brody et al 2000).

The scale of available data is, therefore, fundamental to the effective use of maps in forming policy. Selecting an optimum scale at which to frame policy is a question of balance since too coarse- or too fine-scale data can both impact our ability to interpret spatial patterns (although as we shall see, additional tools are now available to help overcome these issues).

Acquiring detailed data is no mean feat. Even in those countries such as the United Kingdom that have a well-developed data infrastructure, it is rare that datasets such as the Census of Population or Employment Census are fully comprehensive—people often forget or refuse to fill in the forms correctly. Although there is no single definite database containing detailed information on the individual, vast amounts of data about consumers and their lifestyles are collected in the private sector. Geodemographic data suppliers, credit checkers, supermarkets, banks and most of the businesses we come into contact with, in both the real and virtual worlds, assemble huge amounts of data about us (see also Webber and Longley, this volume).

These companies rarely have any motive to share their information without remuneration. Certainly, none of the major geodemographic data suppliers has taken up Charles Booth's example of providing data for the public good. Many would argue that it is the government's job to fulfill this role and indeed, the U.K. Office for National Statistics (ONS) has recently tried to satisfy the demand for detailed socio-economic data by making them available at the ward scale (each typically accommodating around 4000 resident households) available over the Web (see www.neighbourhood.statistics.gov.uk).

At present, these data are not available below the ward threshold because of the population's justifiable concerns about confidentiality and disclosure. The examples described of how such information has been used in the past point to the reasons why the right to privacy remains a cultural issue in Britain. These restrictions apply not only to personal or household data, but also to information on businesses and the economy. The ONS operates under the tight restrictions of the Statistics of Trade Act 1961 which prevents the release of detailed information on particular companies. Indeed, enshrined within the act is the principle that if companies are willing to share their information with ONS, it will not be passed on to other people. Similar statutory obligations restrict the dissemination of information in other government agencies in Britain.

This is not the case in all countries. Detailed information on property, both domestic and commercial, is freely available at the level of the individual building in most states of the United States. For example, the Lucas County State Auditor, based in Toledo, Ohio, actually releases a bespoke GIS tool on CD-ROM to help the public obtain detailed information (including floorspace and valuation data) on all property in the county. This is in contrast to Britain's Land Registry and Valuation Office Agency which are able to release only a limited amount of data from their databases.

There can be no doubt that the technology now exists to easily combine and process data from disparate sources to create detailed profiles on vast numbers of people, businesses and buildings. While confidentiality and ethical constraints protect our privacy to some degree, it is the prohibitive cost and difficulty of combining all these disparate databases that currently prevents the detailed profiling of every citizen. However, even if the data were available at low cost, understanding aggregate pattern in fine-scale data is extremely difficult although, as shall be seen, it is by no means intractable.

6 Seeing the wood for the trees

Confidentiality is not the only concern with fully disaggregated data. When extremely detailed datasets are mapped, the sheer volume of atomised data means that visualising and detecting spatial patterns in the data becomes unmanageable. For Snow, mapping the specific locations of people who had contracted cholera was important, but he was only looking at a small area—a few streets in Soho, a fairly small quarter of London's West End. The policy maker is generally more interested in larger areas, the scale of which is usually determined by a hierarchy of administrative units.

This was borne out when CASA started to model detailed employment and floorspace data for the Office for the Deputy Prime Minister's project on defining town centre boundaries in London (ODPM 2002). Identifying where people worked, where the concentrations of property were, and which parts of London had the most diverse economic base were all central to the success of the project. As discussed by Lloyd et al (this volume), even though the data were aggregated to unit postcode level (and not at their most atomised level, that is individual businesses and buildings), when they were viewed at finer spatial scales, it became more and more difficult to decipher the spatial patterns they made; the granularity of the data was simply too high.

Figures 4A, 4B and 4C show how challenging highly granulated data can be to visualise, especially over a large area. The larger the circle in each of these figures, the more people are employed in retailing at that location. The maps show retail employment data from the ONS Annual Business Inquiry (ABI), one of the two key datasets used in CASA's research for the ODPM. Here the data are aggregated to the level of the unit postcode, the georeference of which is the spatial average of the buildings that fall within it. Those locations where many people are employed in retailing are points shown using proportional circles.

Retailing is one of the most important sectors in the U.K. economy (see Longley et al, this volume). The mapping of retail employment data is one way in which policy makers can understand the sector as a whole and develop policies to adjust the direction of its development. An excellent example of this is PPG6, which makes it much more difficult for major retailers to locate in a non-town centre location, reversing the trend that blighted many town and city centres (DETR 1996). The impact of this and other policies needs to be tracked at a number of different levels, from the nation through to the high street.

With this in mind, retail employment data are mapped at three scales, typical of those at which the policy maker would choose to view the data: the county or subregional level, the district or borough level, and the ward level, the most disaggregated scale at which ONS currently releases data (figures 4, 5 and 6). Three successive scales—country, borough and ward—are shown, repectively in parts A, B and C of each figure. Green rectangles are used to identify the location of the borough shown in figures 4B, 5B and 6B, and the ward shown in figures 4C, 5C and 6C. The Thames and London's road network are also shown with railways being added to the map of Bow ward to give further context.

As can be seen in figure 4, point maps are harder to interpret at coarser scales, such as the county level. Despite the obvious detail, or granularity, of the data, one simply cannot see the wood for the trees. Viewed at the slightly finer scale of the district, the main clusters of retail employment found in the central London towards the west of the map are sharply focused. It is also possible to see how retailers outside the central core tend to cluster along the main radial routes, like the Mile End Road which falls in the centre of the map, or in distinct centres such as Stratford, located in the northeastern part of the map.

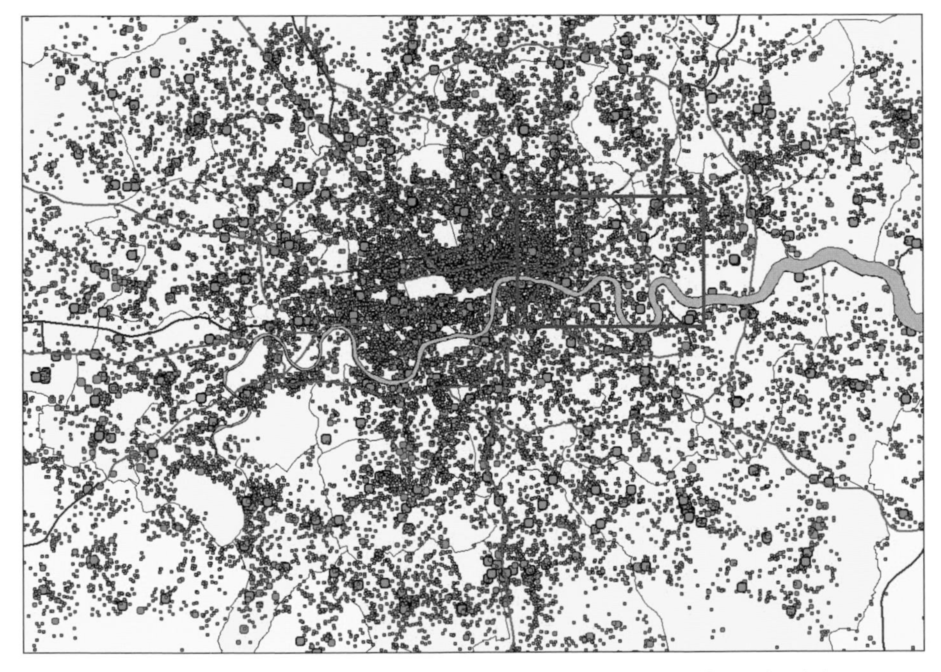

Figure 4A Retail employment data in London presented as points—county (Greater London)

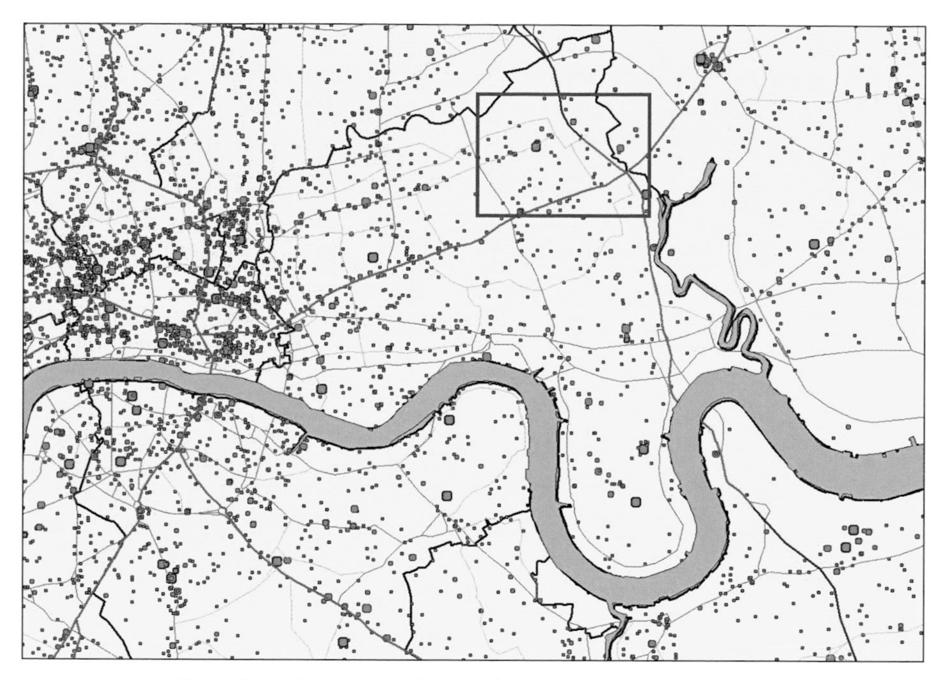

Figure 4B Retail employment data as points at borough level—Tower Hamlet

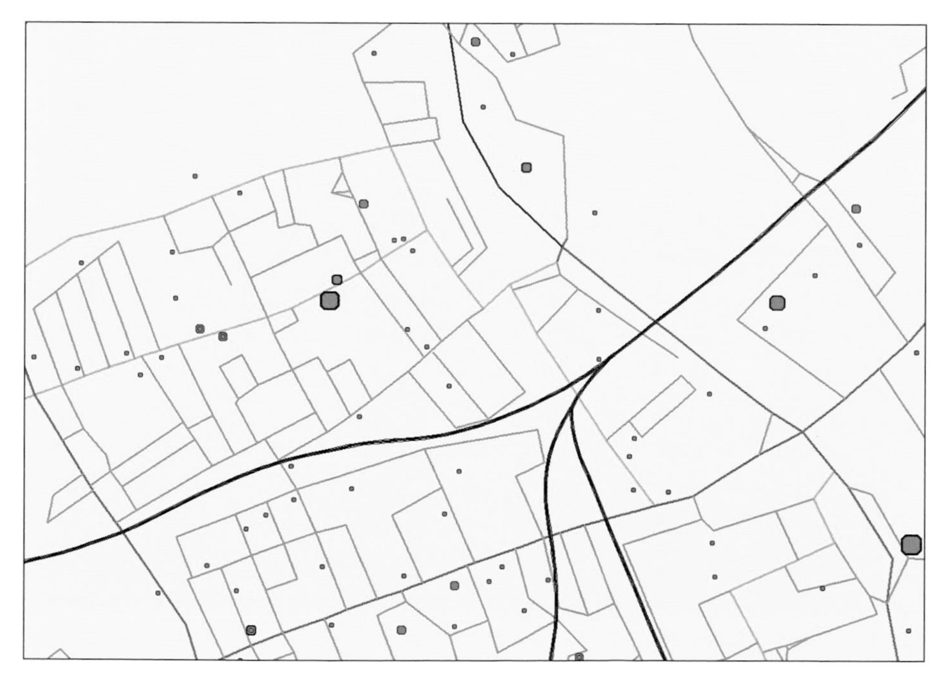

Figure 4C Retail employment data as points at ward level—Bow

Figure 5A Retail employment data presented as zones—county (Greater London)

Figure 5B Retail employment data as zones at borough level—Tower Hamlets

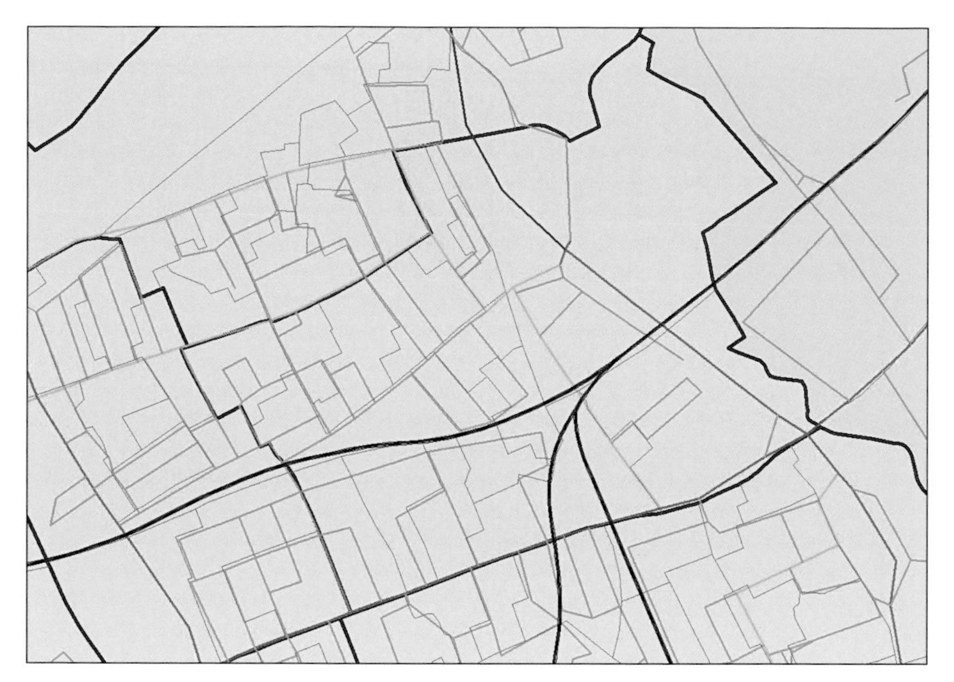

Figure 5C Retail employment data as zones at ward level—Bow

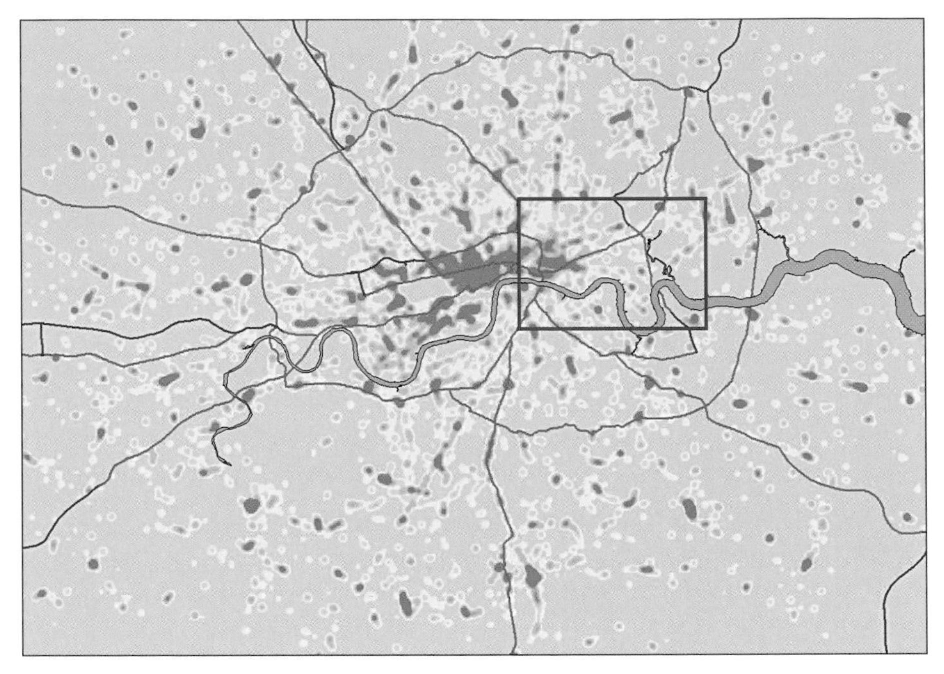

Figure 6A Retail employment data in London presented as a surface—county (Greater London)

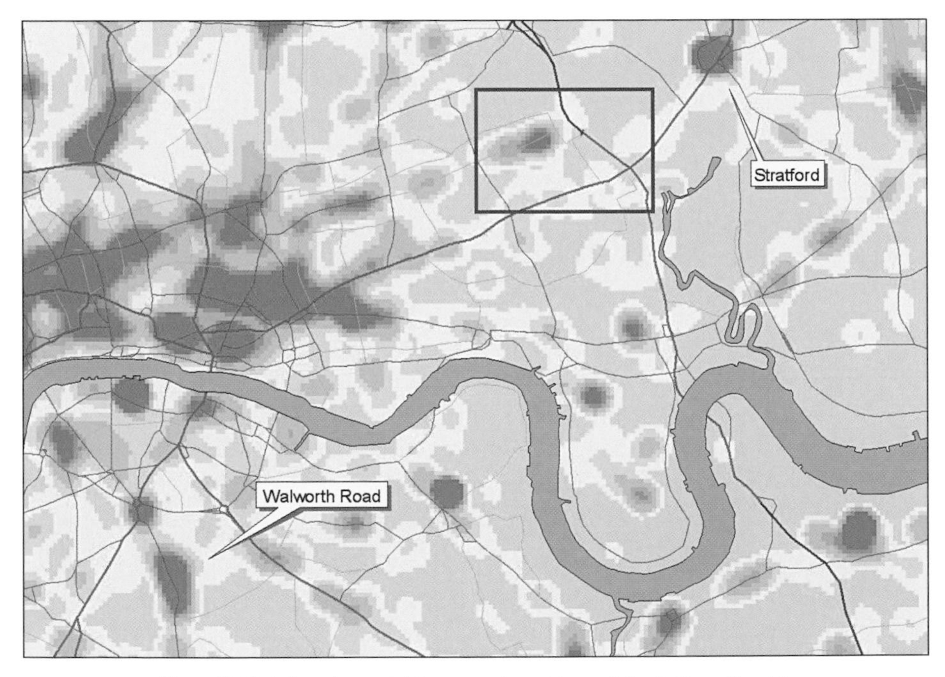

Figure 6B Retail employment data as a surface at borough level—Tower Hamlets

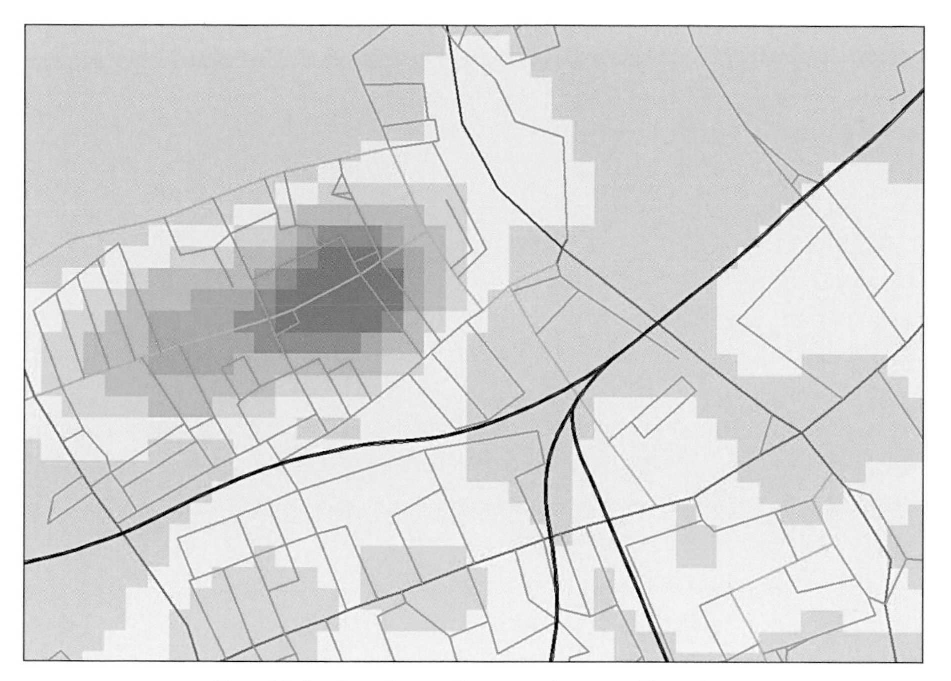

Figure 6C Retail employment data as a surface at ward level—Bow

This demonstrates, for example, why shopkeepers on the Mile End Road have opposed the creation of a no-stop zone along this road, in anticipation of its negative impact upon business. We can also identify the main supermarkets and retail warehouses in and around Tower Hamlets whose competitive strength has forced many of the small independent retailers out of business. The policy maker could use these data to see if, over time, there has been a quantifiable drop in the number of these smaller local retailers, which pepper the local neighbourhoods such as those of the Roman Road in Bow or Brick Lane in Shoreditch.

Of the three scales, it is easiest to detect data patterns at the district scale. The county-level map is too cluttered, and detecting patterns at the ward level is made harder because the data are aggregated at the level of the unit postcode, rather than for individual buildings as in Booth's map. Even if the data were available at the building level, the fundamental heterogeneity of the data at this scale makes detecting overall patterns impractical.

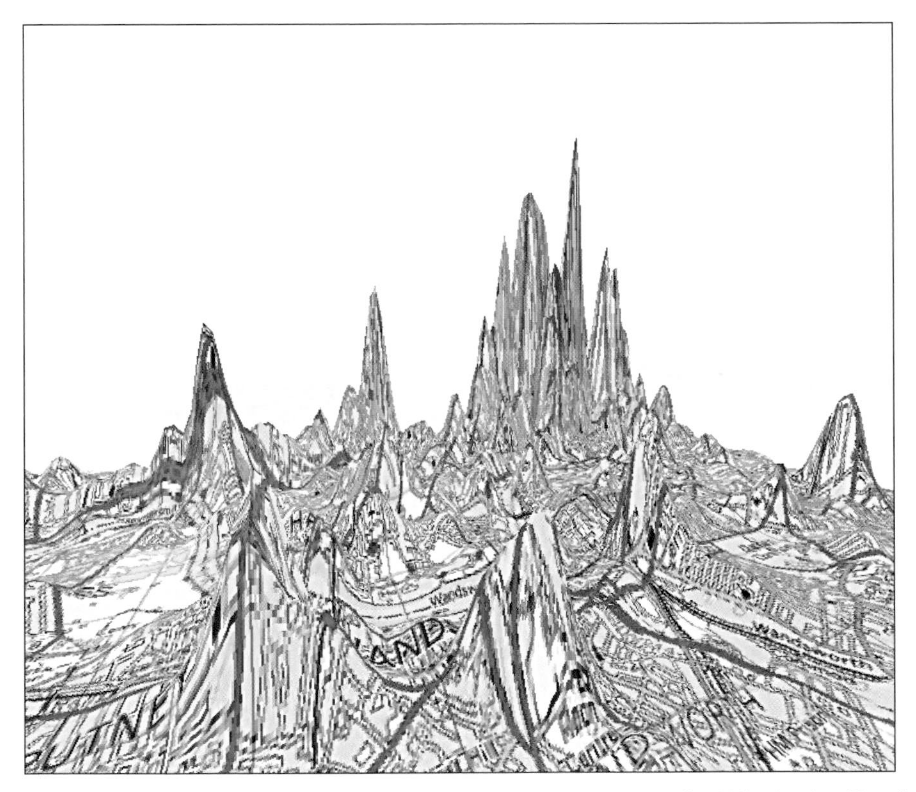

Figure 7 A 3-D perspective of retail employment density in London viewed from the Putney in South West London. (The tallest peak in the distance is Oxford Circus.)

In figure 5, the retail employment data are aggregated into ward units and are presented at the same spatial scales. In order to preserve the distribution of the data, the colour scheme is shaded on a bivariate scale by standard deviation increments: purple wards have a below-average number of people employed in retailing, while the orange wards, increasing in hue, are those with a higher than average retail workforce.

While the pattern of the county scale map (figure 5A) is perhaps easier to interpret than its equivalent (we can now clearly see that retailing is concentrated in the West End of Central London), local variations in pattern are now lost. Those developing parking policy, for example, would find it more difficult to assess the impact of new strategies on local retailers if they viewed the data through the filter of the wards. The data have become homogenised at the level of the ward and little can be deduced save that Bow falls in an area where retail employment density is lower than the London average. As we shall see, were policy makers to see this as a cause for concern, they might end up spending precious resources chasing shadows.

Contrast this with the three maps of figure 6 where the data have been generalised into a data surface (the well-established process of which is explained in more detail in Lloyd et al, this volume) shown at a resolution of 50 metres. Using the same purple to orange shading system used before (with white colouring showing areas with average retail employment density) we can see a terrain map which shows the peaks of retail activity as dark orange, with the retail flat-lands where there is little retail employment shown in purple. The principle is exactly the same as the topographic maps we use to navigate around the mountains. It is only when we look at the data in a 3-D perspective that we see just how concentrated retailing is in Central London as shown in figure 7.

7 Data surfaces: the best of both worlds

Whether viewed as a flat terrain map or in 3-D perspective, the data surface allows us to clearly identify the distribution of retailing at the macro-, meso- and micro-scales. Surface analysis combines the strengths of both the point pattern and the zonal pattern offering both detail and generalisation. Patterns hard to discern through the clutter of the point map are smoothed to reveal a complex retail geography, a geography that is all but obliterated by the crude generalisation of the zonal approach. We can see the importance of district centres throughout the conurbation, and their survival in the face of intense competition from central London.

This hierarchy of retailing becomes even more apparent when the data are seen at the district level—the regional centres of Stratford and the Walworth Road assert themselves some distance from central London. And in Bow, we can see that there is after all an above-average concentration of retailing along the Roman Road; a pattern completely misrepresented in the zonal representation, yet still hard to detect without some generalisation in the point pattern.

The reason for this is that Roman Road retail area is neatly bisected by the boundary separating the Bow and Park wards, which effectively divides the data, making the pattern disappear. Of course, the areas of most interest to the policy maker tend to lie, paradoxically, along boundaries. Even if the data were aggregated to the smaller census enumeration district (the boundaries of which are shown as the light brown lines in the ward scale diagram in figure 5C), their shape bears little resemblance to the point pattern of the more atomised point data.

These two issues expose the Achilles' heel of mapping data with areal units: their connection to the underlying geography of the phenomenon they are trying to explain is often tenuous. They are usually designed either with the ease of data collection in mind (as is the case with census enumeration districts) or are defined by an irrelevant administrative framework, the roots of which may be historical or unrelated to aspects of administration. In contrast, data surfaces are no respecters of boundaries.

The artificiality of ward boundaries could hinder the development of good policy to combat social exclusion in the United Kingdom. The ONS has recently released data aggregated at ward level to help central and local government and the public at large to better understand issues of urban poverty. Furthermore, it is expected that the analyses of these data should support local authority bids to central government for funding to ameliorate social deprivation in their areas. Harper (2002) shows that areas of poverty, inequity and social exclusion do not, however, correspond to the arbitrary geography of wards and districts. This can be clearly seen in figure 8 where pockets of urban poverty are shown in red in the London Borough of Brent.

This means that areas of urban poverty which are either very small relative to the rest of the ward, or which straddle ward boundaries (as did the retail concentration along the Roman Road) will not be detected in the ward data. This means that government policy makers are attempting to tackle the very real issues that

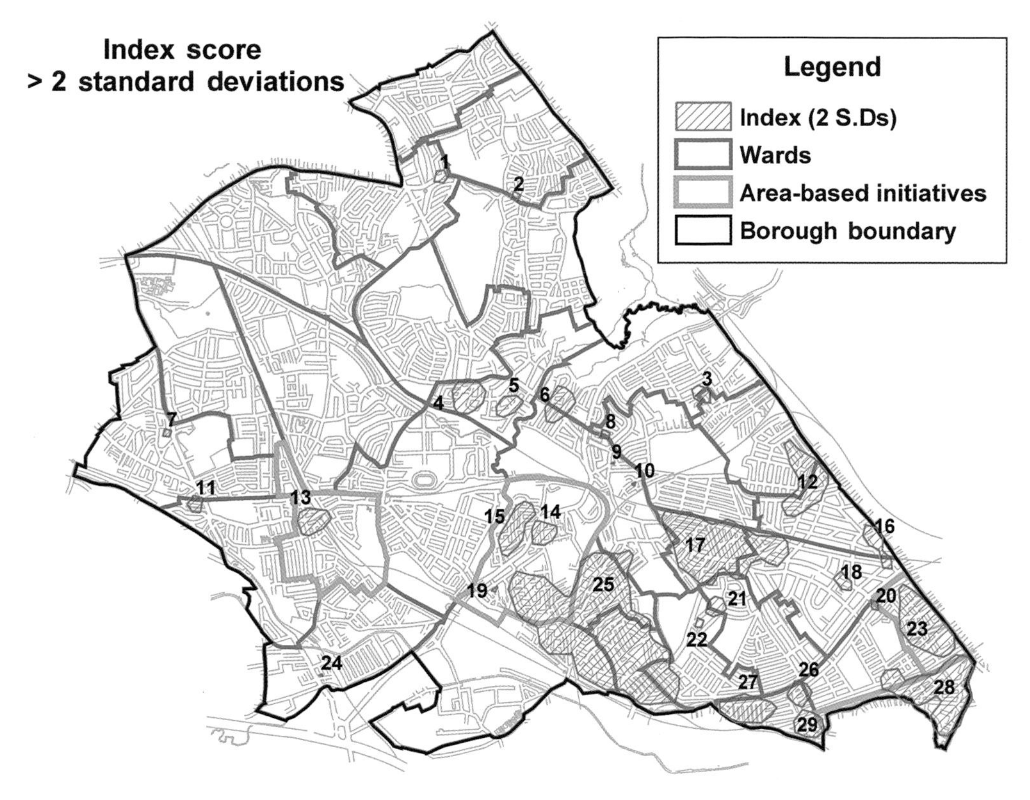

Figure 8 Ward boundaries hiding pockets of urban poverty in Brent

cause and sustain urban poverty with poor and inappropriate measures that affect the quality of analysis and inevitably lead to the formulation of poor policy.

For the reasons of confidentiality and data unmanageability outlined above, the mapping of extremely fine-scale data, whether the univariate poverty measures employed by Booth or the credit worthiness of the population, is not feasible. For example, there would be around 15 times as many points on the maps of figure 4 if every property was shown as a discrete point. Here is another opportunity for a data surface to help us to understand macro- and micro-patterns efficiently. Such an analysis has the important subsidiary benefit of smoothing and summarising data in a grid, rendering the underlying information anonymous.

This ability of the data surface to convey information at both the local and strategic level is vital to the creation of good policy. It enables policy makers to understand the ways in which local, regional and national trends coalesce at different locations. An excellent example of this was the study by Coombes and Raybold who mapped how employment accessibility (defined as the distance that people on average have to travel to secure work) changed between 1981 and 1991. The surface representation enabled the identification of relatively small areas that have witnessed growth despite being located in regions suffering decline, such as Leeds, and areas suffering decline despite being in areas of major growth, such as the nineteenth-century core of Greater London (Coombes and Raybold 2000: 329).

8 A new policy geography?

It is tempting to caricature policy makers universally as impartial government officers quietly meeting society's needs according to obvious findings based on available data. In many contexts, of course, the agendas of planners, businesses, communities and special interest groups clash. This need not imply that data analysis becomes useless. Rather, that its outcomes must help to inform debate and—in an ideal world—encourage maximum involvement of all parties in seeking consensus. This is the context in which issues of sustainable development are commonly contested. We have already seen how the representation of spatial data in surfaces can begin to help us map and visualise some of these issues. If only the creation of policy were that easy.

In the sustainability field, a major problem facing policy makers is the vast number of different indicators advocated by the various proponents. No single, correct set of indicators that should be measured, no 'one size fits all' solution has been agreed upon. The challenge is to combine the different indicators into manageable elements that can form the consensus by which policy is evolved through the various stakeholder groups (Cox, Fell and Thurstain-Goodwin 2002).

In principle, CASA's work for the ODPM in defining town centres achieved this—a range of different indicators were drawn together using a process called indexed overlay (the means by which data surfaces can be mathematically combined) to create a composite indicator that endeavours to index town centre activity (Thurstain-Goodwin and Unwin 2000). There is also promise in using data surfaces as a means of conveying sensitive and confidential datasets at fine resolutions while managing the risks of disclosure. The possibility of using them to create datasets that address particular policy issues, such as Coombes and Raybold's employment accessibility surfaces, is also very exciting.

Throughout history, maps used in policy contexts have been constrained by the basic representations of points, line and polygons—the latter proving to be the most popular and sustained. As we have explored, this has been because of the relative paucity of data and the difficulties of handling and analysing what data were available. The rise of computing and GIS, in parallel with the availability of more data, means (at least in the developed world) more opportunities are suddenly available to policy makers.

There is little point investing in the collection of detailed data if they are not to be used to their full potential. Policy needs no longer only be formed and understood by zonal geographies; data surfaces get closer to ground truth, but remain flexible enough to arrive at a common understanding of these more difficult fields of policy formulation. The only risk, perhaps, is ascribing them too much power. Useful as they are, data surfaces are still maps. They can confuse as well as enlighten and be used to oppress or to liberate. Ultimately, their use, like the policy they help to mould, is a matter of perspective.

Chapter

9

Urban remote sensing: the use of LiDAR in the creation of physical urban models

Sarah L. Smith

The use of remotely sensed data in the creation of physical models of the built environment has increased in recent years, following cost reductions in data and their wider availability. Many previous studies have concentrated on the extraction of urban features from various types of imagery, but the majority have neither assessed nor quantified the accuracy of the various stages of the modelling process. An understanding of these accuracy issues is key to quantifying the propagation of error within the whole modelling process and to assessing the fitness-for-purpose of the city model products. This chapter presents preliminary research investigating the early stages of physical model creation from first return LiDAR data of Southampton, United Kingdom and discusses both the identification and extraction of urban objects from remotely sensed data. The importance of the accuracy of both the imagery and the models derived from it are discussed using specific examples. We focus on urban remote sensing research within CASA, where emphasis is on the creation of city models for use in analysis and predictive modelling. This also complements research into remote-sensing algorithmic methods in geomatics and photogrammetry.

1 Data sources for urban remote sensing

This chapter presents research focusing on LiDAR-derived models and their accuracy, and also reviews aspects of urban remote sensing in general to better put the research at CASA into context.

Urban remote sensing

Urban remote sensing entails the collection and analysis of imagery of heavily populated areas which comprise dense networks of artificial structures. It is an active field of research, partly because of the wide variety of data sources that may be used for creating city models.

Data sources

Traditionally remote sensing of the urban environment has been conducted using aerial photographs. Photogrammetry is the science and technology of obtaining spatial measurements from photographs. Typical procedures involve measuring distances, areas and elevations. Aerial photographs can be used to generate accurate digital elevation models (DEMs), orthophotos, thematic GIS data, and numerous other derivatives (Lillesand and Kieffer 2000). Photogrammetric techniques continue to be used extensively today in many organisations. However, the focus of much urban remote sensing research has shifted in recent years to alternative methods of imaging—both from Earth-orbiting satellites and more recently onto airborne and terrestrial platforms. Donnay, Barnsley and Longley (2001) suggest that this transition can be explained by the wide availability of such imagery, greater frequency of update and falling costs. In addition, the use and fusion of different sources of remote imagery can be used to create much richer models, which may include such characteristics as surface roughness, vegetation indices and thermal properties.

Images from various platforms have been used for many different applications. Data acquired from satellite platforms (such as the Landsat 5 thermal channel) have been specifically used to understand the complexity of the urban environment and in many cases to predict patterns of change and growth. In recent years, airborne platforms have been increasingly used to obtain higher resolution imagery. Such data, often at submetre resolution, have been used to infer urban land use (Barr and Barnsley 1999), population sizes (Doll and Muller 1999), settlement structure (Pesaresi and Bianchin 2001) and to study the effects of pollution in the urban environment on vegetation (Malthus and Younger 1999).

Most recently, vehicle-mounted and human portable scanners have been used to produce very high-resolution data which can be used for street reconstructions and for pedestrian modelling (see Batty, this volume; Zhao and Shibasaki 2001; Kitazawa 2000).

However, one of the most common applications of such high-resolution data in recent years has been in the production of 3-D city models. There exists a huge range of potential applications for 3-D city models including urban regeneration, tourism and visualisation of the urban environment—not least because all of these applications focus on the visualisation and communication of information. More specific uses such as telecommunications cell planning, tracking of noise and air pollution, environmental impact assessment, carbon budget calculations, and flood modelling all require models that are both highly accurate and highly precise (in the range of centimetres rather than metres). The emphasis in these applications is not so much on the aesthetics of the model but on the mode of creation and the reliability of measurements and analysis that may be performed on it.

At CASA there are a number of projects that have looked at the creation of city models for the applications mentioned above. One such project used a city model to perform a visual impact assessment of large buildings on protected views of St Paul's Cathedral in London. The resultant city model is shown in figures 1A and 1B.

A common theme of much of the research conducted at CASA is the development and assessment of a range of novel techniques integral to creating digital city representations (see Smith and Evans, this volume), and to quantifying and evaluating differences between competing techniques in an effort to understand the errors and inaccuracies within models. Many data sources exist for this purpose, and these are increasingly widely available at low cost. Given this, it is imperative that a fuller understanding of the limitations and the subsequent applicability of these data sources is gained. This topic is the focus of the current research reported here. The use of airborne urban remote sensing within multidimensional models is an active area of research, both at CASA and within the wider academic community. Researchers from a wide variety of specialisms, including computer vision, computational geometry, VR, GIS and remote sensing are currently involved, perhaps reflecting the importance of this area of research and the urgency associated with the efficient creation of up-to-date city models. CASA's contribution towards this field of research offers a new approach towards understanding the reliability of the physical models.

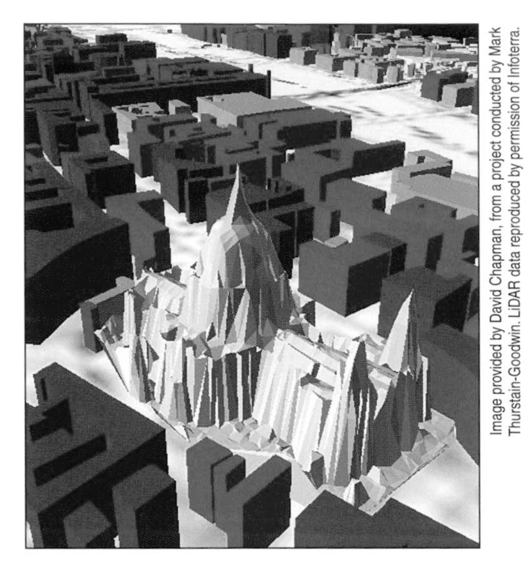
Image provided by David Chapman, from a project conducted by Mark Thurstain-Goodwin. LiDAR data reproduced by permission of Infoterra.

Figure 1A A LiDAR-derived model of St Paul's Cathedral, London.

Image provided by David Chapman, from a project conducted by Mark Thurstain-Goodwin.

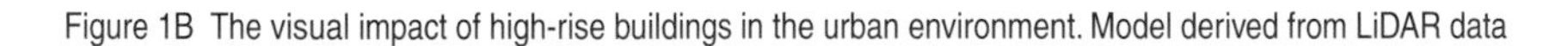

Figure 1B The visual impact of high-rise buildings in the urban environment. Model derived from LiDAR data

There are a great variety of data sources available for creating models of the built environment. The most commonly used are very high-resolution satellite images and those derived from airborne RADAR and laser sensors. This chapter focuses on the utility of airborne LiDAR imagery and places it in context, with a brief discussion of the benefits and suitability of the alternatives.

Very high-resolution satellite data

Of all the satellite products currently available, the two most relevant in terms of urban feature extraction are IKONOS and QuickBird images. IKONOS is now commercially available at a spatial resolution of 1 m and QuickBird at 0.6 m. These data have been used in a variety of studies. Fraser, Baltsavias and Gruen (2002) investigated the potential of IKONOS imagery for submetre 3-D positioning and building extraction. They found that by using only six ground-control points a planimetric accuracy of 0.3 to 0.6 m and a vertical accuracy of 0.5 to 0.9 m could be achieved (using C-C Modeller for reconstruction). However, it should be noted that the ability to capture larger buildings accurately does not necessarily mean that small buildings have been captured—this is important for assessing the overall accuracy of the 3-D city model.

Recent work by Sohn and Dowman (in press) has also reported promising results. However, in that study only large detached buildings were of interest, and as such, the application of this technique for the entire city environment is constrained.

Synthetic aperture RADAR (SAR)/ IfSAR (Interferommetric SAR)

A larger number of studies have been conducted based on extraction from SAR/ IfSAR images, perhaps because of the relative ease of access of this data source (SAR data can be acquired very rapidly) and, of course, its competitive costs. IfSAR imagery is particularly well suited to building reconstruction because these data include measures of elevation in addition to the coherence and intensity of the backscattered radiation. Both the coherence and the intensity measurements can be combined to give specific information about the scattering properties of the viewed surface. The fairly obvious disadvantage of using IfSAR instead of optical aerial photography is that the IfSAR images are generally of lower resolution and they also contain a greater amount of noise than the alternatives. They also contain specific artefacts that must be removed before analysis can be conducted. The removal of these problematic features of IfSAR data was the focus of early work in this area. In addition to the intrinsic IfSAR system noise level, there are also problems of multiple scattering that are attributable to building geometries and to layover effects.

Despite their popularity, radar derived 3-D images are more difficult to analyse than other images (Gamba and Houshmand 2000). These difficulties arise as a result of the stronger interaction of the electromagnetic fields with urban materials in the microwave region than in optical wavelengths. Shadowing and layover effects, caused by line-of-sight masking and multiple reflections from urban surfaces, result in significant problems for the image analyst.

Photogrammetry

A range of techniques for feature/object extraction from aerial photographs is available in the literature. Images may be processed to facilitate the extraction of edges or homogenous regions. These edges are subsequently combined using geometric or perceptual rules in order to complete the object description. Bellman and Shortis (2000) have described how in some cases edges are, in fact, matched to models of generic objects and how, as a consequence, the recognition of these objects occurs in the reconstruction phase. This, in fact, implies a cognitive approach to vision, where the recognition task is largely performed as a cognitive rather than a visual process (see Li and Maguire, this volume for a broader discussion of cognitive issues and representation). Results using aerial photographs are often the most detailed and offer an attractive methodology for urban model creation. However, there is a clear difference in the degree of operator intervention that is required for modelling from different data sources. Aerial photographs, in particular, require a considerable amount of user intervention. This makes it a time consuming, expensive and rather inefficient way of creating models. Having said this, some recent papers such as Gerke, Heike and Straub (2001) have used a more automated approach towards the detection of buildings. In this study knowledge about each building's surroundings was employed in order to support the detection of individual buildings.

LiDAR and its potential

This chapter focuses on the use of first return LiDAR data in the creation of city models. The LiDAR acronym stands for **L**ight **D**etection **A**nd **R**anging (there is another definition of LIDAR as **L**aser **I**nduced **D**etection **A**nd **R**anging, and this refers to the same laser scanning data). It is a particularly useful data source for city modelling for a number of reasons. First, LiDAR is an active remote sensing technique. Pulses of laser light are directed towards the ground. The time taken for these pulses to return to the sensor is measured and processed in order to determine the distance between sensor and the object or surface. Recently, very high resolutions for LiDAR have been obtained from both airborne and ground-level sensors. Several additional advantages of LiDAR have been reported in the literature. Lawless (2001) suggests that digital surface models (DSMs) created by LiDAR are the most suitable for the creation of models of urban form. Furthermore, it has been noted (Smith et al 1997; Day and Muller 1989) that LiDAR does not suffer from the breakdown of image-matching algorithms that is common when using other sources during the creation of stereoscopic models. Finally, DSMs derived from either photogrammetry or IfSAR require intensive operator intervention, making them more expensive (time-wise) to produce. It should be noted that LiDAR differs from the other data sources described above in that essentially it is a technique for measuring the height of objects, while other data sources all derive height from inferences based on the subsequent analysis of the imagery.

In the research presented here, airborne LiDAR is being used to create a three-dimensional city model. Many 3-D city representations have been constructed in the past from a variety of data sources for specific applications. However, this model aims to be more generic and to focus more on understanding the differences between elevation data derived from different image sources. The research focuses on error and uncertainty within the model, and the spatial autocorrelation of errors is also investigated. All of this research aims to aid understanding of elevation models so that they may, in the future, be used most appropriately.

2 Methodology: the first stage of city model creation from LiDAR

The modelling process involves two fundamental stages: first, the detection of urban features above ground height and, second, the identification or differentiation of these features into classes of objects such as buildings, vegetation and street furniture (artefacts). The methodology developed for these two processes in this research is presented below.

Defining the ground height

The first stage of constructing a multidimensional model from remotely-sensed imagery involves ascertaining the height of all features above ground. Raw image data (figures 2 and 3) return height values for pixels that represent the height

Figure 2 Raw LiDAR image of the football (soccer) stadium, Southampton, United Kingdom

Figure 3 Raw LiDAR draped with aerial photography showing the football (soccer) stadium in Southampton, United Kingdom

above datum rather than above the ground. Left in this raw state these height data cannot be used for creating a city model. The true heights of the objects can nevertheless be ascertained if a ground-height elevation model is subtracted from the raw DSM. The subsequent elevation model contains all feature heights and is often termed a normalised digital surface model (nDSM). Clearly the accuracy of the derived nDSM is almost entirely reliant on the accuracy of the ground height model or digital elevation model (DEM) that is used in this process.

Such ground height models can be derived from raw imagery data using a variety of techniques. A new technique has been created here which aims to create a very accurate ground surface model (bare earth DEM). The accuracy of this bare earth model has been assessed against ground heights from Real Time Kinematic (RTK) GPS (taken by the author).

Without a full appreciation of the accuracy and the spatial variation in error of the ground model used, the subsequent accuracy and precision of the derived city model cannot be ascertained. This problem provides much of the motivation for much of the recent research into the creation of an accurate bare earth digital elevation model from LiDAR data at CASA.

Modelling ground height from LiDAR imagery

The identification of elevated objects from a DSM is more difficult than might first be expected. As stated above, raw LiDAR pixel values do not reflect the actual height of objects above ground level, and, as such, classification into sensible objects cannot be conducted on the basis of pixel values alone.

The best method identified within the context of this research applies object-oriented image processing techniques to the DSM in order to classify cells or groups of cells in accordance with the difference between each pixel group and its neighbours. So, for example, if a pixel group is considered to have a mean difference from its neighbours of more than a specified amount, it is classified as being an object above ground level.

The algorithm that has been written as part of this research passes through several iterations of this classification, searching for progressively smaller groups of pixels in order to capture information about small objects such as trees and cars. The image objects are assigned to one of two classes: either above ground objects or bare earth. Then they are used to cookie-cut the original DSM image. This essentially removes any object greater than one square metre in plan form and classified as being above ground.

The resulting DSM contains only ground-level information but is no longer spatially continuous because of gaps left after the objects have been removed. In order to restore continuity across the dataset an interpolation method has been employed. Of course, there are many different methods of interpolation, which are each appropriate to specific circumstances (see Thurstain-Goodwin, this volume; Bailey and Gatrell 1995). The method used here employs simple techniques of spatial prediction using a kernel methodology. In this particular algorithm a kernel of varying size passes over the data. Within the boundary of each kernel the mean value of successive pixels is first calculated and then attributed to any pixel in the gaps between successive windows. This process is performed iteratively to build up sensible values across the gaps and is shown diagrammatically in figure 4 below.

The resultant model thus represents the bare-earth digital ground, as shown in figure 5.

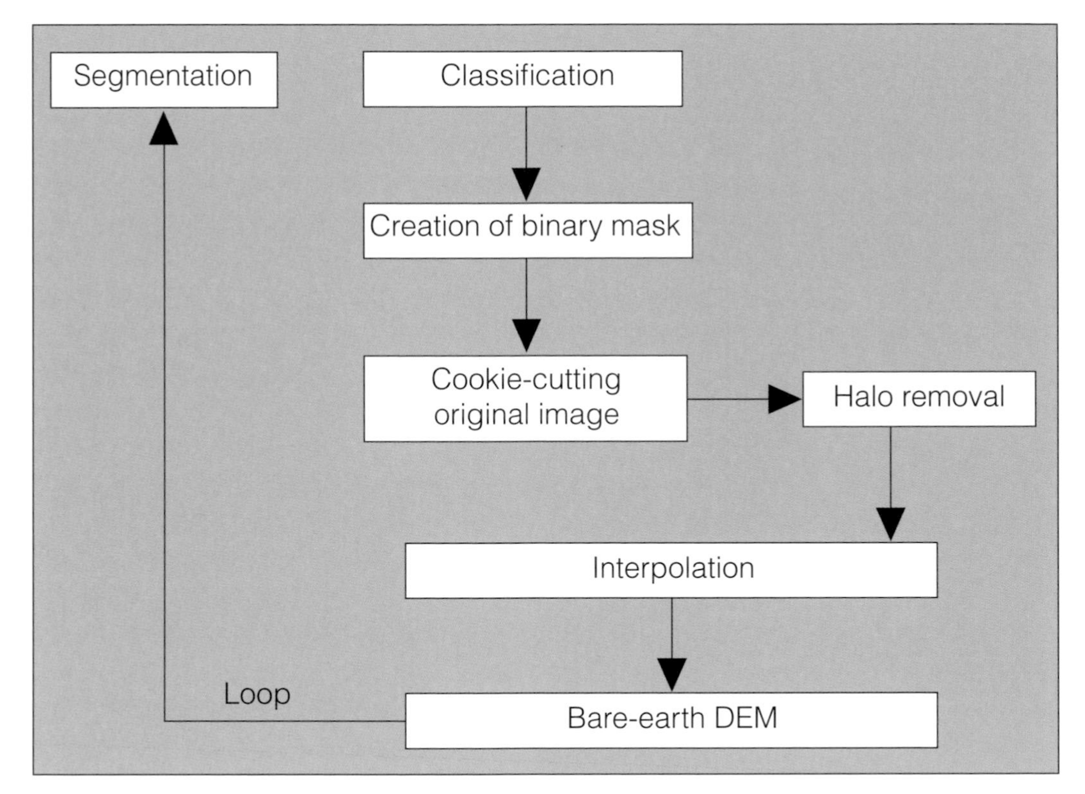

Figure 4 Methodology flowchart

What is the surface of the earth?

However, this methodology assumes that the bare earth is simply the surface model minus any above-surface super-structure, an assumption that essentially treats the surface of the earth as a binary boundary. In other words, objects are either part of the bare-earth surface or they are not. Whilst this is an attractive simplification, there are some fundamental associated problems arising out of the

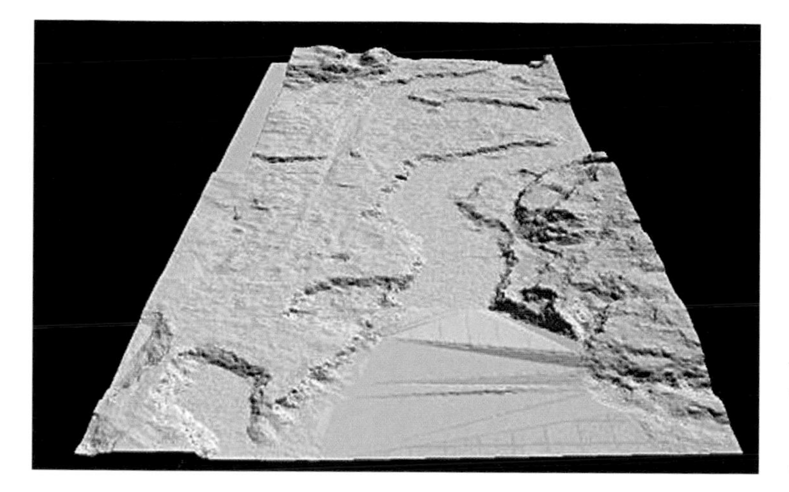

Figure 5 The continuous bare-earth digital elevation model produced from the algorithm described above

inherent uncertainty in all geographical classifications of the real world. A more detailed discussion of such uncertainty is beyond the scope of this chapter (for a fuller discussion, see Zhang and Goodchild 2002).

In many applications the uppermost level of concrete in the urban environment can be taken to be the earth surface. In accordance with this definition the concrete overlying the natural earth surface essentially becomes the most important surface and is referred to as ground level. However, problems arise when the concrete surface does not lie over the earth but is a free-standing structure in its own right. These structures are very common in the urban environment and include bridges and fly-overs at motorway intersections. Here the different levels of the motorway present multiple artificial surface levels, rather than a single Earth surface measurement. Depending upon the application for which the model is to be used, the upper surface may or may not be considered to be bare earth. In this instance the ground height model is used to assign heights to objects for inclusion within a city model. If the case of a road bridge over a railway line is considered, it is obvious that we may wish to include the height of any street furniture along the bridge or the height of vegetation beneath the bridge within the model. In both instances it must, therefore, be assumed that the road surface should be included within the bare-earth model. For this reason the raw LiDAR height of the road network was included within the bare-earth model in a further extension of the algorithm. The result is shown in figure 6.

Applications such as flood modelling, line-of-sight analysis, and particularly telecommunications planning would all require the inclusion of such overlapping transport networks. With this in mind, the heights of all railways were also

Figure 6 The bare-earth model including road heights overlain with aerial photography

included to permit the road network to fly-over. It should be noted here that where a road passes over a railway or another road, only the top surface is modelled. It is the aim of this research to eventually investigate methods of creating a fully 3-D model, although research to date has not attempted this.

Creating the object height model and assessing its accuracy

As already discussed, bare-earth elevation models are used in the creation of 3-D models from remotely sensed imagery. For such cases, the bare-earth models are subtracted from the surface model (DSM), which is obtained from the raw remotely sensed LiDAR data. This operation produces a normalised digital surface model (nDSM) that contains the elevations of objects above the surface of the earth: in other words, this process returns relative object heights rather than absolute heights above datum.

Clearly, if there are significant errors in the bare-earth model, the subsequent nDSM will depict the incorrect heights of objects. This may have severe implications for analysis based on such 3-D models. For this reason, considerable research has been undertaken to quantify the accuracy of the model.

The accuracy and error of DEMs has been a subject of much discussion in recent years. However, there has been a general paucity of research into the accuracy characteristics of the elevation data. Indeed, metadata supplied with DEMs often give an indication of how the cell elevation values have been calculated, but rarely does this indicate how accuracies vary across the surface, or indeed what the surface represents (or in other words how the bare-earth surface has been defined).

As part of this research, some preliminary investigation was undertaken to ascertain whether the bare-earth model (DEM) accurately represents the bare surface of the earth. The DEM designed here has been validated against RTK GPS readings that are accurate to 2 to 4 cms in the z dimension. Over one hundred readings were used in this validation process. Initial analysis has indicated that the model is more accurate than some alternatives. The model created here has a lower root mean squared error (RMSE) value from the GPS results than models derived from photogrammetric methods (table 1). This suggests that the use of LiDAR for the creation of the bare-earth DEM may reduce subsequent error in 3-D city models.

RMSE bare-earth model produced here	10.33
RMSE environment agency model	11.22
RMSE photogrammetric model	9.38
RMSE ordnance survey profile	13.16

Table 1 Root mean square error estimates of four models

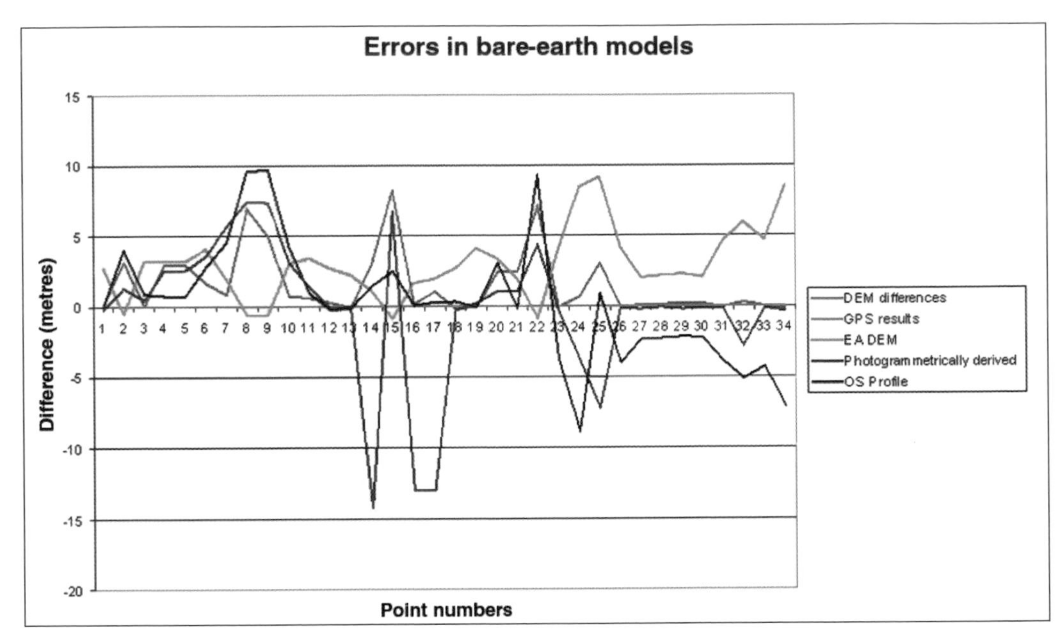

Figure 7A Comparison of the accuracy of four bare-earth models, produced from different data sources

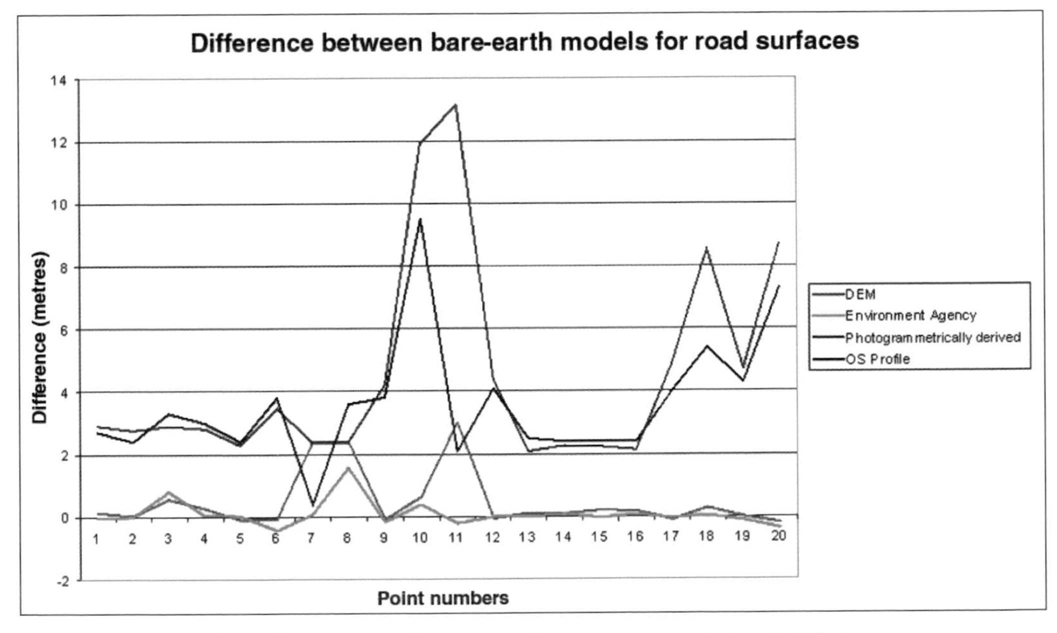

Figure 7B Comparison of the accuracy of four bare-earth models for road surfaces

The RMSE values shown in table 1 highlight the lower values for the LiDAR models. However, it should be noted that these values are all still rather high. Indeed, when comparing models of the bare earth, it is interesting to look not only at the overall level of accuracy within the model but also at the spatial variation of error. Figures 7A and 7B show the differences between the GPS readings and four bare-earth models for 35 points in the dataset. It can clearly be seen that there is some correlation between the results, in that in some instances all four models have a large error for the same point. This seems to suggest that the terrain at this point was difficult to model.

The 35 points used previously were simply classified in accordance with the terrain they represented. Attributes of points were classified as road, water edge, vegetation or part of a bridge structure. The models all performed much better on road structures, and within this investigation it was found that the DEMs produced from LiDAR were more similar to the GPS readings than those produced from other data sources. Further research will investigate methods of modelling water-edge environments from LiDAR with greater accuracy.

This process of quantifying the spatial variation of errors within the bare-earth model aids in our understanding of the uncertainties inherent in the derived height model (nDSM). The accuracy of these object heights may be compared with photogrammetrically derived heights, such as those of buildings, for further validation. This uncertainty can then itself be modelled, as Zhang and Goodchild (2002) have described. These authors have used uncertainty descriptors such as surfaces of standard error that have been derived as by-products of Kriging (see Burrough and McDonnell 1998 and Longley et al 2001 for overviews of this technique). Indeed, the importance of using measures such as these cannot be underestimated, as errors within digital elevation datasets reveal crucial information about the elevation represented in the model.

Thus, quantification of the degree of accuracy of the bare earth and the image objects can be used to assess whether the results are suitable for the next stage of analysis. Assuming that the predicted bare-earth and object heights fall within a suitable tolerance threshold, the object height pixels may be classified, or segmented, for inclusion within the city model.

Segmentation of the images

Segmentation, or differentiation as it is often called, involves the classification of the surface model into object classes based on a number of possible criteria. These

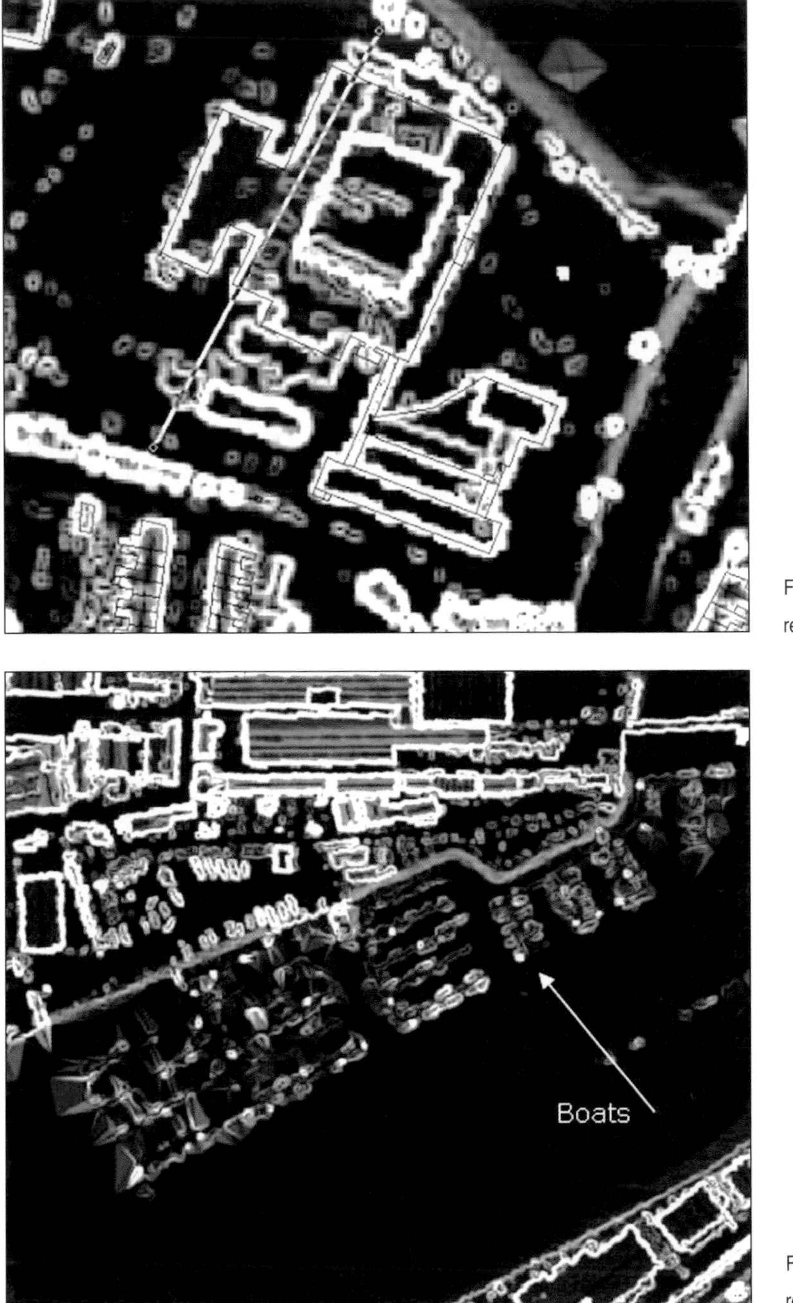

Figure 8A Gradient image revealing buildings

Figure 8B Gradient image revealing boats

may include spatial, geometric or spectral differentiations (depending upon the type of imagery used).

This research is focused on the potential of LiDAR for the creation of city models, and as such it is most appropriate to use spatial and geometric properties of the image as a basis for segmentation. Several geometric properties have been used for image segmentation, including gradients of pixels, the shape of image objects, the compactness of objects, mean difference between neighbours and the mean height of objects. All of this segmentation has been achieved using simple object-oriented techniques within the E-Cognition™ environment. In this way, object recognition has been achieved using context and the proximity of neighbouring objects in order to produce the best classifications.

One of the most useful parameters for segmentation is that of gradient. A gradient image, such as that shown in figure 8A, can be used to detect the edges of the buildings and vegetation. The gradient value for each pixel can be further used to differentiate between different classes of buildings and vegetation. Furthermore, as can be seen in figure 8B, the gradient image is a particularly effective way of picking out small objects such as cars and boats. It is important to identify such objects when creating a model from LiDAR data, as these need to be excluded

Figure 9 A simple level of segmentation of the nDSM into image objects. Blue pixels represent the bare earth, green pixels depict vegetation and red pixels are classified as buildings

from the final model. The fact that they can often be identified according to their pixel gradient values is particularly useful.

Some promising results have been achieved using constraints such as the gradient value of pixels in conjunction with other spatial parameters. Buildings, vegetation and bare earth have been differentiated reasonably successfully. Furthermore, some initial research has also shown that different types of buildings can also be differentiated in this manner. Figure 9 shows some preliminary object differentiation.

The segmented and classified image objects can subsequently be reconstructed. This reconstruction phase is an exciting and very active area of research, particularly in a number of institutions in Germany (Technical Universities of Dresden and Munich) and the Netherlands (TU Delft). Recent work has focused on the level of automation that may be achieved from different processes and from different image sources. In many cases the goal of research in this field is the creation of a fully automated algorithm that can reconstruct the 3-D object vectors, create the topology, and verify the accuracy with which this has been achieved. However, despite this being an active field, an immediate solution is not yet on the horizon, though many current developments show considerable promise. The recent focus on LiDAR data for the creation of city models perhaps reflects its enormous promise and potential.

Figure 10 depicts some 3-D building structures that were derived photogrammetrically. It is envisaged that a similar level of detail will be achieved from LiDAR

Figure 10 Bare-earth DEM draped with aerial photograph of area, overlaid with 3-D buildings

in the next stage of research. This reconstruction phase and its validation is seen to be the final product of the research within the current project.

3 Prospects

Work in the next phase of the research will focus on the extraction of roof morphologies from LiDAR data in order to increase the detail and the usability of the final city model. Accurate roof morphologies are required in many applications (see Smith and Evans, this volume), most specifically by the telecommunications industry and for many detailed rain run-off models that need to know not only the shape of roofs, but also the gradient of the roof slope, the surface material, and potentially, information about artificial drainage networks (gutters) that form part of the roof. Extracting such information accurately from remotely sensed data has enormous implications for the type of analysis that may be conducted in 3-D GIS environments and would open up a range of potential future GIS applications.

Promising work using a random sampling technique has been noted for the extraction of roof morphologies from LiDAR. Instead of using the LiDAR dataset for detecting edges, the dataset can also be used for detecting roof components, which Lawless (2001) claims can provide a more flexible, although also more arbitrary surface reconstruction of buildings. The acquisition of roof components may be achieved through the dual processes of primitive fitting and primitive extraction (see Lawless 2001 for more detail). Such an approach represents an exciting approach to LiDAR modelling which requires further investigation.

In addition, the increasing availability of point data of greater density per unit area and the supply of first and last return data together with intensity images means that this area of research is currently expanding rapidly as opportunities for development and analysis open up.

The work presented in this chapter has been the result of the initial stages of CASA research looking at optimal methods for the extraction and reconstruction of objects. Further work will look at a comparison of methodologies for the recognition of objects within the height model. Such methodologies will include the use of wavelet coefficients, neural networks and model-based methods such as the Hough Transformation. This comparison will describe the errors and uncertainty within each result produced from the different methodologies and thus will represent an exciting new area of research in this field. Indeed, most studies have concentrated on the derivation of one very detailed algorithm that focuses on one part of the model creation process. This research, in contrast, attempts to provide a more holistic view of the modelling process and to ascertain how uncertainty is created and propagated to the final result. Only with such an understanding can we really begin to use the elevation and city models for analysis with any degree of confidence. Perhaps the most important point to make about this research is if we want to follow the transition from data (raw LiDAR) to information (the city model) to knowledge (the analysis on the model), then we must understand the spatial pattern of uncertainty within the original data and how this propagates through the modelling process. We can only begin to move from data to information when we have the combination of data plus quality metadata.

The research at CASA builds on previous work from fields, including computational geometry, remote sensing, mathematics and computer science, but it offers a different approach to the use and analysis of the models. It is anticipated that this research will facilitate a wider understanding of the utility of LiDAR for city modelling. It was identified earlier in this chapter that 3-D city models are used by telecommunications companies and environmental planners for prediction and planning in analysis that often has fundamental environmental and legislative consequences. Accordingly, this research represents a key stage in the development of both the creation and the use of models of the built environment.

Acknowledgements

The author wishes to thank Ordnance Survey (Great Britain) who financed this research.

Section *III*

Location in physical and socio-economic space

GIS provides critical context to decision making whenever we ask the question *where*? Chao Li and David Maguire make a measured appraisal of the ways in which handheld technologies allow us to both ask and answer this question in an ever-wider range of settings, and then examine some implications of the ways in which we might use them to understand more about human behaviour. This kind of understanding, coupled with clearer thinking about store formats and adaptation to change, is of crucial importance to the retail industry, as described by Paul Longley, Charles Boulton, Ian Greatbatch and Michael Batty. Retailers remain staunch advocates of the use of geodemographics, which provides one of the cornerstones to the GI data industry. Here, Richard Webber and Paul Longley describe how this approach is being extended from core areas in retailing to encompass public service planning, and in so doing, geodemographics are being used to reveal much more about the difference that place makes. Locational sensitivity is a theme that is developed by Daryl Lloyd, Muki Haklay, Mark Thurstain-Goodwin and Carolina Tobón in their discussion of the design of GIS to visualise structure in urban data.

Chapter

[Location in physical and socio-economic space]

The handheld revolution: towards ubiquitous GIS

Chao Li and David Maguire

Over the last decade, information and communication technologies (ICTs) have developed rapidly, and there has been remarkable convergence between what previously have been considered quite separate developments. Recent innovations in wireless telecommunications have led to rapidly increasing numbers of mobile device users. Further diversity in the range of GIS applications will be an inevitable consequence. This chapter begins with a discussion of the current technological setting, with reference to developments in handheld GIS, mobile communication networks, positioning technologies and location-based services. A series of emergent new research issues is then identified and discussed in relation to mobile location-based services and spatial knowledge acquisition. Interest in such research issues is seen as a further stimulus to the growing ubiquity of GIS technology.

1 Introduction

Since the early 1990s, information and communication technologies have continued to develop rapidly. This decade saw the innovation of the Internet, the advent of the World Wide Web as a mass communications medium, the near ubiquitous application of networked PCs in business and in recreation, the growing use of the Global Positioning System (GPS), and the maturation of geographic information systems (GIS) for the management, analysis and visualisation of geographic data. There have also recently been rapid developments in wireless telecommunication networks, the advent of high-level communication protocols, and mass production of handheld wireless devices. GIS have converged with other information and communication technologies such that the potential now exists for GIS to become ubiquitous as mobile geographic services.

The number of mobile device users increased dramatically during the late 1990s. More than 40 per cent of the population of Europe and nearly 70 per cent in the United Kingdom own a mobile device (see Shiode and Batty, this volume). There are nearly 800 million mobile-phone subscribers worldwide. This number is predicted to reach one billion by 2003 and 1.3 billion by the year 2005 (source: www.idc.com). In the United States, further usage growth will be stimulated by the U.S. Federal Communications Commission enhanced 911 (E911) mandate. This requires reporting the location of mobile phones used for emergency calls and provides the basis for a wide range of additional, value-added location-based services (LBS). There is also considerable market drive for providing such services in Europe. The new application areas for providing such location-related information and services in real time are more focused on networks of individuals and in mobile contexts. In view of this prospect, there are substantial research opportunities concerning communication and interaction between individuals through such applications.

In this chapter, we describe developments which could herald a new age in behavioural geography (Golledge and Stimson 1987), particularly with regard to measuring and monitoring individuals' cognitive abilities, their spatial awareness and the ways in which they acquire spatial knowledge. Current projects at CASA are at the forefront of this revitalised approach to behavioural geography (see Clarke 1998 for the general case of field computing in physical geography). We begin by considering how handheld devices and mobile location services offer the prospect of making GIS a near ubiquitous technology. We then go on to discuss the spatial cognitive processes that are important in acquiring spatial knowledge using technology. This leads us finally to focus on a number of research issues concerning location-based services applications in current research at CASA. There are also links between the material discussed here and some of the other drivers towards ubiquitous GIS that are covered elsewhere in this book (for example, Hudson-Smith and Evans, this volume). Some of the related hardware developments not discussed here include wearable GIS, with which people have wearable hands-free input devices and direct display into the human vision field, and transportation telematics such as in-car navigation devices.

2 Handheld GIS and mobile location services

Technological developments in the further miniaturisation of computer components have led to the wide availability of handheld devices. This, in turn, has stimulated GIS solutions that utilise such devices and the rise of mobile location services. This section develops the theme that the current technological environment will provide the potential for ubiquitous GIS and also pose important research challenges in the area.

Handheld GIS

Software for mobile mapping and GIS for handheld computers and latterly personal digital assistants (PDA) has been available since the late 1990s. Such mobile GIS usually comprise software with a restricted command set and lightweight hardware. For real-time positioning, this configuration is coupled with GPS, and real-time communication is achieved through wireless technology. This is bringing about changes in the way geographical information is collected and used. For data collection, geographical databases can be compiled directly or checked and amended using a handheld GIS device in the field. For users, GIS have become easily portable and can be queried during work, recreation or social activities.

Although many people focus on the handheld component of a mobile GIS solution for good reason, of equal if not greater importance is the back-end support system (Maguire 2001). Current handheld devices do not have the processing capabilities, screen real estate, or software engineering tools to create and run sophisticated applications. As such, they are best seen as complementary mobile field tools. Desktop systems are useful for data preparation and integration tasks such as data format conversion, projection and data selection. Enterprise servers can be used to manage multi-user access to large databases and to host sophisticated GIS application services (for example, complex land suitability or traffic assignment models: see Evans and Steadman, this volume).

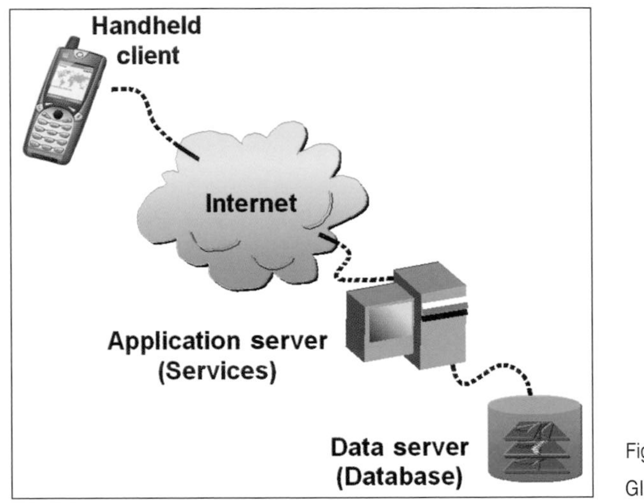

Figure 1 A generalised architecture for handheld GIS solutions

A generalised architecture for a handheld GIS is shown in figure 1. The key components are:

- multi-user database(s) able to serve and, for read-write mobile applications, receive data. Large databases should be built and managed using a commercial off-the-shelf database management system (DBMS) technology (Longley et al 2001: chapter 7).
- application server(s) able to provide geographic information and processing on demand to distributed handheld, mobile clients either directly or via a desktop interface (see the following). In modern implementations of this architecture, the geographic data and processing capabilities are exposed as Web services. In this context, a Web service is a self-contained, self-describing software unit that serves data (for example, addresses, streets or land-cover areas) and processing (for example, mapping, geocoding, routing and land-suitability assessments) on request to handheld GIS.
- network connectivity. This may be a wired local, wide-area or Internet network if the connectivity to the handheld device is via a desktop system. Alternatively, if there is no GIS intermediary involved, then an interface to a wireless network (most usefully a cellular telephone network) is required.

Bandwidths vary from about 10 kbps (wireless 2G telephone) to over 100 mbps (ethernet).

- handheld GIS application. The handheld device should be able to use a wireless link direct to Web services running on one or more application server(s) or be connected to a desktop system. The unit may run all software locally (thick client), or else access data and applications over a network (thin client).

One of the interesting aspects of mobile and handheld solutions is their application and technological diversity. Applications range from agriculture (for example, agricultural census field checking in Iowa) to zoology (for example, mapping elephant distributions in Kenya): see figure 2. Technologies range from, for example, very lightweight locationally sensitive telephones able to perform simple tasks like displaying image maps to comparatively heavyweight laptops weighing 10 pounds or more that are able to run highly sophisticated, full-featured GIS applications used for utility work order management. Table 1 classifies the

Figure 2 A handheld GIS (ArcPad) with GPS device used for mapping fire perimeters

Characteristic	Consumer	Simple professional	Advanced professional
Weight	< 4 oz	< 1 lb	< 10 lb
Screen	Colour or mono, 200 × 180 pixels	Colour, 320 × 240 pixels	Colour, 1024 × 768 pixels
Cost	<$400 (often free with phone plan)	< $700	< $2000
User Interface	Voice, simple keyboard	Pen, simple keyboard or voice	Pen, full keyboard or voice
Location determining technology	Low accuracy, network determined, or GPS	Medium to high accuracy, GPS	High accuracy, differential GPS
Battery	> 24 Hours	> 8 hours	> 3 hours
Location of business logic and data	Server	Client, and possibly server	Client
Applications	Very easy to use and highly focused	Medium complexity, highly focused	Advanced, general purpose
Expandability	None	Some	Extensive

Table 1 Generalised characteristics of current handheld GIS solutions

main dimensions of the handheld market into three main groups: consumer systems, simple professional and advanced professional. The entries in the table are generalised to emphasise the variations in systems, although in practice, there is considerable overlap.

Consumer systems are typically characterised by their limited capabilities and are used to perform simple applications such as image mapping, point-to-point routing, and locating the closest facility. Consumer systems are essentially mobile phones enhanced to support geographic applications (usually termed 'location-based services', LBS, in this market). Location is usually ascertained by the telecommunications operator and is provided as an inherent part of the phone's service plan. Both the business logic and the data are maintained centrally and are delivered to the handheld device as a service. Simple professional handheld applications are based on PDAs or pocket PC hardware. Typical applications include mobile mapping and simple field data entry (both of geometry and attributes). The hardware is relatively lightweight, fairly low in cost, and moderately easy to use. The business logic and data can be stored on the handheld device,

Figure 3 Some devices and software/communication protocols used for handheld GIS applications (WAP = wireless application protocol, MMS = multimedia messaging, J2ME = Java2 Micro Edition, PC = personal computer)

but data and processing services may be accessed over a wireless network connection. Advanced professional handheld applications utilise standard notebooks. It is expected that newly released Tablet PCs (November 2002) will provide the platform for a significant part of this market in the near future. The focus here is on supporting advanced applications, rather than lightweight field portability. Example applications include utility work order management and sophisticated field-based data entry. Typically, all the data and business logic are held on the local machine; the data and processing required in professional applications typically exceeds the wireless bandwidth currently available.

An interesting trend was the emergence in late 2002 of hybrid systems that combine several of the capabilities previously outlined. For example, a recent innovation is a single system that combines the PDA and mobile phone.

Mobile communication networks and positioning technologies

The anticipated availability of broadband communications for mobile devices, the development of positioning technologies, and the growing volumes of available location-specific information will inevitably lead to the demand for services that deliver to individuals location-related information that is customised to their current location. Such services are generally known as location-based services (LBS). Many believe that GIS will play a central role, as they form the foundation for a large part of LBS developments. GIS has been considered a key area in LBS for managing, processing and delivering spatial information in meaningful ways.

Mobile telecommunication networks have developed dramatically (Dornan 2001) from

- second generation, 2G, such as Global System for Mobile Communication (GSM), Code Division Multiple Access (CDMA) and Time Division Multiple Access (TDMA) to
- 2.5 generation such as General Packet Radio Service (GPRS) and Enhanced Data Rates for Global Evolution (EDGE) and soon to become
- third generation, 3G, such as Universal Mobile Telecommunication System (UMTS), Wideband Code Division Multiple Access (W-CDMA).

The bandwidth of the networks is moving from 9.6 kbps to 115 kbps, and will support bandwidth of 2 mbps with UMTS. The development of mobile telecommunications has also transformed service status from focusing on transmission of voice data to various applications of transmitting multimedia information. Handheld, mobile and small size wireless devices such as PDAs and mobile phones have been enhanced and can be connected to wireless networks via infrared, GSM modem, or radio signal. With the development of location awareness technologies and the increasing number of mobile device users, services providing location-related information are likely to become major applications of the new technologies.

The key to LBS is sensitivity to location in service provision or the position-fixing capability of the mobile communication devices. There are several main location-determining technologies plus combination hybrids:

- Network Cell Identification (NCI) or Cell ID—this network-based method identifies the area around the mobile network base station used by a mobile communication device. The accuracy varies from about 250 metres in urban areas to less than 10 km accuracy in rural areas.
- Time of Difference Of Arrival (TDOA)—this network-based method calculates the time difference of the transmitted signal from the handset arriving at three separate base stations. The accuracy threshold can reach about 125 metres, although the cost is rather high because of the need for high synchronisation within the network of base stations. This technique can be used with existing generation handsets.
- Angle of Arrival (AOA)—this network-based method measures the angle of the same signal arriving from at least two base stations. The accuracy of this method is not reported but is likely to be similar to that of TDOA.

- Global Positioning System (GPS)—this device-based method works by including a satellite navigation receiver in the handheld device. The accuracy can be within 20 metres. Assisted-GPS (a second supplement signal) is required for covering situations such as indoor areas and tunnels.
- Enhanced-Observed Time Difference (E-OTD)—this is a device-based method that calculates the time taken for a signal to arrive from at least three base stations. The method requires installing software on the handheld device. The accuracy is about 50 metres. It can also be a network-based solution as a modification of the TDOA method previously explained.
- Bluetooth and Wireless Fidelity (WiFi)—these methods are mainly used for ascertaining location indoors and in dense urban areas. They can be deployed to complement other positioning techniques.

Mobile location services

In recent years, LBS have been depicted in various ways. Some regard them as 'an information service that exploits the ability of technology to know where it is and to modify the information it presents accordingly' (CSISS 2001). It can also be viewed as a new breed of mobile geographic application in the GIS area that provides up-to-date locationally sensitive information through wireless networks. In general, LBS can be considered the provision of location-relevant information to individuals via a mobile telecommunication network. Such information can be delivered to individuals irrespective of where they are, yet the location of individuals can be vital to the customisation and provision of such information. The convergence of new information and communication technologies, such as mobile telecommunication systems, locationally aware technologies and handheld devices with the Internet, GIS, and spatial databases, provides the foundation of LBS (Brimicombe 2002).

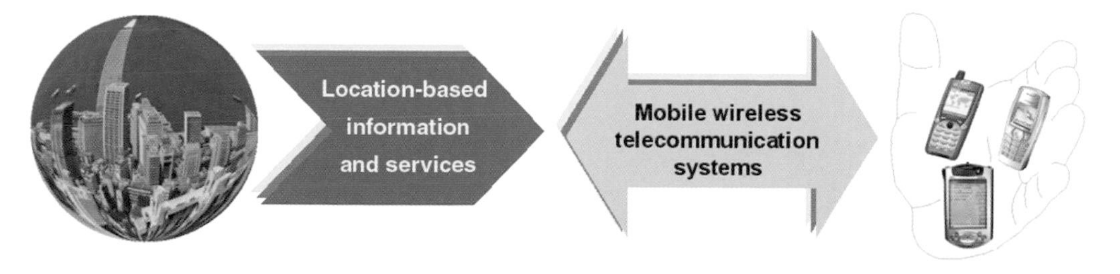

Figure 4 The concept of location-based services (LBS)

LBS have considerable growth potential. The typical applications can be assigned to the following categories (Hunter 2001), some of which are currently under development and some of which are already available:

- pushed online advertising
- user-solicited information, such as local traffic information, weather forecasts, and local services
- instant messaging for communicating with people within the same or nearby localities
- real-time tracking
- mapping/route guidance, directing people to reach their destination
- emergency services for a stationary location (see Schietzelt and Densham, this volume)
- location-based tariffs

A range of input and output methods has been proposed and offered in LBS applications in an attempt to identify the most desirable and effective way to deliver information. Voice has been suggested as a user-friendly method for inputting requests and giving instructions. Speech delivery via a mobile phone as navigating instructions during driving is one such example. A phonetic transcription of city names, street names, and points of interest to the digital maps is another example for inputting information through voice. Text is another appealing method for certain individuals and situations. Different types of graphics and images are also being studied according to their visualisation characteristics. The ways in which these technologies come to be used may well depend on individuals' spatial cognitive processes and familiarity with the type of environment in which they find themselves. For this reason, theories of spatial information acquisition become important.

3 Spatial knowledge acquisition

At this point, it is appropriate to review our understanding of spatial cognition, since this is likely to be a particularly important dimension to the design of mobile and handheld devices in which navigation and GIS are likely to be central. People's spatial knowledge structures, as embodied in their cognitive facilities, are generally viewed as providing the basis for interpreting places in the environment and are a subset of their total knowledge of an environment. In this section, we begin with a review of the typology of spatial knowledge before considering the processes of spatial knowledge acquisition.

A typology of spatial knowledge

Spatial knowledge has been classified into different types over the years, including egocentric and domi-centric knowledge (Trowbridge 1913), strip map and comprehensive map knowledge (Tolman 1948), and topological/projective/Euclidean knowledge (Piaget and Inhelder 1956). The concepts of route, survey landmark, and configurational knowledge can been seen in the research of Shemyakin (1962) and Siegel and White (1975). Kuipers (1978) suggests separate classes of sensorimotor, topological and metrical knowledge. Thorndyke and Hayes-Roth (1982) draw a contrast between procedural and survey knowledge. Stern and Leiser (1988) identify three levels of spatial knowledge known as landmark, route and survey knowledge. Golledge and Stimson (1987) further contend that spatial knowledge comprises declarative, relational/configurational and procedural components.

Declarative knowledge refers to those objects or places that have meaning or significance attached to them. It is also referred to as landmark knowledge, frames of reference or cue knowledge. Procedural knowledge concerns an individual's understanding of the process of how to travel or navigate from one locality to another and, hence, can also be defined as route knowledge. Route knowledge

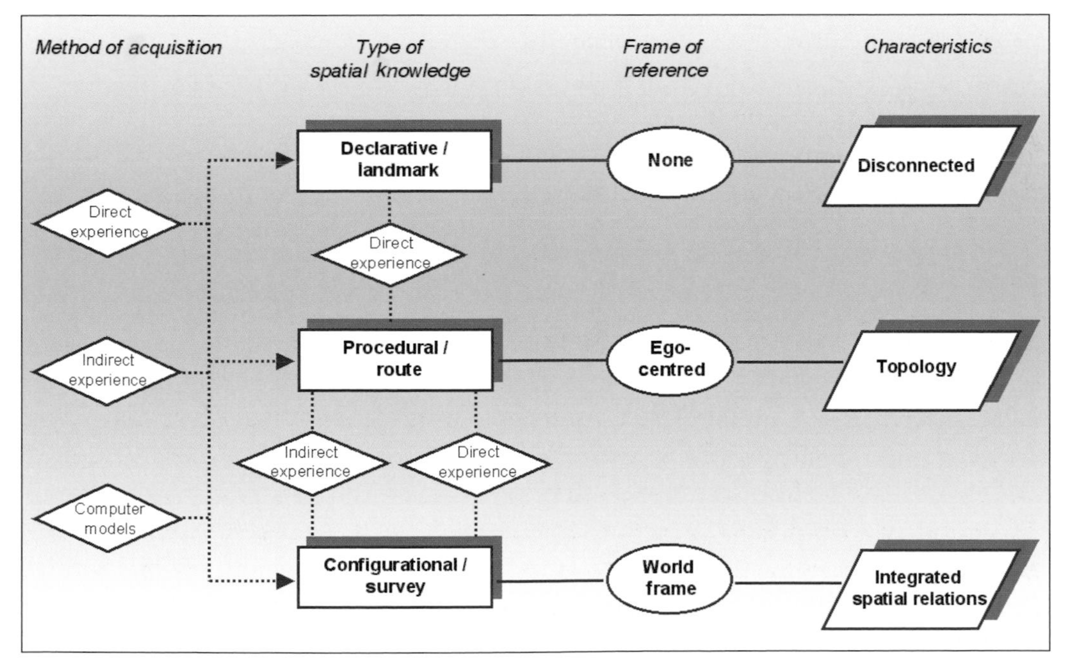

Figure 5 The typology of spatial knowledge as summarised from the literature

typically refers to knowledge about movements and mostly consists of procedural descriptions, some landmarks, and path elements. The term 'configurational knowledge' generally refers to integrated knowledge of the layout of a space and the interrelationship of the elements within it along with people's ability to traverse that space in complex configurations of paths and nodes within some external frame of reference. This knowledge is considered to comprise not only visual and geometric, relational, perceptive and descriptive information, but also spatial relations. Survey knowledge, relational knowledge and metric knowledge are generally regarded to be in this category. Configurational or relational knowledge, in particular, refers to knowledge about spatial relationships between objects or places. Figure 5 summarises the types of spatial knowledge and their characteristics.

Although this is a widely accepted classification, knowledge relating to areal information, denoted as areal or map-like knowledge, has been differentiated from configurational/survey knowledge in some of the literature (Aitken and Prosser 1990). Survey knowledge refers to the concept of 'sense of direction' (Kozlowski and Bryant 1977), while areal knowledge focuses more on the familiarity with places and routes in an area. The exact distinction between configurational or survey knowledge and areal or map-like knowledge seems unclear in most of the literature.

Spatial knowledge acquisition

People acquire and develop spatial knowledge through various experiences and processes, which may include recognising and understanding the characteristics of objects and localities, the interrelationship between elements in environments, and so on.

Shemyakin's theory (Shemyakin 1962) suggests that spatial knowledge can be acquired by a process of starting with landmark knowledge, progressing to route knowledge, and finally to survey knowledge. As knowledge is accumulated, so its accuracy in terms of angularity, direction and proximity improves. According to Piaget and Inhelder's development theories (Piaget and Inhelder 1956), spatial knowledge development progresses over four phases of one's life: the sensorimotor period covering infancy; the pre-operational period covering preschool age; the concrete-operational period covering middle childhood; and the formal operational period covering the age from adolescence onwards. An individual proceeds from an egocentric, prerepresentational space, to topological, projective and finally to a Euclidean metric relational structure through these development phrases. Stern and Leiser (1988) further suggest that spatial knowledge progresses from landmark knowledge through route knowledge and to survey knowledge through accumulated direct navigation experience or map-centric learning at different stages.

Siegel and White (1975) suggest that the process of spatial knowledge acquisition entails recognising landmarks, finding routes connecting the salient landmarks, and then developing complex and general configurational survey representations. Thus, spatial knowledge begins with landmarks and develops into route knowledge by the process of joining up known landmarks. This route knowledge progresses from being topological to metric. Groups of landmarks and routes are then organised (or chunked) into clusters based on their metric relationships. The topological relationships between each cluster are also retained. In the final stage, a coordinating frame of reference develops and results in survey knowledge. Kuipers (1978) also contends that knowledge of an external environment is hierarchically organised with landmarks, routes and configurations assembled into a coherent structure organised around the relative location of landmarks. The anchorpoint theory of Golledge (1978) is similar with a hierarchical ordering of locations, paths and areas. He suggests that some locations become primary nodes or anchorpoints and that a skeletal structure develops outwards from these nodes as spatial knowledge develops. Through this spread effect, survey knowledge develops. Montello (1995), in contrast, suggests that metric knowledge, such as survey knowledge, is acquired and accumulated from the onset of exposure to an environment. He also contends that non-metric knowledge, such as landmark and route knowledge, is not necessarily a precursor to configurational knowledge and suggests that it may exist concurrently with metric knowledge. Thus, non-metric and relatively pure topological knowledge are described as being used in linguistic systems for storing and communicating spatial knowledge about places. Acquiring spatial knowledge in large-scale environments is, therefore, regarded as an incremental accumulation with consequent refinement of metric information instead of a qualitative change from non-metric to metric knowledge. In other words, types of spatial knowledge are developed concurrently rather than sequentially.

In the above theories of spatial knowledge acquisition, one has to further consider whether knowledge is obtained via direct experience or is acquired indirectly. Direct experience refers to that gained through activities in a real environment, while indirect and conceptual experience is that gained through exposure to simplified and symbolised representation rather than real environments. Direct experience also refers to active learning modes, in which people can view or experience an environment via perceptual focusing using head and body movement. Indirect experience can be described as a passive learning mode, mostly involving only one modality such as vision, without direct sensory contact with the environment. From the perspective of spatial knowledge acquisition, direct experiences include route-based learning in a spatial environment and indirect experiences include map study and verbal instructions. Acquiring information through maps delivered over the Internet or by mobile device is deemed to be indirect experience. Spatial knowledge can also be acquired through computer models that are structured to simulate real environments. Some point out that the spatial knowledge acquired via the survey method, such as map reading, is usually assumed to be the most advanced level of spatial knowledge (Hart and Moore 1973).

A number of studies (Thorndyke and Hayes-Roth 1982; Golledge, Dougherty and Bell 1995) suggest that spatial knowledge acquired through direct experience enables better distance estimation and orientation, while indirect experience (for example, map learning) facilitates better Euclidean distance and object judgements. For performing object location tasks, people with direct navigation experience have to transform route knowledge into survey knowledge. They tend to perform the tasks with lower accuracy and in a longer time than people who have conceptual map-reading experience. The ability of those with direct experience to locate objects also shows little improvement after repeating the task. On the other hand, people acquiring spatial knowledge through direct navigational experience in an environment are more likely to give route-oriented descriptions. The study by Golledge, Dougherty and Bell (1995) further suggests that people's knowledge acquired via map learning is more effective than that acquired via direct route learning for understanding spatial relations, and suggests that people who are new to an environment will acquire more spatial knowledge if they learn from maps. It is commonly accepted that people acquire better survey knowledge through indirect conceptual experience, particularly through map study, and acquire better route knowledge through direct experience of navigating through environments. These studies also raise the question as to whether direct experience does lead to survey knowledge or whether this depends upon the complexity of the environment.

Computer model experience, as a means of environmental exposure, shares many common characteristics with direct experience, despite subtle differences. A number of studies provide evidence for similarities in direct experience and computer model experience. For example, Tlauka and Wilson (1996) show that spatial knowledge acquired through direct and computer model experience is orientation-free when compared with map learning experience. Studies also point out that people who acquire spatial knowledge in virtual reality often have similar capabilities to those who acquire knowledge via direct experience. They can produce extensive and accurate route knowledge but exhibit less well-developed survey knowledge (Witmer et al 1996; Wilson 1997; Ruddle, Payne and Jones 1997). Other studies show evidence that survey knowledge may be acquired more quickly using computer representations. The study by Rossano et al (1999), with regard to spatial knowledge acquired via computer models, reports some tendency towards a better performance compared with map learning for route knowledge and poorer performance for survey knowledge. Computer model experience shows little evidence of orientation specificity but no improvement on survey knowledge, particularly the accuracy of object relationships in the environment (Rossano et al 1999). Nevertheless, the study suggests that both route and survey knowledge can be acquired through computer model experience. It should be noted, however, that the experiments carried out in this study required participants to be largely passive rather than engaging in active movement. This would be like viewing a computer screen as opposed to being in an immersive virtual environment. Such differences could have had an important impact on performance in spatial knowledge acquisition.

4 New research questions for mobile geographic services

The fast-emerging fields of mobile location-based services, allied to GIS, provide the technological setting for enhanced abilities to access and utilise geographically related information. New geographical information services are being introduced to the market in both a traditional mapping sense, through map Web sites, and at a conceptual level, using next-generation technology and mobile telecommunication systems. These GIS applications provide the ability to deliver data, computing capability and integrated functionality to individuals at distributed locations and thus reduce the traditional limitations of spatial and temporal separation. The rapid developments of technologies, the potential for ubiquitous GIS as mobile geographic services, and the focus on individuals in mobile contexts pose research challenges in cognition, spatial knowledge acquisition and the human/computer interface for geographical information science.

There has as yet been too little systematic research on the interaction between individuals and the new information and communication technologies that deliver

LBS. We still understand too little about how these technologies can assist our understanding of the processes by which knowledge is acquired, valued, communicated and applied. Furthermore, there is the need to understand how new mobile technologies can be used to complete particular tasks and to derive generalisable measures of performance (see Tobón and Haklay, this volume for the case of desktop systems). Taking the example of navigation, humans are very skilled at giving directions, and computer emulation of these skills would be useful in the design of handheld or in-vehicle navigation systems. Yet the kinds of instructions generated by the route-finding analysis functions of GIS are often quite different to those that might be given by a human. An individual's directions are typically given in familiar and colloquial terms and use many more landmarks and hints designed to make the user's task less error-prone.

Information necessary to complete location-specific tasks using new mobile technologies

Given that the demand for real-time, fast-changing information is increasing, people will be able to receive more relevant, timely information in various forms. But what information do people need in the new 'mobile age', and, therefore, what kind of data are needed? Personal needs are being emphasized. There are related questions on how to provide access to large numbers of people in high transaction environments in LBS. The dynamics and complexity of the urban arena are another consideration, particularly with regard to selecting points of interest (POIs) and landmarks for giving wayfinding instructions to mobile users. What constitutes relevant and meaningful information? How will people use new mobile technologies to perform wayfinding tasks? For example, what are the preferred landmarks that should be available via a mobile handheld device to allow people to make wayfinding decisions?

Delivering geographical information with regard to spatial awareness and cognition

LBS applications need the ability to integrate address location functionality, searching, mapping and routing. A primary requirement is identifying user-specified content. Location information can be provided in a range of formats such as graphical, textual and voice. Interest will become focused on the capability and the diversity of devices to deliver and access location information. There are also discussions based on the effectiveness and usability in how location-based information is delivered, such as 'speak and listen', maps and text. However, research issues are concerned with the means of communication by which task-oriented geographical information is delivered to, and is most easily assimilated by, the recipient. There is also a need to investigate the relationship between individuals' spatial abilities and the means most appropriate for them to acquire spatial information. The conventional means of communication of spatial information has been the map. How might the necessary evolution in the means of communication enhance people's ability to acquire spatial knowledge?

The semantics of communication between individual and database with or without intermediary

Another area requiring research is the semantics of communication between individuals and a database with or without the intermediary of a call operator. In navigation LBS applications, instructions are given to people by means of maps, spoken word and text. It is envisaged that various landmarks, points of interest (POIs), and key features of neighbourhood environments might be provided as spatial cues via mobile devices. On the other hand, people's ability to answer spatial questions may be limited either through lack of awareness or confusion over vocabulary and map reading. This strikes at the traditional core principles of GIS such as data modelling, data handling, generalisation, cartography and navigation. Can objects be described in terms of neighbouring features in exactly the same way that people might describe them after looking at a map? The precise relation of cognition to language remains controversial (Mark 1999). Can we arrive at a common semantic for spatial descriptive terms between systems and people when they are receiving information through handheld mobile devices in the field?

Figure 6 An illustrative virtual reality environment for studying urban wayfinding using mobile devices

Many of the assumptions concerning usage and behaviour in handheld GIS and mobile location-based service applications are untested and should not be taken for granted. The research being carried out at CASA in this area is aimed at studying some of these fundamental research questions.

Ongoing and future research

We propose that our research agenda will move to test the interaction between individuals and mobile communication devices in a controlled virtual reality environment (figure 6). This will entail representing a range of urban morphologies and the landmarks within them. A series of urban wayfinding tasks under different contexts will be choreographed. We hope that this will reveal how spatial knowledge is acquired, valued, communicated and applied with the assistance of handheld devices and their applications. In this way, a clearer understanding of key issues previously discussed and which underscore LBS implementation can be formed. This research is ongoing.

5 Conclusion

In this chapter, we have discussed the current technological setting for provision of mobile geographical information services, including handheld GIS, mobile communication networks, positioning technologies, and location-based services. The relevant research on spatial knowledge acquisition has been reviewed. New research focuses have been identified and discussed in relation to the development of mobile location-based services, spatial knowledge acquisition, and the potential for ubiquitous GIS. This research is still at an early stage. Mobile location-based services are likely to grow to become part of our everyday social and business activities. From the dual perspectives of, first, handheld GIS becoming available and affordable through mobile devices and, second, the reliance of LBS on GIS to underpin spatial data management and query, we are likely to see GIS becoming a still more ubiquitous technology in the near future.

Chapter 11

Strategies for integrated retail management using GIS

Paul A. Longley, Charles Boulton, Ian Greatbatch and Michael Batty

The retail system is central to the organisation of the modern economy. It has multiple manifestations on the geographical structure of cities and regions. However, the extent to which the system can be studied in isolation from non-spatial and related organisational factors complicates spatial analysis in ways that make it difficult to cut through its independencies with other systems. In this chapter, we make an initial analysis of this complexity through a series of interviews with key United Kingdom high street retailers, focusing on their perceived needs for augmenting and extending traditional spatial analysis by incorporating new issues not currently addressed by technologies such as GIS. We conclude by identifying four areas where research in GIS needs to be extended: the mobile consumer, functional and spatial networks of relationships between retailers, new formats for stores based on new kinds of geodemographics affecting niche markets, and the influence of behavioural and cognitive landscapes affecting the shopping experience, not previously considered important in determining retail location.

1 Retail management, systems integration and GIS

The retail sector is a large and vital part of all advanced economies, with increasingly strong interdependencies within the global economic system. Yet it is also one of the central structures around which more local urban and regional economies and city systems are built. Today the retail hierarchy is often thought of as the essential determinant of the status and prosperity of an urban centre, whereas previously it would merely have reflected available markets that were themselves largely a function of the primary and secondary sectors. Growth in the volume of activity in the retailing sector means that its continuing success, structure and organisation are fundamental to the international economy, at a range of scales from the local to the international trading bloc. Its organisation has explicity spatial consequences, with traditional geographies along the classical lines of Central Place theory, for example, playing a less significant role.

In the United Kingdom, the retail sector accounts for at least 16 per cent of GDP, and it is now the largest distinct sector in the economy—but it is also becoming one of the most volatile. The advent of e-tailing, increased cross-border shopping in the Single European Market, and new loci of retail growth (in unconventional outlets such as airports) have also brought an international dimension to shopping behaviour with new competitive pressures upon the retail sector. Access to retail opportunities remains pivotal to the quality of life of individual households, and thus the spatial dynamics of retail organisation is a matter of considerable importance, not only for the companies concerned but also for the spatial structure of society as a whole. In the United Kingdom and the rest of Europe, traditional high streets and town centres retain an important role in civic life and still account for considerable infrastructure spending and planning (Thurstain-Goodwin and Batty 2001).

The future of retailing is thus a pivotal issue, both from the point of view of the firms involved and from that of the customer and the community. GIS has been central to retailers' own analysis of market performance, as well as in their tactical and strategic decision making. GIS has been adapted to anticipate external changes to the demographic, transportation and technological environment, as well as traditional internal issues of efficiency, supply chain management, capital markets, management style, and response to market needs (Birkin, Clarke and Clarke 2002; Wrigley and Lowe 2002). Yet caricaturing the remit of GIS analysis as dealing with exogenous change in the retail system, or endogenous management of retailer activity masks the interdependencies that exist between the activities of retailers and the health of the retail system at the local and regional levels. Retailer decisions affect the urban system in which they are embedded and thus affect the markets that the retail organisation is responding to. This situation can be thought of as a co-evolutionary process consistent with the emerging theory of complexity in which the impact of change affects the very way change is perceived, and these changes in view affect future impacts (Allen 1997). In short, changes in the internal structure of the organisation affect its environment, which in turn influences internal changes in its structure (Holland 1995).

Such effects are of considerable importance for the planning system, particularly in settings where regional or national market share is concentrated in the hands of a few retailers. In the United Kingdom, for example, the largest four grocery retail corporations control over half of all food sales and an increasing proportion of non-food sales such as clothes and drugs; similar concentrations occur in retail banking, drugs and clothing. Casual observation confirms that the pattern of much of the retail sector no longer corresponds to the classical models of locational analysis. Moreover, Allen (1997) suggests that their dynamics are not well represented through static models or strategies for short-term profit maximisation. The danger for the planning system is that there are no apparent rules to guide the evolution of the retail system. For the retailer, the danger is that, over time, retail firms may find themselves in evolutionary dead-ends from which escape may be difficult.

For example, the demise of the clothing retailer C&A in the United Kingdom reflected a strategy that produced a distribution of stores and costs based on a particular supply and delivery system. When this was coupled with a particular market strategy based on quality and style of products, the firm found itself at an impasse, which could not have been anticipated through traditional methods of analysis and management. Arguably, a similar situation faced U.S. retail giant, K-Mart. Following major restructing during 2002 to 2003 (including a major programme of store closures), K-Mart is (as of May 2003) emerging from the protection against creditors afforded by chapter 11 bankruptcy legislation. Viewed more prospectively, U.S. and U.K. retail grocery corporations alike continue to ponder the relative merits of specialised picking centres to establish and serve e-tailing, versus those of serving home delivery demand through established store networks. There is now some accumulated international evidence to suggest how existing retailers should not adapt to these challenges (Birkin, Clarke and Clarke 2002), but there are few generalised rules to guide decision structures towards successful adaptation. Moreover, whatever strategy a retailer pieces together will co-evolve and compete with the wider environment based on other stores and on volatile trading arrangements. Long-term success and sustainability of a company thus requires a series of factors and mechanisms that enable it to identify threats and opportunities in its environment (on which it relies), and to respond to these with changes of product range, service delivery and marketing that fits its technology. Indeed, it is interesting to think of a retailing operation as a particular technology that aims to deliver the right kind of commodities and service to the appropriate people. The location, unit size, and the quality, reputation and style of its offerings must be adapted to those who are accessing them, and any system that is able to cope with this must be able to anticipate those who will access its offerings in the future.

This implies an integrated systems view of the retailing sector. It reaches back from customers and markets through buyers, distributors and producers to the supply chains and the design processes at their base. Failure in the retail sector reflects either a lack of response or a response to changes in the environment that is too slow. Either the threat is not perceived or the response is not forthcoming, but, whichever the case, the result is the same. Successful co-evolution with the environment (technology, demography, changes in residential patterns, transport access, fashion, changing tastes, changes in competition) requires an integrated systems view that can anticipate problems and opportunities and allow possible strategies and ideas to be rapidly explored and tested. Successful spatial analysis thus requires representation of a retailer in a setting where it co-evolves with its changing surroundings.

This, in turn, implies successful linkage of regional and urban models and databases concerning changing patterns of lifestyles, geodemography and income distribution with knowledge of changing local and regional accessibilities, local and regional opportunities, and planning restrictions regarding retail development and evolution. This is more clearly the domain of GIS-based analysis, and the dynamic, spatial representations of urban and regional change described elsewhere in this book (for example, Batty, this volume; Evans and Steadman, this volume) are also relevant in providing valuable context and strategic tools for retailing. Successful representation of an integrated system requires that fast-changing patterns of population, income and accessibility be related not just to one another but also to decisions concerning market development, store format and store location. The broader challenge in most European settings is to devise an exploratory tool that makes it possible for a user to investigate scenarios concerning the development of out-of-town shopping, of high street shopping, and of the spatial retail hierarchy. By extension, it should also be possible to explore changes in the global environment, in e-tailing and in the impacts of the adaptive strategies of particular retailers to changing situations.

The impact and organisation of Internet shopping provides a particular challenge. Some limited successes in grocery retailing aside (Birkin, Clarke and Clarke 2002: 89-108), the main beneficiaries of e-tailing to date have been parcel deliverers such as FedEx where the business is dominated by routine sales and where physical delivery patterns are well established (for example, see Longley et al 2001: 45-6). The most high-profile companies such as Amazon.com are only just turning a profit on e-sales alone, and we would argue that it is the wider competitive environment that is least understood in enabling these online businesses to meet supply and demand and effect market clearing. There would also appear to be a need to specify integrated systems that would be useful in understanding the extent of the impact of Internet economies, not just in order to suggest optimal ways of managing new delivery systems, but also to suggest strategies for combining them with existing physical selling infrastructure (the so-called clicks and mortar strategies). In particular, the largest U.K. grocery retail corporations are currently mired in this problem with respect to their own schemes for online grocery ordering and, thence, physical delivery to the customer.

Last but not least, there is a need to help assess and hence improve the capacity of retailers to adapt and change successfully in the future. This implies an ability to perform learning simulations to help identify and choose between strategies. These strategies might integrate location, type of outlet, style and quality of goods, and age of prospective customers with the anticipated changes occurring in the urban and regional system. Integral to this would be a holistic appraisal of the diversity of household, social, economic, demographic and behavioural characteristics at a range of scales, and the dynamics of change in such characteristics. Together this might create a picture of the dynamics of retail demand.

2 The complexity of contemporary retailing

The retailing literature documents the many recent systemic difficulties and failures that have occurred in the retail sector—and that also concern service providers in the public sector (Birkin, Clarke and Clarke 2002; Longley and Clarke 1995). The difficulties encountered by service providers and retailers are of interest to the firms themselves, to those that advise them, to the local authorities, high streets and communities that are affected by any decline, and to developers and city planners interested in creating successful and sustainable urban development. Many well-known retail chains face the spectre of hostile takeovers, branch rationalisation, or, indeed, complete withdrawal. While at a macro-scale this can simply be seen as the normal workings of a market economy, at the urban and regional scale it often involves great human stress and a huge waste of assets, value, employment and community infrastructure (specifically in the high streets of town centres) and access. Such losses might be avoided if the strategies of the retail chains and other service providers were able to adapt better and faster to changing lifestyles, demography, technology and consumer preferences (Allen 1998). Taken together, there is a need in both the retail and planning sectors to understand the sustainability of a retailer, which we might define as its successful co-evolution with its environment (Allen 1998).

This in turn suggests that an appropriate way to understand what is going on in retail systems is through the concept of the sustainability audit. This concept, which is currently popular in complexity theory and systems thinking, focuses on three key areas:

- the ability of an organisation to monitor and model its environment, allowing the anticipation of changed conditions;
- the ability to deduce the implications of these changes for the organisation, and to perceive threats and opportunities for the future;
- the ability of an organisation to adapt and transform itself and its products/services, supply chain and delivery systems in an appropriate way, within the context of a coherent over-arching strategy.

Retailers each have their own constituent systems of supply, product design, buyer strategy, style and quality of goods and advertising, and branch locations. The decline or failure of many well-known retail names reflects their inability to match their own constituent systems to the tastes of the consumers who have access to their stores. Yet there remains no consensus as to what precisely is needed to improve the capacity of these firms to co-evolve successfully with their changing environment.

Allen (1998) suggests that there are four sets of questions that retailers should ask themselves in the framework of a sustainability audit:

- What is required for an organisation to monitor and model its environment? What factors, processes and indicators are relevant, how can they be defined and measured, and how can this be achieved?
- What is required to translate the output coming from the perceived future environment into threats and opportunities for the company? How could this be enhanced?
- What is required for an organisation to be able to adapt and transform itself in an appropriate way, within the context of a coherent strategy? How could this be achieved, and what mechanisms, factors and indicators would be relevant?
- How, technically, can these capabilities be integrated to provide a rapid and useful tool? What are the techniques and models needed to provide this integration?

The first point of monitoring and modelling the environment itself breaks down into a number of additional questions that are applicable to GIS:

- How may we develop measures of the heterogeneity of customer bases, and the individual and aggregate propensities of households to adopt new retail innovations?
- How may retailers adapt to the new imperatives of e-tailing, through conventional e-tailing and hybrid (clicks and mortar) responses?
- What are the planning and infrastructural implications for retail hierarchies?

Of course, the environment is far more than simply a changing spatial distribution of different cohorts of customers. It reflects the changing macro-economic situation and the changing technologies and strategic issues that affect service and product delivery. It is thus important to develop a view of these factors as part of an overall service delivery strategy.

A joint project between CASA and the Complex Systems Management Centre at Cranfield University set out to establish what is required for an organisation to monitor and model its environment. In the context of the problem as already defined, we set out to identify the mechanisms that exist in a representative range of retailers for considering possible future scenarios, and how they can be translated into possible threats and opportunities for retailer operations. Thus, we hoped to form a view of the way in which each retailer perceives itself in relation to its environment. This in turn would allow us to understand better the factors and assumptions that each retailer believes accounts for its place in the larger system. We anticipated that this might guide us to the variables and factors that need to be represented in our integrated system, might also reveal weaknesses in the completeness of a company's perceptions and might provide wider lessons for retail planning.

The related purpose of the sustainability audit concerns the ability of an organisation to adapt and transform itself in an appropriate way within the context of a coherent strategy. This entails attempting to establish how the interaction of different component systems within a retailer generate innovative and adaptive responses to the changing environment. This would allow us to understand the qualitative structure to be specified as part of our integrated system. We also sought to consider how, technically, these capabilities (anticipating environmental changes, generating an appropriate response) might be integrated to provide a rapid and useful tool. This was achieved through two workshops, held towards the end of our feasibility study.

Finally, we mounted a conceptual exploration of different kinds of spatial decision support tools with an emphasis on flexibility and integration. As a result of various surveys of retail organisations, we reached an important conclusion: retailers need flexibility in developing geographical tools and integrating them with different non-spatial approaches. To progress these ideas which emerged from the retailer researchers and analysts involved, we developed certain notions about how we might extend GIS to embrace dynamic trends in the retail and related sectors and to relate locational aspects to more functional and organisational issues within the structure of the retailing operation.

3 Research hypotheses, objectives and methodology

Failures in the retail sector and in service provision must result in large part from deficiencies in strategic management and the consequent failure to redeploy resources effectively at an appropriate range of organisational levels and spatial scales. Thus, our first research hypothesis was that such deficiencies arise out of an inability to anticipate, and respond sufficiently rapidly to, changes in the environment. This is not to say that any component or actor within the system is necessarily at fault, but rather that a tool to provide this rapid vision does not exist. Our second, related hypothesis was that this situation could be improved by a tool that integrates and updates information about the environment with that of possible options for company strategy involving supply-chains, product/service design and development, and structures for delivery systems across different spatial hierarchies.

Our ultimate objective is, therefore, to develop an integrated strategic management tool that will link the product service design strategy and supply chain management to the strategy concerning retail location, spatial scale and hierarchy, thus allowing faster, better-directed responses and adaptations to a changing situation. However, the objectives of the feasibility study that we have carried out to date are to ascertain what would be needed to develop such an integrated system. Therefore, our initial objectives were to:

- conduct research to explore with potential users the validity of our research hypotheses, taking account of their comments to precisely define a project that could develop such a system and test our overall hypothesis.
- establish clearly what would be required to facilitate a faster response to change compared to the tools that exist. We, therefore, gathered evidence from experts and from potential users about the factors, processes and difficulties that the retail sector faces.
- analyse this information to establish a clear view of the factors, processes and issues that a successful system must be able to include and address.

Agents in most areas of retail activity think of the retail environment in broadly similar ways, yet the language that they use to describe it may differ, as may the ways in which they represent it in formal systems like GIS. These differences in language and technique may arise out of differences between companies with respect to the nature of their operations, their corporate histories, or the nature of their legacy GI systems. In this context, our feasibility study set out to pursue our initial objectives by investigating the commonalities and differences in retailer perceptions of the retail environment.

In the first phase, we carried out a series of in-depth interviews in order to ascertain how retail managers perceive, understand and measure the two principal connotations of the retail environment: first, on the supply side, the structure and organisation of the business in space and, second, the demand side, the consumers. Our approach was to work with the senior professionals responsible for modelling retail interactions in virtual space and to compare and contrast the ways in which they resolve commercial problems. In this formative stage of our work, we decided to use a series of face-to-face semi-structured interviews as a means of identifying commonalities and differences without imposing our own preconceived structures. A problem of some previous retail research, as well as GIS applications in retailing, is that technique has been applied without due sensitivity to spatial and organisational context. In our own research, we were seeking to develop a closer partnership and shared mission with the retailers that agreed to

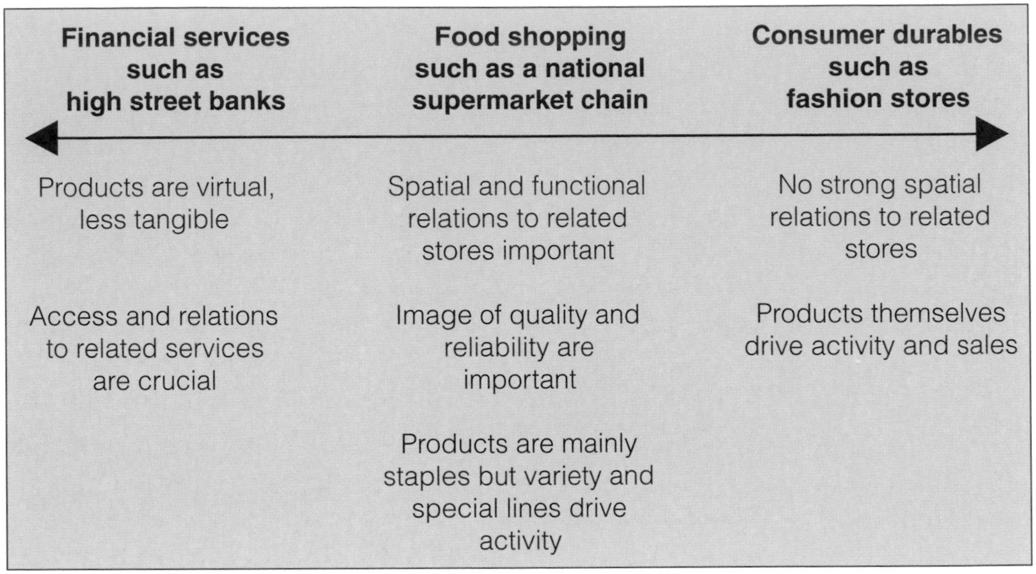

Financial services such as high street banks	Food shopping such as a national supermarket chain	Consumer durables such as fashion stores
Products are virtual, less tangible	Spatial and functional relations to related stores important	No strong spatial relations to related stores
Access and relations to related services are crucial	Image of quality and reliability are important	Products themselves drive activity and sales
	Products are mainly staples but variety and special lines drive activity	

Figure 1 A continuum of retail activity. We set out to recruit one retailer from each extremity, and others at spaced intervals along the continuum

be part of our study (see Clarke and Clarke 2001 and Birkin, Clarke and Clarke 2002 for a discussion of the merits of this approach).

Securing the cooperation of senior representatives of the retailer community is by no means straightforward in an industry in which participants are more concerned with short-term competitive advantage than any longer term commonalities of interest. We wanted the participants in our survey to span the full spectrum of retailing activity illustrated in figure 1, from banking to supermarket chains. We also wanted the participating retailers to have national coverage throughout the United Kingdom. Our research was greatly facilitated by our collaboration with the Demographics User Group (DUG), a retail industry group that, amongst other objectives, lobbies government for greater provision of access to spatially referenced public sector data (www.demographic.co.uk). This community is, of course, largely self-selecting and, as such, disproportionately represents the more sophisticated retail multiples. However, the concerns that its members articulate are likely to be very representative of the wider concerns of the other main U.K. retail players. DUG member representatives are 'spatially aware professionals' (Longley et al 2001: 18-20), who are typically GIS literate and who report on tactical and strategic geographic/locational issues to board level. Some organisations and representatives not involved in DUG but known to our team were also

Retailer A	High street provider of personal care products. A range of formats and layouts.
Retailer B	High street clothes, grocery and household store. Some diversification into specialist micro-stores format.
Retailer C	High street bank.
Retailer D	Nationwide department store selling clothes and household items, often providing restaurant/café facilities. Often incorporates other high street or designer franchises in store
Retailer E	Nationwide supermarket chain. A range of store sizes from out-of-town hypermarkets through city-centre stores. Also sells electrical goods, clothes, snack food, newspapers, tobacco, etc.

Table 1 Some salient characteristics of the five participating retailers

contacted through other channels, and they agreed to participate in the research. Some of the characteristics of the five companies that agreed to take part in the interview process are summarised in table 1.

The interviews were carried out during the period from February to April 2002, each was conducted at the retailer's offices by two interviewers and was typically two to three hours in duration. Although the interviews were not tape recorded, detailed note-taking made it possible to review the interview data with reference to our initial problem structures and to develop new insights. Where appropriate, we recontacted interviewees to refine our understanding or to extend the discussions in the light of later insights

4 The interview results: commonalities of retailer perspectives

We identified a series of common messages from our interviews, but with slightly different orientations that were specific to individual retailers. All of our retailers had planning horizons beyond five years and in some cases out to 15 years. This reflects the time frames over which large capital investments are made, over which demographics and urban land-use structures are deemed to change, and over which the retailers see quite fundamental changes in consumer behaviour patterns. Most retailers are actively concerned, not only with the direct impact of planning policies, but also with unintended consequences that have been seen to arise or could be expected to arise in future.

There was considerable common ground in the different retailer assessments of demographics, spatial structure, and the evolution of the retail system. Location is of crucial importance to retailers and will remain so for the foreseeable future, even given the Internet 'revolution'. But the importance of location is not straightforward: over the last decade retailers have adopted a series of subtle strategies that has led to the creation of different formats of outlet for different consumer purposes, be they the weekly 'big shop' or supplementary 'top-up' trips. The innovation of multiple formats makes it possible to make more focused use of locations in a wider range of sites. However, the corollary is that multiple formats create the need to understand how to optimise the different locations, matching them to consumer needs and to consumer behaviour patterns, and then designing and stocking the outlets as appropriate. This challenge is thrown into sharp relief in retail banking where a single format may serve multiple purposes: on most occasions the customer desires to complete a routine transaction quickly and efficiently, but on others they may be seeking sophisticated understanding of a complex product. This obviously requires quite different interfaces with the consumer.

A perennial GIS-able problem recognised by the retailers that we interviewed centres on accommodating regional variations in consumer behaviour, expectations and local competition through outlet format and location. Additionally, it became apparent through the interviews that retailers have different cost bases, and they also have different concerns about the flexibility that is open to them in seeking to adapt or change company strategy with regard to location and format.

However, as with most complex systems, as soon as one component changes its behaviour, the other interacting components adapt. There was a general perception amongst our retailers that lifestyles have changed in recent years, and that this has led consumers to adapt their shopping patterns to different formats and locations. There was clear recognition that retailers need to keep abreast of these secular changes and that, in responding to them, there were particular risks of cannibalisation of their own stores from locational and format adaptation. We were particularly interested to learn that there was no clear consensus amongst retailers as to whether they were essentially responding to changes in customer lifestyles or whether there was a significant sense that they were causing them. It was not apparent whether consumers respond to retailer initiatives and learn how to shop the new networks of stores they are seeing, or whether retailers respond to observed changes in consumer expectations and new expressions of demand. There was, nevertheless, clear consensus on two matters: first, that the marketplace is becoming increasingly dynamic and, second, that moves by competitors are becoming more significant, reinforcing the notion that all retail stores are becoming integrated more closely into a wider system.

A key part of our discussions centred on the internal decision-making processes of our retailers. We were particularly interested in the use of data and data-manipulation tools, the extent and use of modelling, and the integration of such tool-derived insights with the experience of the retailers' managers. Specialist teams invariably take the lead on issues of location and outlet design. Beyond this commonality of approach, it became apparent that the teams employed by particular retailers work in slightly different ways and with different degrees of autonomy. For example, some are tightly integrated with the overall marketing of the organisation, while others work in parallel to it but with clearly defined boundaries to marketing and, for example, to the merchandising groups.

All of the retailers commented that the volume of data available to retailers is less of an issue than the use made of such data. Store loyalty cards potentially provide retailers with a huge number of insights about the individuals who patronise their outlets, but they know much less about those who do not. This means that, on the one hand, there are continuing concerns about how customer data can be anchored to characteristics of the population at large and, on the other, that retailers have common problems identifying what value can be derived from their own data. In the context of GIS, all of our retailers were seeking tighter integration between data manipulation, modelling, and the decision-making process. All felt that they integrate quantitative information well, and that they are similarly adept in assimilating past experience in decision making, though a range of different processes was used in order to reach those decisions.

The majority of those we spoke to felt that their GIS infrastructure was well suited to their needs, and that the commercial GIS providers are able to generate appropriate tools as and when required. There were, nevertheless, several caveats expressed about using the systems without careful interpretation of the messages that might be extracted from the data and the tools (see Tobón and Haklay, this volume). A key finding of our work was that there is little appetite for a tightly integrated modelling suite. The complexity of the models in use and the variety of issues being addressed, coupled with the sheer uncertainty of what will be modelled next in pursuit of a new insight, indicates that flexibility is valued more than integration. This almost certainly reflects the dynamic nature of the marketplace and the search for competitive advantage through insights derived from new models. What emerged was the clear need for flexible tools that might be used quickly to generate the parameters of how the retail system in general might

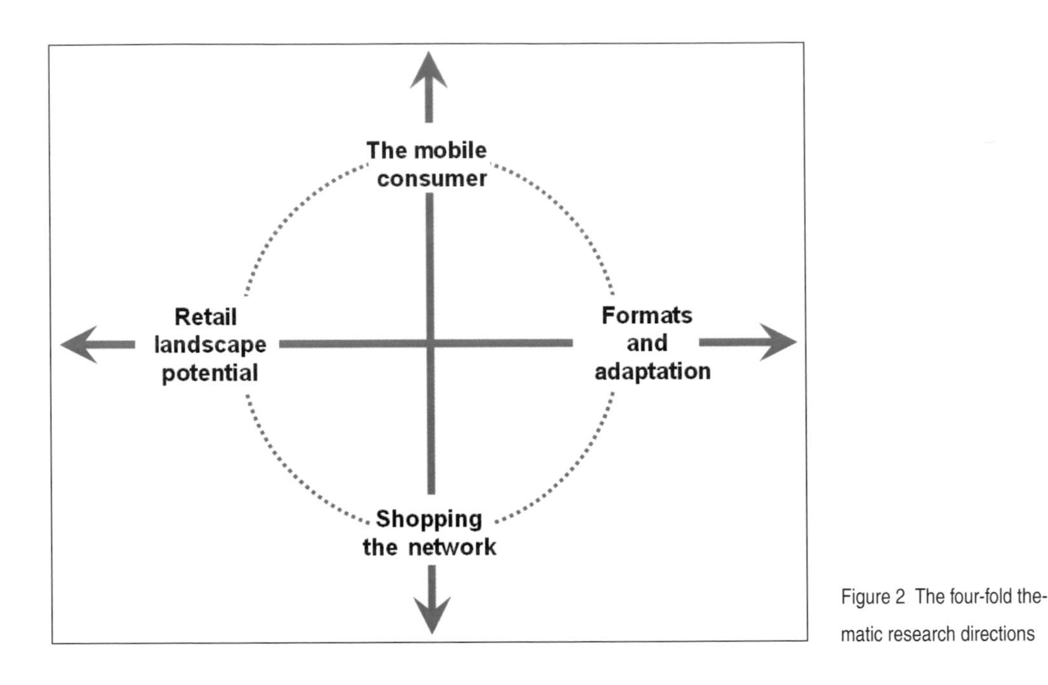

Figure 2 The four-fold thematic research directions

behave. These parameters need to be linked to the specific strategies relevant to particular stores or users working to maximise the commercial advantage and sustainability of their particular store or chain.

5 Interim GIS-able themes

As a result of the interviews, we have identified four themes as candidates for future research that represent the combined interests of the selected retailers and CASA; we show these in figure 2. Each may be enabled using GIS as presently configured, but each also invites further research and exploratory analysis.

The mobile consumer

It is apparent that, contrary to traditional concepts of trip and trip-type, shopping trips are not necessarily initiated from home. The increasing mobility of the consumer, coupled with work commuting, makes a proportion of shopping strongly conditioned by mobility patterns. These include commuting, shopping based from the workplace(s), or shopping events that combine different missions but depend on mobility. Since each of these factors can lead to consumers shopping by using different travel modes, there is a need for retailers to build a deeper understanding of such modes and behaviours—probably using different sorts of data.

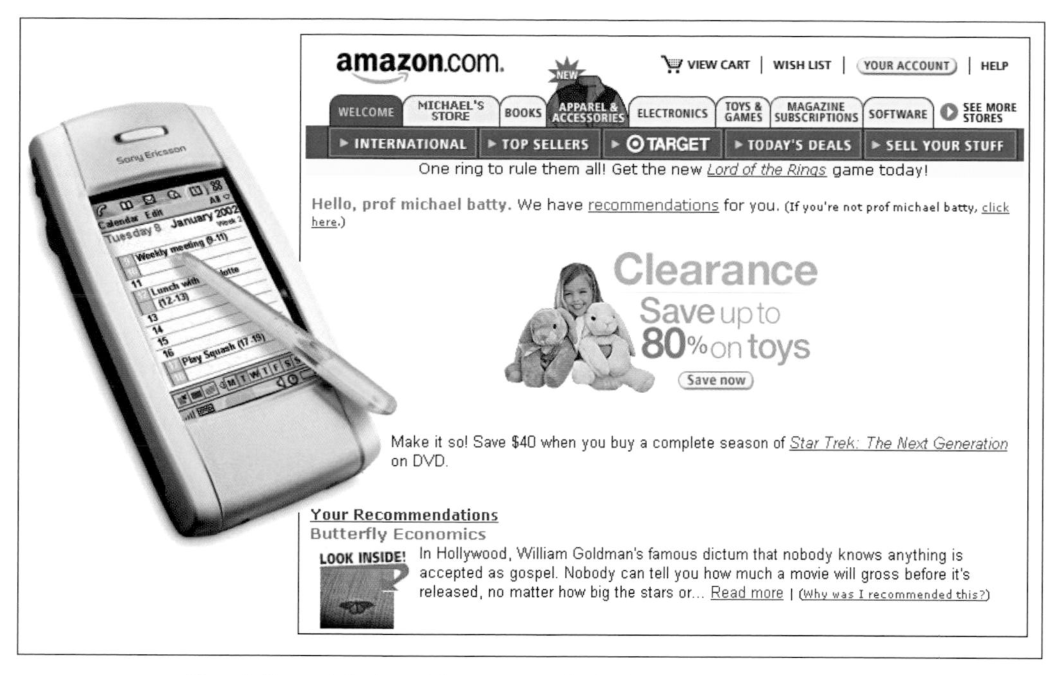

Figure 3 Prospects for the mobile consumer: online buying and selling while on the road

There is considerable interest in an increasingly complex definition and description of shopper-types and trip-types. If we can identify clear behavioural patterns and temporal signatures of either type, we can make bolder, more predictive, and more accurate statements about the kind of shop that needs to be in place. Essentially, this is a question of developing and applying new techniques to the long-recognised problem of developing an adequate understanding of the customer and the customer mix. This theme implies taking understanding of customers to an unprecedented level: understanding their motivations, spatiotemporal behaviour, and the purchasing patterns associated with these behaviours. In terms of spatial scale, this means that mobility must be linked to the global as well as the most local scale. This suggests that new kinds of models of local behaviour at the level of the store where customers move within stores and within shopping streets, malls and town centres need to be available (see Batty, this volume). At the more macro level, mobility involving shopping abroad or over the Internet needs to be integrated with the demand for goods in any particular store or outlet (Dodge 1999). Radical changes in new ways of shopping which make use of these technologies, linking stores to consumers through fixed and wireless networks, are almost upon us as we illustrate in figure 3.

Shopping the network

Retailers are developing different formats of store and, hence, consumers are faced with a network of different sorts of stores available to them, from which they can shop in different ways. The retail system can be thought of as a network, in which there are brands, stores, products and companies that build together in a hierarchical way. Each company contributes a network of stores, comprising different store types, each selling different mixes of products to this system. Taking the big picture, the retail industry is concerned with identifying how consumers will respond when faced with a given retailer's network. There is a clear and demonstrable need to monitor and model consumer interaction with retailer networks—in other words, to identify how consumers shop the network of different formats and whether this behaviour changes between various retailers and retail sectors.

In fact, there is currently substantial research into networks which suggests that patterns of supply and demand have a much more coherent structure than has been thought hitherto. For example, customers shopping in particular locations have relatively well-structured networks of where they shop individually and, if the focus turns to measuring such networks at the individual level, these might be tied to consumer demographics in a way that would give new insights into multiple shopping trips and related movement patterns. In terms of supply chains, there are also relatively well-structured networks that are nevertheless volatile to the price structure of delivery. Better understanding of the relationship between such networks and locational patterns now seems possible, and this is likely to provide a much stronger set of tools for exploring the locational dynamics of both demand and supply. Some suggest that such networks are central to the way in which supermarket chains, for example, supply products to their customers in particular store settings. For example, J Sainsbury plc have developed an experimental SimStore model, which shows the way in which retailers might be able to integrate networks of various kinds with consumer data and product inventories to create much closer matches between what customers want and what the store provides (Venables and Bilge 1998). We show the outline of this model in figure 4.

Formats and adaptation: niche locations, brand loyalty, and the new demographics

As we have discussed, retailers have recently developed a multiformat approach to store placement and type. The layout, product lines, and location of stores are

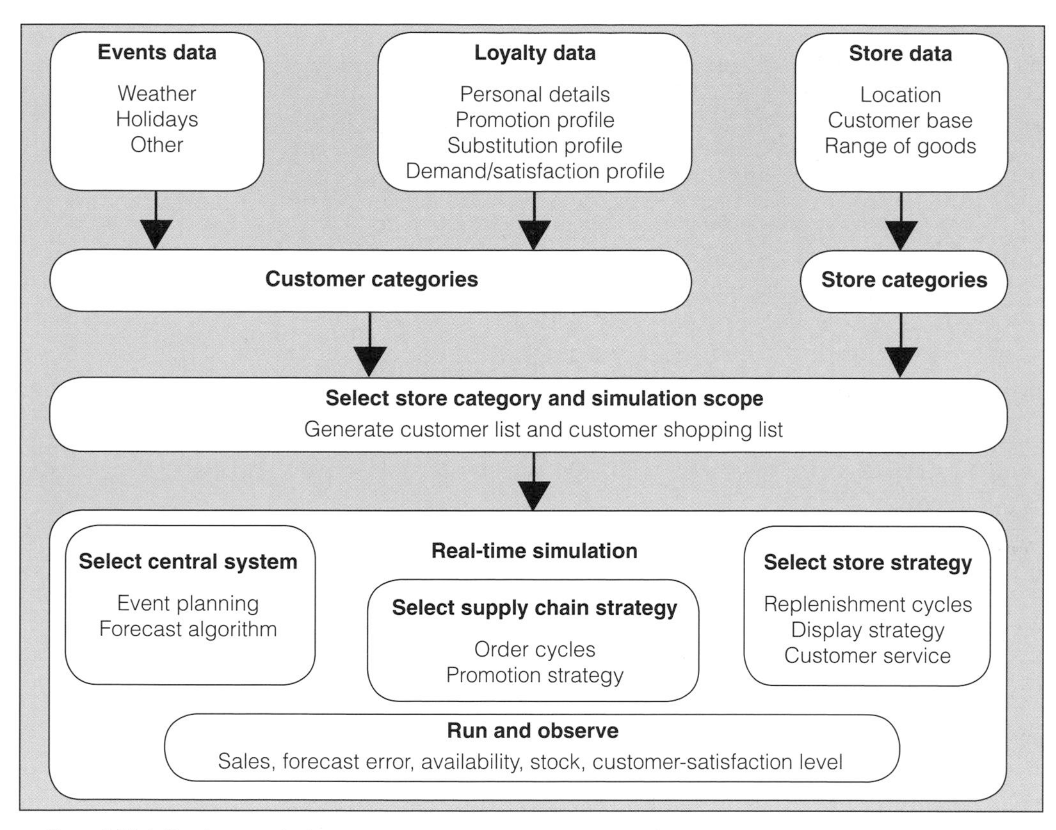

Figure 4 Modelling the network of supply chains into the consumer's domain of the store itself (from Venables and Bilge 1998)

increasingly tailored to suit the perceived market—the archetypical example being small metro style shops which sell a range of sandwiches, ready meals, cosmetics and toiletries and may provide some other service such as photographic developing, medical prescriptions, toys or clothes. Moreover, these are not necessarily provided in small up-market shopping centres but may be found everywhere from city centres to neighbourhood strips to out-of-town malls, as we illustrate in figure 5.

There is a general need to research the number of formats a retailer needs in order to address a target market. Particular attention should be paid to addressing the interrelationships between designing the formats, choosing the categories in each type of store, and identifying the extent to which a given store brand can be stretched. Research needs to address whether it is better for a retailer to offer a definitive and tightly prescribed suite of store types, the most appropriate of which can be placed within a given geographical market, or whether the retailer

Figure 5 Niche marketing and new formats

should think of itself as offering just one store type model that is adapted to the particular market. Alternatively, under what circumstances should a retailer rely on a scale economies argument and have one store format, selling a quality product at a low price, and making this the mainstay of success?

This research agenda also begs questions about the most appropriate network configuration for a retailer: macro-scale (the choice of categories within formats), micro-scale (the choice of products within categories), and nano-scale (the choice of shelf-layouts by product) (see the 'Shopping the network' section). We believe retailers would like the ability to create models that work across or through these scales, thus understanding to what scale an individual problem belongs and creating a communication link that runs throughout these scales. This implies that there is a need for linking the sorts of models described by Evans and Steadman (this volume) with those described by Batty (this volume).

The retail landscape potential

Our interviews suggested anecdotal evidence of growing interest in new ways of evaluating place and spatial relationships. We uncovered dissatisfaction with some of the functionality and approaches of current GIS technology. In particular, many cases have emerged where stores are performing either better or worse than predicted because of factors not typically taken into account in traditional models of the retail hinterland. One example concerned a store that was considered a small convenience store located in a small town, with neither a large residential nor a large workplace population. From the very start of its commercial life, this store out-performed many of its competitors to a completely unexpected level. After an investigation, it was postulated that this performance arose because of the store's location at the intersection of several routes to work, leisure hubs and home.

The implication for future research is the need to devise more sophisticated measures of retail landscape potential, in order to take into account factors other than residential demographic composition and the location of workplace populations. The agenda thus entails finding new ways of valuing market potential in the retail landscape by evaluating the widest range of other (synergistic or antagonistic) factors. The particular ways in which customers perceive the complexities of this landscape and navigate to make the most effective decisions suggest that our models should not only address the problem of spatial integration but should also take account of behavioural, perceptual and cognitive factors to a much greater extent than we have done so far.

Themes and co-evolution

This research agenda implies a need to model the actions of retail organisations as agents that are subject to system-wide structures. The framework to the research is provided by soft-systems thinking as well as the emerging ideas in the theory of co-evolution. Our own medium-term research objectives are set in this framework and are to:

- develop a representation of the mechanisms of adaptive response within the retail organisation itself. This will entail appraisal of the ways in which companies formulate adaptive responses, through the 'sustainability audit' (section 2) framework. In particular, this will establish the existence or absence of key connections of various types of environmental change to the generation of an adaptive response by the company.
- elicit the different dimensions of manoeuvrability that organisations see as available to them and develop a tool to explore the consequences of implementing different possible options.
- understand the criteria of evaluation that are used by an organisation to assess different possible option outcomes.
- decide how to bring together the representations of an organisation's adaptive mechanisms and integrate them into a dynamic, spatial representation of the environment.
- examine what changes would be required to allow such a tool to address the needs of other possible users such as the U.K. National Health Service, the local government and planning sector, and developers.

6 Consolidation and conclusions

To quote Caroline Gye (Strategic Management Group, J Sainsbury plc: personal communication) ' . . . all major retail companies are grappling with a range of complex issues resulting from the speed of change arising from both the dynamic market and the Internet revolution'. Yet she continues: ' . . . no-one in the retail sector . . has the skills and resources needed to prepare a fully integrated model which will aid decision making'. The retail sector is a complex system, and there is a clear need for individual firms in retailing to become aware of the 'big picture', but they are unlikely to be able to do this alone. Much the same holds true from a town planning perspective—most planners have not been trained to view retailing as a complex system, and few planning authorities are geared up to benchmarking the experience of their own local areas against what is going on elsewhere in the retail system. Few planners have experienced formal training in the likely implications of new information and communication technologies (NICTS) for retailing.

The main goal of our ongoing research is to gain a better understanding of the retail environment through developing an integrated systems approach which will be built around the qualitative themes that are central to current system thinking and the emerging theories of development based on co-evolution. We see such integrated tools as being accessible though computer environments and including a range of tools that enable qualitative insights to be gained which are then useful to strategic decision making.

New and improved models of integrated systems cannot be calibrated using only outmoded sources such as the census, particularly when they contain few behavioural correlates of consumer behaviour. Many other conventional sources of data on retail demand and supply have been restricted to overly coarse and inappropriate zonal units (see Lloyd et al, this volume; Goodchild and Longley 1999). There have been no sustained attempts to reconcile rich sources of retailer data with system-wide retail data, such as that used in CASA's Town Centres projects (see Thurstain-Goodwin, this volume). There is much to be gained from using GIS to link rich customer data to system-wide representations of other aspects of the retail environment. Local authorities also stand to benefit from thinking of the task at hand as planning complex integrated systems.

Acknowledgements

This research was funded by the U.K. Engineering and Physical Sciences Research Council (EPSRC) grant 'Integrated systems for retail company management strategy: a feasibility study of demand-led aspects'.

Chapter 12

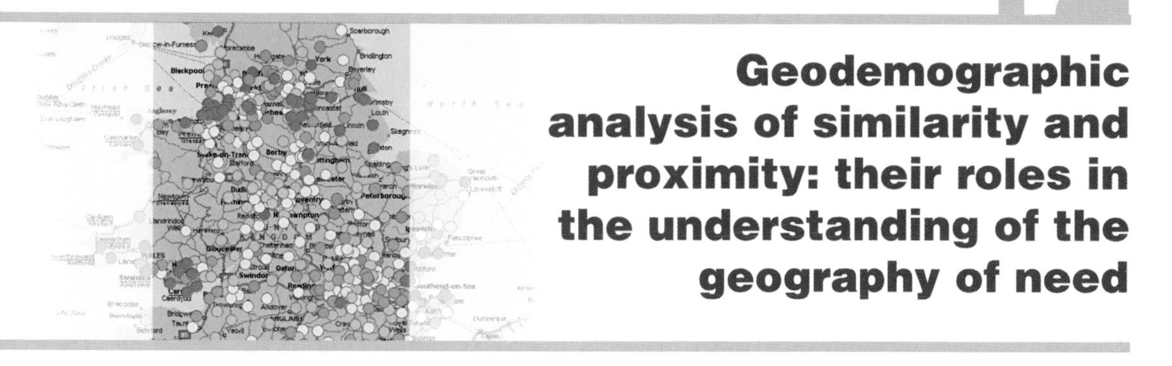

Geodemographic analysis of similarity and proximity: their roles in the understanding of the geography of need

Richard Webber and Paul A. Longley

This chapter considers the very different methods that are traditionally used by government and business when they construct reliable estimates of local demand or need. It considers the hypothesis that for such estimates to be reliable, the estimation process may need to take into account aggregate information held at a variety of different levels of geography. According to this hypothesis, the probability of a household experiencing a particular demand may be influenced not just by attributes of the household itself and of the immediate neighbourhood in which the household lives (for example, the geodemographic classification of its postcode) but also by attributes of the local community and indeed the wider region.

The multilevel approach developed here seeks to combine two contrasting practices whereby information for small areas is summarised for reporting purposes and analysis. The one favoured principally by business involves the summary of information for sets of small areas that have been grouped together on the basis of their social similarity into a restricted number of geodemographic clusters and uses mean values at the cluster level to make inferences about each of their constituent areas. The other, favoured principally by government, involves the summary of information for sets of small areas that have been grouped together on the basis of their locational proximity into a set of administrative units large enough to provide robust population samples. This approach uses mean values for these administrative units to make inferences about each of their constituent areas. Combining these approaches and using geodemographic clusters as one of a number of levels, this chapter develops a statistical approach that can be used to quantify the relative contribution of each different level in accounting for the variation in demand at the individual household level. This approach is applied to a number of commonly used indicators of affluence or deprivation and the results are compared. It is found that the relative contribution of each level of influence varies quite considerably, even between indicators that are highly correlated with each other. The chapter concludes with a discussion of how one might predict, in advance, the levels that might need to be taken into account when estimating local variation in different phenomena and of the selection of social indicators appropriate for identifying different aspects of social need in different areas.

1 Introduction: requirements for estimating local demand

For government, the accurate measurement of local demand for particular public services is a critical input to resource allocation. In the United Kingdom, the proportion of public expenditure that is allocated to services operating on a local area basis such as health, education, law enforcement and social regeneration has been increasing consistently since 1997, and the cost effectiveness of the delivery of these services has now usurped taxation as the primary concern among the United Kingdom electorate. Whilst local agencies remain responsible for the administration of these services, the share of their cost burden that is assumed by central government has also grown relentlessly over this period. A third noticeable trend is that an increasing share of the funding of these programmes comes from discretionary programmes which require local or regional authorities to justify statistically the appropriateness of the programme for the areas for which money is sought. The criteria used to estimate the level of local demand for these services clearly have a critical impact on the level of investment and quality of life in local areas and, therefore, present an increasing source of tension, debate and conflict within the political process.

In the commercial sector, the estimation of local demand for goods and services is a critical input into the financial appraisal processes of the developers of new shopping centres and of the investors who fund them, into the site selection processes of retail multiples, and into the negotiation of rents by landlords, agents and retail occupants (see Longley et al, this volume; Birkin, Clarke and Clarke 2002). Small area estimation is a critical input into the process whereby media owners justify the cost-effectiveness of their advertising rates, whether on regional television, in local and regional newspapers, on local radio, on poster sites, or using door-to-door distribution. Most field sales forces selling to consumers are also targeted on the basis of some form of small-area estimation (Sleight 1997). Motor manufacturers are just one of many types of commercial organisations that use small-area estimation to set appropriate sales targets and to evaluate the performance of local franchises (Birkin, Clarke and Clarke 2002).

Small-area estimation models are used in many communications programmes that involve the targeting of individuals and households. For example small-area estimation is a major input into the process whereby names and addresses are selected for cold prospect mailings as well as for customer communications. General insurers, credit granters and even life assurance companies rely on small-area estimation to set prices or business terms. Call centres and Internet sites increasingly use geodemographic information about the postcodes of callers or visitors as well as their proximity to distribution outlets to determine the scripts or banners that will be most appropriate for them. The aim of broadening access to higher education by awarding institutions additional funding according to the proportion of students from disadvantaged neighbourhoods is an example of how government is also beginning to employ local-area estimation techniques to target at the level of individual (Ainley et al 2002; Webber and Farr 2001)

The research reported here is part of a wider theme at CASA, which is seeking to investigate the properties of spatial dependence and spatial heterogeneity in a range of applied settings. Our applications include the creation of fine-scale representations of income (Harris and Longley 2002) and appraisal of the scale and aggregation effects inherent in representing patterns of urban deprivation (Longley and Tobón 2002). The interrogative approach developed here also has much in common with the CASA theme of exploratory spatial data analysis (Tobón and Haklay, this volume).

2 Properties of data available for local area estimation

'Target group' is commonly used as a generic term to describe those people who exhibit whatever behaviour for which an analyst wants to generate a reliable estimate on a small-area basis. The incidence of this target group is expressed as a rate which is constructed by dividing a target population (for example, people who read a particular newspaper such as the *Daily Telegraph*) by a base population (for example, 'All Adults'). The output of the estimation process is typically the best possible estimate of the incidence of this target group (Sleight 1997). To construct these estimates most analysts will make use of one or more proxies, a proxy being a variable which is expressed in rate form and thought to correlate well with the incidence of the target group (see Harris and Longley 2002). Some such indicators may be direct (for example, measures of household income provide a fairly direct indicator of household expenditure on consumer goods), while others are indirect (as in the use of automobile ownership or ownership of consumer durables such as freezers as surrogate income measures when the latter are not available). In both cases there are likely to be imperfections in the way in which an indicator represents a phenomenon of interest, and such imperfections may have a geographic dimension. For example, there may be a geography to household savings or enforced savings in high-cost housing areas where property purchase is a major component of household expenditure. Similarly car ownership rates may under-represent income levels in areas of high-income metropolitan singles. Low levels of ownership of freezers are likely to be a more pertinent indicator of deprivation in rural areas where the availability of top-up convenience stores (with long opening hours) may be very limited. Nevertheless, the percentage of workers in the socio-economic group 'professional and managerial', for example, might be a useful proxy for the target group readers of the *Daily Telegraph*. The level of recorded crime is usually a good, but never perfect, proxy for the target group 'victims of crime'. Last year's unemployment rate is usually a good, but never perfect, proxy for the target group 'currently unemployed'.

The manner in which proxies are used to construct the local distribution of target groups varies, not so much because different methods have their own adherents, but rather because different proxies and different target groups vary with respect to a number of important properties. The great strength of the decennial census in Great Britain, both as a source of proxy and target group information, is its coverage, notwithstanding the missing millions that some commentators have argued were not enumerated in the census (for example, Champion 1995; White 2002). Information on all but a very few households is recorded. By contrast, information on many of the target groups of interest to commercial organisations can only be cost effectively recorded for the 40 000 or so households that a syndicated market research survey will typically interview in any one year.

One strength, shared by the census and by market research surveys, is that their respondents are representative of the population at large. This is in comparison with consumer lifestyle surveys which, in comparison with conventional market research surveys, are more comprehensive but less representative. There are also further elements to the comparison that we will not dwell upon here (but see Longley and Tobón 2002). The fact that census statistics are usually only refreshed every ten years can be a major impediment when representing fast-changing or rapidly developing aspects of social systems but is less significant in the analysis of neighbourhood processes where filtering results in individuals being replaced over time with others that share similar characteristics and lifestyles. These various problems conspire to create difficulties when specifying multifaceted, multi-attribute phenomena—as in the connotations of deprivation with poor physical as well as poor social conditions. Such multi-attribute indices may be inherently ambiguous and their specifications contentious (Fisher 1999).

Another key consideration, particularly important where administrative data are involved, is the distinction between benchmark and performance measures.

Administrative data will on many occasions measure not so much the absolute level of demand for a service (the benchmark) as the effectiveness (or performance) of the delivery organisation in catering for or responding to that demand. For example, we can draw a distinction on the one hand between the extent to which a school experiences truancy and on the other hand the level of truancy one would expect in a school given the nature of the streets from which it draws its pupils. Likewise, the effectiveness of a police force is measured not just by the number of criminals it catches per 1000 inhabitants, but whether this ratio is high or low by comparison with rates in forces operating in areas of similar demographics. Often the objective of a small-area estimation model is to identify the difference between the two and thereby to evaluate the performance of the delivery organisation in relation to a reasonable benchmark based on the population characteristics of the small area that it serves.

The size or frequency of the target group will also affect the suitability of the estimation method. For example, if a target group, such as owner-occupying households, has a penetration of close to 50 per cent, then the size of the geographical units for which it will be possible to create current or future estimates based on previous information could be quite small. If, on the other hand, the target group is a very rare one, such as the risk of a death through a house fire, it will require the accumulation of data for a very large and potentially diverse area before the numerical size of the target population becomes statistically sufficient to compare with that of the base population.

Target groups vary considerably in terms of their volatility. The gender and date of birth of most people is fixed for life. Attributes related to their housing tend to remain constant for an average of ten years between house moves. Attributes of a person's employment also tend to remain constant until they change their job or employer, which might be each five years on average. Such attributes are ongoing conditions. For these target groups, previous data usually provide a reliable guide to current and likely future levels of the target group. By contrast, the estimation of the current and future incidence of events, such as burglaries or car accidents, is more difficult for, as with stocks and shares, the level of previous performance is not necessarily a reliable guide to the future or, in these cases, even to the present. This is because the subjects of burglaries and car accidents change on a daily basis.

Target groups also vary considerably in terms of their variability on a geographic basis. The proportion of married women who are in employment varies only within narrow limits at postcode sector or Parliamentary (political)

Constituency level, by contrast with the distribution of people employed in agriculture or born in Bangladesh, target groups that are highly concentrated in a very few localities.

Finally, the distribution of a number of target groups is highly influenced by accessibility. Accessibility may be a function of geographic proximity, as in the cases of people shopping at a particular department store or citing sailing as a leisure activity in a lifestyle survey (see figures 1 and 2). In these examples, the local incidence of the target groups will clearly be a function of proximity to the chain's shops or to a sheltered estuary. In other instances, such as products promoted by mail-order companies or advertised in niche media, the distribution of the target group is frequently affected very significantly by the decisions that marketers make regarding the types of person and small area to which they target their promotional offers.

Figure 1 Accessibility: a key consideration in estimating local usage of specific fashion retail outlets

Figure 2 Sailing: a leisure activity whose local popularity is largely determined by opportunity

3 Why the choice of estimation strategy must respect the properties of the target group

We have already noted a significant difference in the methods used for local-area estimation between the government and academic sector on the one hand and commercial organisations on the other. Culture may contribute a partial explanation for this difference, but this is likely to be secondary to the properties of the different proxies to which each sector typically has access and to the different target groups whose local distribution it wants to measure.

Government, in the main, has relied on the decennial census, on administrative records and on a small number of sample surveys to meet its information needs for small-area estimation. It has used the coverage of the census and the fact that its administrative datasets have complete coverage to meet most of its needs for local-area estimation for those of its services which are targeted on a local-area basis. So, for example, in determining the allocation of funds to health authorities, it is able to make use of various measures of deprivation that may be derived from the census as well as standardised mortality rates for the areas covered by individual U.K. National Health Service (NHS) trusts. In many domains government is either the sole provider of administrative data or is in a position to insist on their supply where the service is supplied privately (as for example in education). This makes it comparatively easy for the government to access statistics that are comprehensive and to design, implement and regularly update its neighbourhood statistics dataset. Commercial businesses, which by their nature tend to operate in competitive markets and not to have legal rights to insist on accessing information from competing suppliers, find it much more difficult, and indeed in most cases impossible, to access administrative data which would provide statistics on the total level of demand whether locally or nationally.

For policies that are applied at the household level rather than the area level, the United Kingdom government has typically relied on sample survey data such as the General Household Survey (GHS) and Family Expenditure Survey (FES) to build estimation models operating at the household level. Only in exceptional cases have these models incorporated data attributes for postcodes or any higher levels of geography (in the United Kingdom, one exception concerns monitoring of student applications to all universities). Not having a requirement to selectively target promotional material at their clients, most government agencies have tended to have less reason to purchase geodemographic classifications than have commercial organisations. Perhaps as a result, government statisticians have had fewer occasions and opportunities to investigate the extent to which information at postcode or census enumeration district level contributes incremental explanatory power over equivalent information at the person or household level in analytic models.

In the cases both of administrative data and sample survey data, information has been collected and, where appropriate, published for the smallest level of geography (that is, the finest spatial scale: Longley et al 2001: 102) for which there are sufficient numbers of respondents within the target group to satisfy the twin requirements of statistical reliability and respondent confidentiality. Where the proportion of the population falling into these groups is high and the sample size is large, as for instance with the condition-type (see previous section for definition) questions typically included in censuses, these may be as low as the census output area (enumeration districts in the United Kingdom and blocks in the United States typically comprise 150 to 250 households). In the case of administrative datasets such as those relating to vehicle registration, property transfer, data, crime and unemployment, data are aggregated to and published at either postcode sector (areal units not dissimilar in population size to five-digit U.S. ZIP Codes) or ward level (U.K. wards being slightly larger than postcode sectors and being the electoral divisions used for elections to local councils).

Commercial organisations, by contrast, do not have the luxury of having their target groups included in the census questionnaire. They also operate in a competitive and dynamic market where their own administrative records cover only a minority of the consumers in whom they are interested. For these reasons they have typically relied to a much greater extent on syndicated market research surveys such as Market Research Bureau International's (MRBI) Target Group Index, National Opinion Poll's (NOP) Financial Research Surveys, and Market Opinion Research Institute's (MORI) Financial Survey market research surveys whose comprehensiveness in terms of questions asked is very good even if the coverage, in terms of sample size, is relatively weak.

Because of the limitations of sample size, it has therefore been impossible for commercial organisations to obtain estimates of market demand from national surveys for regions smaller than either the eleven standard government administrative regions or the twelve TV broadcasting regions for which advertising slots can be purchased. One way to construct estimates of local demand from such sources would be to construct models at the household level and then multiply the results by the numbers and percentages of households of specified types within any given area. By comparison with government, which can target at the household level by using eligibility criteria that can be recorded when the household fills up a benefit application form, commercial marketers could only deploy such estimation methods at a household level if the fields collected on research surveys could be matched with attributes which could be known about everyone in the population at large. Unfortunately no such attributes exist at the household or person level, so such analysis has not proved practical.

As a consequence, it is understandable that the commercial sector has preferred to use geodemographic classifications as a bridge to link the results of thin national market research samples with the demographic profiles of ad hoc local areas. The two most widely used geodemographic classifications in the United Kingdom are Experian's MOSAIC™ and ESRI Business Information Solutions' ACORN™, systems broadly similar to the PRIZM® system promoted in the U.S. market by Claritas® and to the systems developed by Experian and others for most European markets, South Africa, Australia, New Zealand, Japan and Hong Kong. Using statistics from the 1991 Census at census enumeration district level supplemented with additional information from administrative datasets such as the electoral register at the finer postcode level, these systems use cluster analysis techniques to group each of the United Kingdom's 1.4 million unit postcodes into one of 50/60 clusters. These clusters are designed to be as homogeneous as possible in respect of the age, housing and socio-economic profiles of their residents. Table 1 lists the 12 groups and 52 types used in the U.K. MOSAIC system. Table 2 presents the verbal pen portraits of two of the 52 types and figures 3A and 3B the illustrations of each type that are used in the MOSAIC brochure. These materials are found useful by marketing people when trying to understand the attitudes and values of prospective customers as well as their product preferences.

Using a directory containing the appropriate cluster code for each unit postcode, market research companies and other end users then import into their research databases the cluster which best describes the postcode where each survey respondent lives, thereby allowing behaviour covered by the survey to be cross analysed by type of neighbourhood. The same directory is also used by companies with large customer databases to analyse in which types of neighbourhood their customers are most likely to live and to identify the variations in patterns and levels of customer purchasing that characterise each of the different geodemographic clusters. The classifications are subjective, in that the profiles depicted in MOSAIC are the outcome of the selection of a particular mix of census and other variables, but the approach is widely recognised to provide a valuable means of discriminating between the purchasing patterns of different consumer groups. Figure 4 shows the spatial distribution of the 12 summary MOSAIC groups in Bristol, United Kingdom. Figure 5 shows the extent to which the 52 MOSAIC categories vary in terms of the proportions of their resident who, according to social surveys, (a) claim to visit church and who (b) do not work due to long-term sickness (a score of 100 indicates an average cluster propensity to engage in or experience this behaviour or condition).

To generate local-area estimates from national sample surveys, the first step typically involves importing into a relevant market research survey the geodemographic cluster in which each respondent lives. The particular target group is then analysed to identify the frequencies with which it occurs in each of the geodemographic categories (the penetration rate). These cluster-specific penetration rates are then applied to the numbers or percentages of households within a local area that are known to belong in each geodemographic category. Multiplied out and accumulated to user-required geographic areas, the results generate estimates in the form of a quantified number (for example, 23 455 owners of a PC), a per household or per capita rate (for example, 48.2 per cent of households) and an index value indicating the extent to which the per household or per capita rate exceeds or fails to exceed the national average, which is benchmarked to 100 (for example, 124). Birkin (1995) provides further details and worked examples.

		MOSAIC Types	MOSAIC Groups	Unskilled workers	Unemployed
High-income families	1	Clever capitalists	High-income families	25	47
	2	Rising materialists	High-income families	49	52
	3	Corporate careerists	High-income families	28	40
	4	Ageing professionals	High-income families	25	47
	5	Small-time business	High-income families	37	47
Suburban semis	6	Green belt expansion	Suburban semis	63	54
	7	Suburban mock tudor	Suburban semis	57	60
	8	Pebble dash subtopia	Suburban semis	55	56
Blue collar owners	9	Affluent blue collar	Blue collar owners	67	53
	10	30s industrial spec	Blue collar owners	85	64
	11	Low-rise right to buy	Blue collar owners	99	78
	12	Smokestack shiftwork	Blue collar owners	132	105
Low-rise council	13	Coalfield legacy	Low-rise council	198	150
	14	Better off council	Low-rise council	174	126
	15	Low-rise pensioners	Low-rise council	163	137
	16	Low-rise subsistence	Low-rise council	231	201
	17	Peripheral poverty	Low-rise council	187	208
Council flats	18	Families in the sky	Council flats	198	318
	19	Victims of clearance	Council flats	988	250
	20	Small town industry	Council flats	145	176
	21	Mid-rise overspill	Council flats	123	223
	22	Flats for the aged	Council flats	144	184
	23	Inner-city towers	Council flats	173	244
Victorian low status	24	Bohemian melting pot	Victorian low status	99	162
	25	Victorian tenements	Victorian low status	69	124
	26	Rootless renters	Victorian low status	119	148
	27	Asian heartland	Victorian low status	126	233
	28	Depopulated terraces	Victorian low status	127	129
Town houses & flats	29	Rejuvenated terraces	Town houses & flats	114	109
	30	Bijou homemakers	Town houses & flats	85	80
	31	Market town mixture	Town houses & flats	125	84
Stylish singles	32	Town centre singles	Stylish singles	66	89
	33	Bedsits and shop flats	Stylish singles	77	168
	34	Studio singles	Stylish singles	54	121
	35	College & communal	Stylish singles	81	149
	36	Chattering classes	Stylish singles	35	83
Independent elders	37	Solo pensioners	Independent elders	125	108
	38	High-spending greys	Independent elders	61	68
	39	Aged owner occupiers	Independent elders	77	63
	40	Elderly in own flats	Independent elders	82	96
Mortgaged families	41	Brand new areas	Mortgaged families	94	95
	42	Prenuptual owners	Mortgaged families	109	101
	43	Nestmaking families	Mortgaged families	47	56
	44	Maturing mortgagees	Mortgaged families	59	50
Country dwellers	45	Gentrified villages	Country dwellers	39	45
	46	Rural retirement mix	Country dwellers	91	73
	47	Lowland agribusiness	Country dwellers	61	52
	48	Rural disadvantage	Country dwellers	127	68
	49	Tied/tenant farmers	Country dwellers	50	52
	50	Upland & small farms	Country dwellers	63	59
Institutional areas	51	Military bases	Institutional areas	74	54
	52	Non-private housing	Institutional areas	82	79

Note: GB = Great Britain (the United Kingdom minus Northern Ireland)

Table 1 Types and groups used in the MOSAIC geodemographic system, and concentrations of advantaged/disadvantaged groups by geodemographic cluster type (penetration rates are expressed as a percentage of the GB average: GB mean rate =100)

	Single parents	Long-term sick	Professional managers	Detached houses	Two-car households	Degree
	30	45	216	259	209	257
	61	56	150	263	182	149
	33	42	181	332	226	208
	32	52	257	242	190	330
	29	62	159	277	163	173
	44	64	129	241	168	107
	59	63	134	110	129	146
	43	66	117	70	133	109
	49	68	93	115	149	66
	46	92	83	98	110	49
	81	90	75	76	99	45
	96	142	54	24	56	29
	135	176	40	21	49	14
	140	146	44	14	62	19
	115	168	46	14	39	19
	264	178	43	9	35	16
	328	158	39	13	42	15
	587	194	33	4	17	18
	257	247	19	5	15	12
	172	191	38	17	35	14
	269	225	44	5	15	38
	94	199	51	14	19	25
	329	158	54	3	21	54
	182	106	93	6	46	140
	107	109	90	12	27	172
	152	124	82	31	53	63
	123	144	54	9	37	48
	93	153	71	69	71	46
	112	95	72	14	54	69
	78	78	88	25	95	73
	79	103	82	72	84	56
	64	93	141	52	67	135
	102	114	125	14	46	206
	110	82	162	10	48	286
	98	90	132	12	49	319
	69	55	209	22	84	435
	97	120	79	51	64	51
	27	77	148	239	91	109
	45	85	97	138	104	71
	75	94	124	51	69	117
	96	100	107	127	113	108
	109	110	84	91	94	61
	86	42	119	86	120	126
	76	52	100	128	143	85
	28	52	177	327	226	198
	52	91	93	189	115	86
	41	62	103	266	190	119
	62	82	86	180	135	85
	36	67	48	319	184	92
	37	79	85	326	169	111
	97	25	29	51	74	63
	68	95	122	136	115	192

Pebble Dash Subtopia

Pebble Dash Subtopia consists of housing that was built during the owner-occupier boom of the 1920s and early 1930s, when low-density estates of three bedroom semis mushroomed across new London suburbs such as Wembley, Surbiton, Bexleyheath and Hornchurch. This was a period when 'pebble dash' was preferred to traditional brick as a material for the outer facing of such houses.

With bow windows and half-timbered gables, these houses were originally built to meet the demand of white collar workers for a place of their own, close to the fresh air and delights of the country. With gardens and a space for a car, such estates represented escape from the grime, congestion and confusion of the inner city, and were advertised by the speculative builders who built them as a kind of suburban utopia offering benefits of both town and country living.

Today these suburbs are rather further than they were from the delights of the country and are often suffering from a lack of investment, not just in the dwellings themselves but in schools, shops and public services. Absent are the trendy young singles who are attracted by the more lively and cosmopolitan atmosphere of the Victorian and Edwardian terraces close to town or wish to escape the city altogether to live in modern estates in rural villages.

The ageing population of Pebble Dash Subtopia makes for a rather conservative and unfashionable lifestyle, but some of these areas are now becoming significantly younger and more cosmopolitan than they were. The absence of the old community spirit may be less of a problem for the younger generation and new fashions don't have to be bought locally. In areas where people need to make the most of limited incomes, people will know the latest market value of their houses, the comparative prices of petrol at different supermarket filling stations or the recent retail promotions advertised in the local newspaper.

Pebble Dash Subtopia comprises people who are careful and reliable about their money; they have practical good sense, keep their wilder emotions under control and find glamour in the cinema, the TV or the weekly magazine.

Depopulated terraces

Depopulated terraces occur in areas of bygone Victorian industry where cheaply built older terraced housing often remains in very poor condition. These are neighbourhoods where even today some houses are still reliant on outside wcs and where many people lack modern bathrooms or central heating systems. Roads made from cinders and abandoned pits and railways contribute to an often unkempt and bleak environment.

Many depopulated terraces are found amid romantic moorlands, among the slate mines of Snowdonia, the china clay and ex-tin mining communities of West Cornwall, the old mining villages on the Pennine fringes of County Durham and the crofting communities of the Hebrides. Many of the remoter valleys in South Wales also fit into this category, as does much of inner Liverpool.

Perhaps because of their extreme poverty, these areas retain a very strong sense of local community, fuelled by a shared history of hardship and economic struggle. Such communities do not accommodate newcomers or new ideas with any enthusiasm and remain residual markets for products and brands no longer fashionable in more prosperous areas.

Table 2 Pen portraits of two U.K. MOSAIC Types: (A) Pebble Dash Subtopia; and (B) depopulated terraces

Figure 3A MOSAIC type: Pebble Dash Subtopia

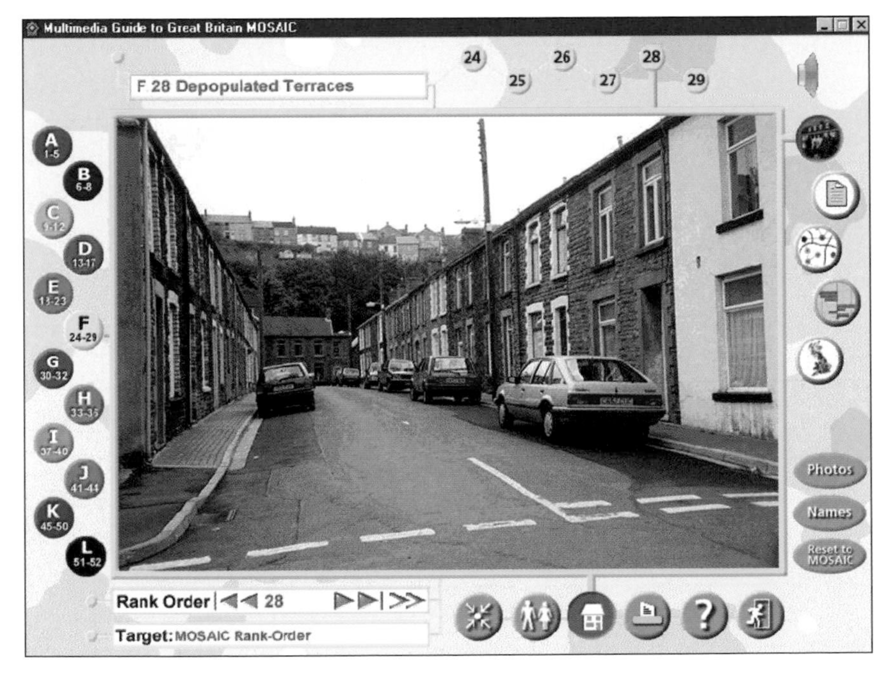

Figure 3B MOSAIC type: depopulated terraces

The geodemographic modelling method has the benefits of being simple to understand, simple to execute, versatile and reasonably robust. It is also a tried and tested technology, with very high levels of repeat purchasing by industry providing testimony to its effectiveness (Birkin 1995). Compared with the use of household models it has the advantage of incorporating area effects. This means, for example, that it is effective in modelling demands that are influenced by area characteristics, for example, the demand for conservatories and lawn mowers or the ownership of dogs, all of which are higher in rural areas, or the absence of a garage, which is characteristic of neighbourhood types dominated by Victorian terraces. It also incorporates a sufficiently wide range of demographic variables and dimensions that make it possible to differentiate between areas on virtually any type of demand, whatever the underlying factors that determine differences on a local basis, not least because recently built geodemographic classifications increasingly incorporate some variables—in particular those relating to employment—for units of a coarser level of granularity that the units which are being classified.

4 The influence of levels

As a result of the recent interest in multilevel modelling, there is a substantial body of evidence to support the contention that the location in which an individual lives has a significant effect on his or her behaviour, and that this is incremental or additional to differences which can be explained using the person or household level only (Goldstein 1995). Furthermore, there is reason to believe that these area effects probably operate at more than one level of granularity, and apply to different degrees at each of these different levels according to the nature of the behaviour.

For example, one might suppose that two households, identically matched in terms of demographics, might nevertheless display appreciably different consumption profiles if, for example, one lived in a low-rise peripheral council (public housing) estate in Sunderland, another in a small estate of privately owned houses in the seaside resort of Torbay. (A similar example might be developed with respect to a comparison between a household living in a rust belt city in the industrial northeast of the United States and one resident in the Californian Sunbelt.) Contributory reasons for such area effects might include one or more of the following: that life in a public housing estate would of itself generate a different mix of needs and expectations to life in a private estate, for example, less

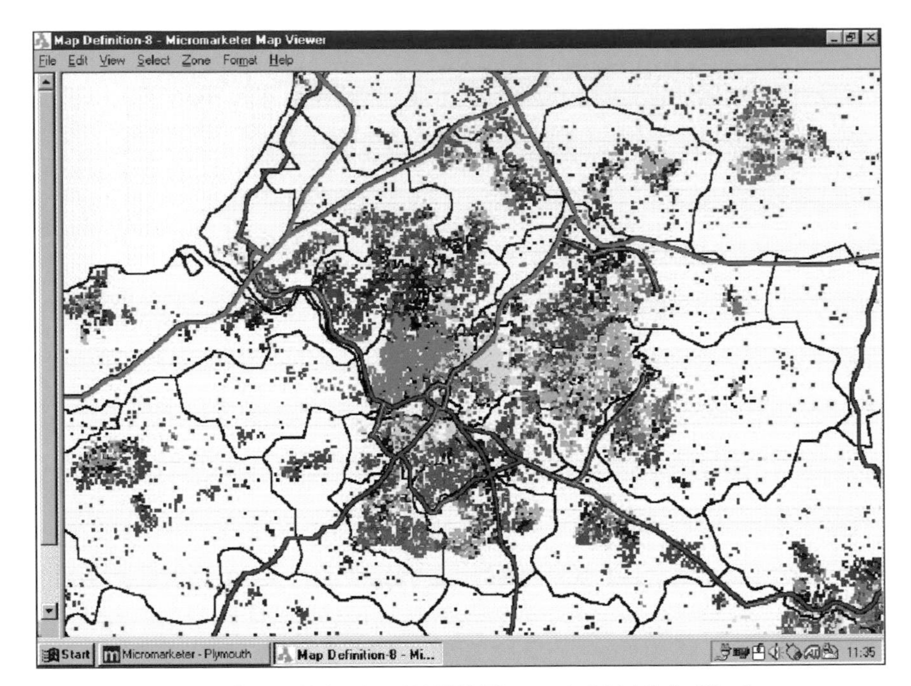

Figure 4 Geographic location of 12 MOSAIC groups in Bristol, United Kingdom

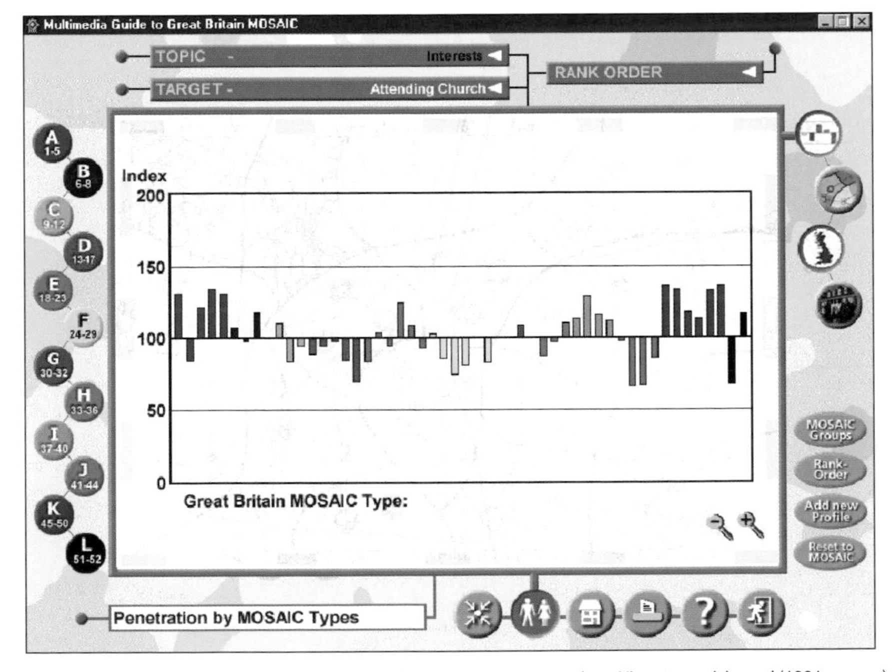

Figure 5 MOSAIC profiles of 'church attendees' (lifestyle survey sources) and 'longterm sickness' (1991 census)

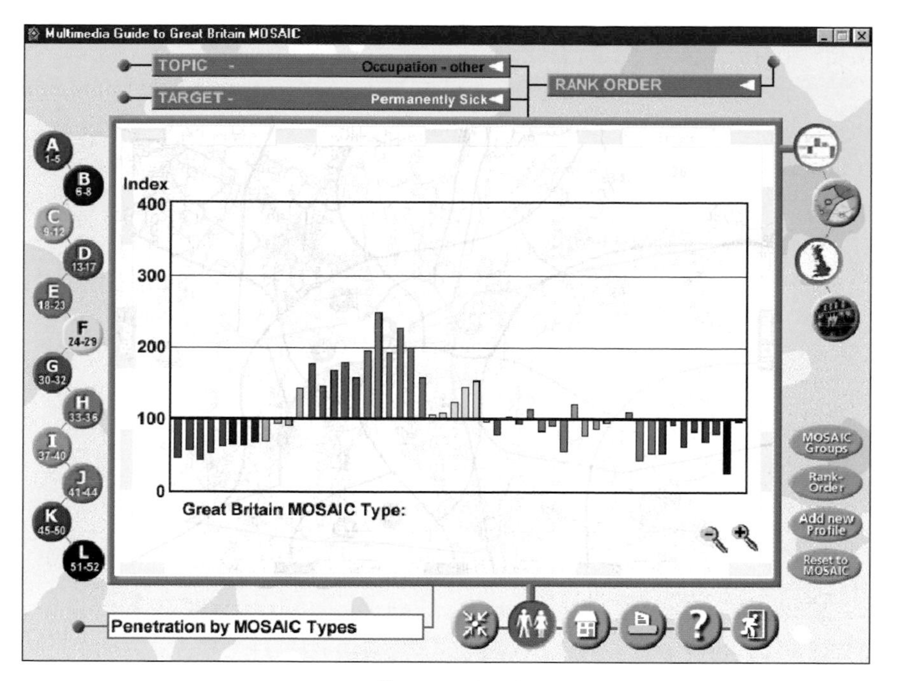

Figure 5 continued

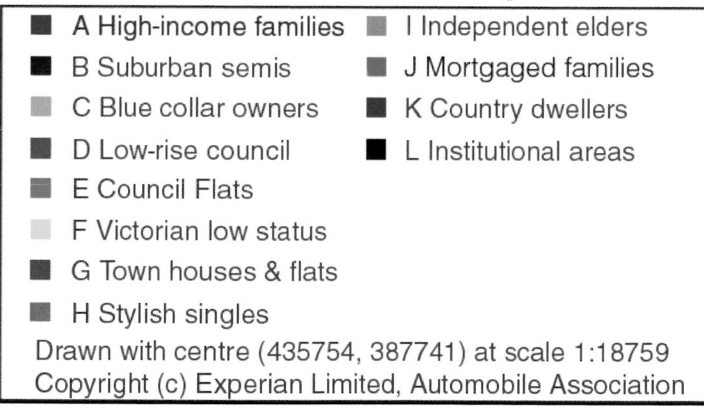

Legend for figures 4 and 5

opportunity to maintain a pleasant garden and hence lower expenditure at garden centres; that the social mores of the community network in Sunderland engender different values to those of Torbay, perhaps resulting in higher use of clubs and pubs; that the traditions of the Northeast lead to a very different set of product and brand preferences than the traditions of the Southwest, for example, a higher consumption of beer and a lower consumption of clotted cream.

To test these hypotheses, we have taken from the 1991 U.K. Census of Population eight separate proxies for affluence and social deprivation and developed a methodology to establish how much of their variance at the person or household level can be attributed to neighbourhood effects, how much to community effects, and how much to regional effects. These eight variables include four measures of social deprivation (single parents, long-term sick, unemployed, unskilled) and four measures of social advantage (persons with a degree, households with two or more cars, detached houses, and economically active persons who are either professionals or managers).

5 Quantifying the effect of neighbourhood

Table 3 shows Great Britain (the United Kingdom minus Northern Ireland) national mean values for these eight proxies (column 1) and the standard deviations that are obtained when the responses to these questions are analysed at the individual respondent level, whether person or household. Essentially the variance represented by these standard deviations is what is normally referred to as standard error. However, it is useful to consider these as standard deviations since this enables us to compare them with the standard deviations of aggregates of these data records, whether clusters based on the social similarity of the census enumeration districts in which they fall or higher order administrative areas such as wards or Parliamentary Constituencies. These differences are shown in column 2 of the table.

To quantify the effect of neighbourhood, tabulations were undertaken to establish the profile of each of these proxies by each of the 52 neighbourhood categories (table 1) used in Experian's MOSAIC geodemographic classification system. The classification was selected because the unit postcode (a level of geography averaging 17 dwellings) would seem to be a reasonably effective level for capturing and describing neighbourhood effects. The average penetration rate of each variable in each of Experian's 52 geodemographic categories is expressed as a percentage of the national average, so that an index of 100 indicates an average penetration

Proxy for affluence / deprivation	GB average	GB standard deviation x individuals	GB standard deviation x MOSAIC cluster	GB standard deviation x constituencies
	%	%	%	%
Unskilled occupation	5.7	23.3	7.8	1.9
Unemployed	9.9	29.8	6.5	4.5
Single parent	2.0	13.8	2.0	1.0
Off work through long-term illness	4.1	19.7	6.7	2.0
Professional/ manager	19.4	35.9	10.1	5.8
Detached house	20.4	40.3	21.9	13.6
Two-car households	23.3	44.3	14.1	9.0
Persons with degree	7.3	26.0	6.7	3.7
Unweighted average of eight	11.5	29.1	9.5	5.2
As % of respondent st. dev.		100.0	32.6	17.9
Number of GB divi-sions		55,000,000	52	642

Note: GB = Great Britain (the United Kingdom minus Northern Ireland)

Table 3 Proxy variables: GB means and standard deviations at individual, MOSAIC and Parliamentary Constituency levels

level of that variable in the MOSAIC type compared with the national average, an index of 200 indicates a level twice the national average, 50 indicates half the national average, and so on. The variance (or standard deviation) of these eight variables is also calculated.

To establish how much of the variance at household level is still retained at the geodemographic cluster level, the standard deviations that were calculated for each of these variables at the person or household level were calculated for a second time, this time at the MOSAIC geodemographic cluster level (see column three of table 3). This statistic measures the extent of variation in the variable at the cluster level. To ensure consistency, this standard deviation was calculated not in raw form but by adjusting to take into account the unequal percentages of the national population in each MOSAIC cluster. This is necessary because the average percentage unemployed (for example) in each of the 52 clusters is not necessarily identical to the average percentage unemployed across the country as a whole and because deviations from the national average tend to be greater in less populous clusters than in those with a higher share of the national population. By comparing the average standard deviation at the respondent level (29.1 per cent) with the average standard deviation at the 52 MOSAIC level (9.5 per cent), one can see that just under one ninth of the variance at the census respondent level is retained simply by knowing which of the 52 geodemographic clusters a respondent lives in.

6 Quantifying the effect of community

To quantify the effect of community the eight variables were next aggregated to the level of each British (GB) Parliamentary Constituency. There are 642 of these units in Great Britain. They are equivalent in function to U.S. congressional districts, though smaller in population size. These units are convenient in that like local authority districts but unlike labour market areas or retail spheres of influence, they tile the country in a comprehensive set of non-overlapping zones. However, unlike local authorities, they are of relatively uniform population size and give a reasonable definition of the broader social community whose values and traditions influence the behaviours of individual consumers. Whilst no Parliamentary Constituency is wholly uniform in terms of its social mix, the U.K. Boundary Commissioners, when reviewing constituency boundaries, are required to organise them around communities of common social and political interests as far as practical (see Schietzelt and Densham, this volume and Thurstain-Goodwin, this volume for discussions of zone design and the delineation of political boundaries).

We can also compare the standard deviations of the eight variables at the respondent and MOSAIC levels (table 3). Column five of this table identifies the standard deviation at the Parliamentary Constituency level and thereby the proportion of the original variance of the eight variables at the individual level that is lost when individual data are summarised at the level of Parliamentary Constituency. In each case it is evident that a greater share of the variance is lost when census output areas are grouped together into spatially contiguous Parliamentary Constituencies than is the case when census output areas are grouped together aspatially but on the basis of their social similarity into geodemographic clusters. Whereas nearly a ninth of the original respondent variance is retained at the 52 MOSAIC type level, less than 3 per cent of variance is retained at the 642 Constituency level.

This analysis begs two interesting questions. First, if knowing which geodemographic cluster a person lives in provides better predictive value than knowing what Parliamentary Constituency they live in, is there any evidence to suggest that knowing both pieces of information contributes incremental predictive value to knowing the geodemographic cluster of a person only? Second, do residents in particular geodemographic clusters have an equal propensity to experience deprivation in whichever Parliamentary Constituency they may reside?

To answer these questions, we built a simple category share model for each of the eight variables and for each of the 642 Parliamentary Constituencies. The constituency percentages of the population in each MOSAIC type were each multiplied by the national average level of each MOSAIC type for each of the eight indicators of affluence and social deprivation. The results of these calculations were then accumulated across all 52 types for each constituency, in order to generate an estimate of the levels of advantage and disadvantage that one might have expected in each constituency, based solely on the distribution of its population by geodemographic cluster. In this way a dataset was constructed for each constituency from which it was possible to establish the extent to which the expected level of the eight variables based solely on its mix of MOSAIC types differed from the actual levels of these eight variables observed in the census.

Were the constituency (or indeed the region) in which a census respondent lives to have no incremental influence on their experience of social exclusion over and beyond that experienced as a result of the type of neighbourhood in which they lived, then the scores of each constituency on the eight estimates of affluence and social deprivation based on MOSAIC would be identical to the actual levels of affluence and social deprivation that result from adding together the data for all the census enumeration districts in the constituency.

Discrepancies between the actual values of the constituencies on the deprivation indices and the values estimated on the basis of geodemographics can provide valuable evidence in a number of respects, as are set out in our conclusions. In particular they can be used to establish the extent to which any area effect operates at the micro level (that is, at the street/postcode/neighbourhood level), or whether it is best considered to operate at a coarse geographical and social level. We can also establish which deprivation indicators are partly or wholly distributed as the result of labour market/economic characteristics, as distinct from factors which vary at a postcode level only and which, by contrast, one might reasonably assume to relate solely to neighbourhood and personal circumstances. Conversely discrepancies can reveal the extent to which the MOSAIC typology fails to pick up subregional or broader regional spatial variations whether on account of its overly fine level of geographical resolution or perhaps because of the limited mix of variables available at the postcode and individual levels to construct it.

Table 4 shows that on each measure of affluence and social deprivation the standard deviations of the actual levels of affluence and social deprivation at constituency level are on average 50 per cent greater than the standard deviations of the expected levels of affluence and social deprivation. The biggest difference is in the level of long-term sickness. By contrast, variations at constituency level in the number of persons with degrees and workers in unskilled occupations could largely be predicted from the MOSAIC mix of the constituency. Dividing the square of the standard deviations of the estimates by the square of the standard deviations of the actuals we can now establish what proportion of the variance at constituency level is accounted for by the mix of neighbourhoods and what proportion by factors that are peculiar to that constituency itself.

GB standard deviation			
Proxy for affluence / deprivation	**Constituencies (A) Observed %**	**Constituencies (B) Expected %**	**(C) Ratio (B) as % of (A)**
Unskilled occupation	1.88	1.48	127
Unemployed	4.46	2.79	160
Single parent	1.05	0.70	149
Off work through long-term illness	2.01	0.86	235
Professional/manager	5.79	4.26	136
Detached house	13.64	10.17	134
Two-car households	9.00	6.39	141
Persons with degree	3.65	2.98	122
Average of all categories	**5.19**	**3.70**	**151**

Table 4 Proxy variables: observed variation and variation expected on the basis of MOSAIC mix

When the observed levels of affluence and social deprivation are scatter-plotted against the expected levels of affluence and social deprivation (as, for example, has been done with unemployment in figure 6), a significant pattern appears which is consistent for each of the eight proxies. There is a clear propensity for those constituencies that were estimated to have high levels of social deprivation to have even higher levels than were predicted on the basis of the mix of the population by MOSAIC type. By contrast, the constituencies which, on the basis of their MOSAIC mix, were expected to have the lowest levels of social deprivation turn out to have even lower levels of social deprivation than predicted. The corresponding pattern is equally apparent for measures of affluence. When regression is used, it is impossible, by definition, for the error difference between actual and predicted values to have any degree of correlation. The fact that the direction of error is positively correlated with the actual level, as is possible using category modelling, shows that the concentration of disadvantage at a constituency level is contributed to not just by the neighbourhood but also by factors which are operating at a much broader geographical scale (or level of granularity) than at the neighbourhood level.

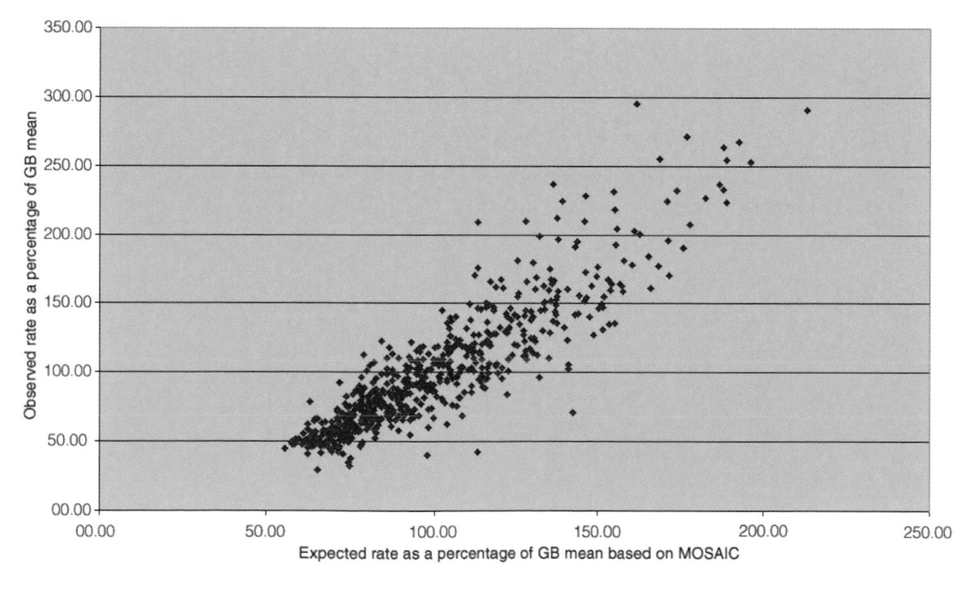

Figure 6 Observed versus expected unemployment rates

This pattern allows us to construct a simple set of eight two-way correlations for each of the eight proxies. For each variable, this correlation coefficient measures the level of association between the expected values on that variable for each of the 642 constituencies and the actual level of that variable as observed at the time of the census. It is by means of this correlation that we can establish what is the aggregate share of variance explained by the combined use of neighbourhood and community influences. It may be helpful to consider this model as quantifying the multiplier effect resulting from the localised concentration of people living in neighbourhoods with a high vulnerability to social exclusion or a high level of affluence.

The results of this model are shown in table 5. Here we can judge not only the share of the original respondent level variance that is retained after data are accumulated to postcode level but also the incremental share of that variance that is retained as a result of the concentration or multiplier effects of people in disadvantaged neighbourhoods experiencing even higher levels of disadvantage where surrounded by high concentrations of other disadvantaged neighbourhoods.

GB standard deviation			
Proxy for affluence / deprivation	**MOSAIC (type of postcode) %**	**Multiplier effect (constituency) %**	**Other unexplained (constituency) %**
Unskilled occupation	11.22	0.18	0.19
Unemployed	4.78	1.10	0.43
Single parent	2.03	0.28	0.06
Off work through long-term illness	11.69	0.55	0.37
Professional/manager	7.86	1.06	0.30
Detached house	29.65	4.65	0.97
Two-car households	10.13	1.81	0.49
Persons with degree	6.71	0.56	0.27
Average of all categories	**7.76**	**1.42**	**0.39**

Table 5 Proxy variables: Percentage of respondent variance explained by various effects

The multiplier effects in absolute terms are largest among those proxies for which MOSAIC works best. But it is significant that the multiplier effects are weak for both unemployment and sickness. By contrast with single parenthood and lack of work skills, differences in levels of unemployment and sickness are much more likely to be the result of particular local circumstances than just the demographic or geodemographic composition of the constituency's population. One likely explanation for this difference is that whilst other measures of disadvantage are largely a function of social circumstance alone, there are causes of unemployment and illness that reflect both historical (in the case of sickness) and recent (in the case of unemployment) patterns of economic change as well as cause relating to the social status and income of an area for which MOSAIC neighbourhood acts as a highly predictive proxy.

7 Quantifying the effect of regionality

The final part of this analysis addresses the issue of what, if anything, explains the constituency level residuals—that is, the differences between the observed and predicted values for each individual constituency. Having taken out the effect of neighbourhood and the incremental effect of the community, is there a regional element evident in the residual that remains? And if so, at what scale does this regionality occur?

When the residual variation left after the MOSAIC mix and concentration effects are mapped at constituency level, a number of clear regional and subregional patterns emerge, whichever of the eight measures one considers. The clearest regional variations in the resulting estimates occur when one considers unemployment and long-term sickness, as can been seen from the map of residuals included in figure 7.

To analyse and quantify these regional differences, an exercise was undertaken to establish and then analyse differences between the nine locationally closest neighbours of each of the 642 constituencies. The first step in this exercise was to establish for each constituency the number of Postal Delivery Points and the geographic centroids of each of its constituent postcodes. These grid references were then averaged, weighted by number of delivery points, to create a population-weighted centre of gravity for each constituency. Straight-line intercentroid distances were then calculated between every possible pair of constituencies, and using these measures the nine geographically closest neighbours to each constituency were identified. In instances where the closest neighbour constituencies were more than 150 kilometres apart, as occurred in the north of Scotland or where travel between the pairs involved crossing the Bristol Channel, the lower Thames estuary, or the Moray Firth, these pairs were removed from the database and not used in subsequent analysis and, as a result, for a small number of constituencies fewer than nine pairs were used in the subsequent analysis.

To establish the incremental impact of regionality on the variation in the eight variables that was not already explained by the neighbourhood and the community level data, we then examined for each remaining pair of constituencies the

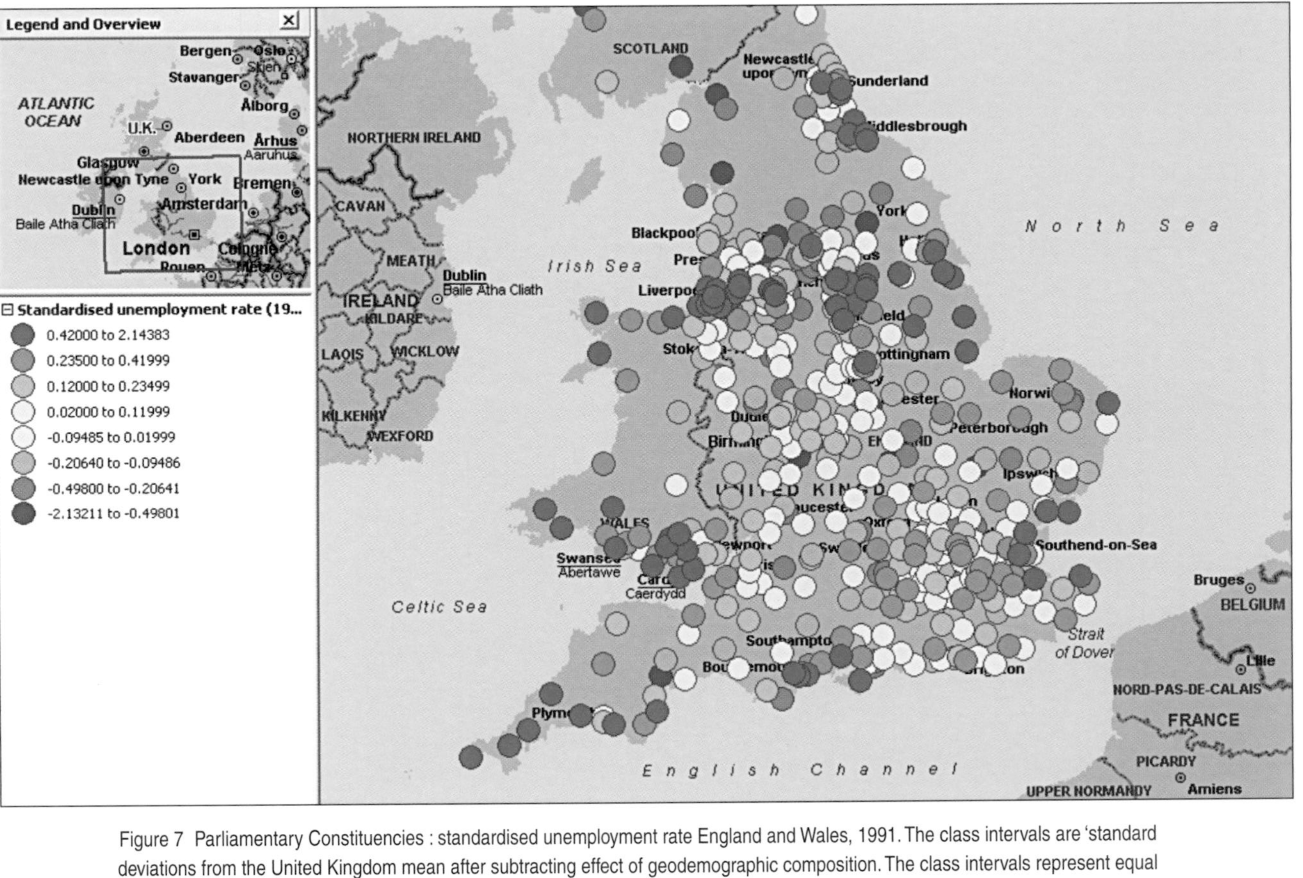

Figure 7 Parliamentary Constituencies : standardised unemployment rate England and Wales, 1991. The class intervals are 'standard deviations from the United Kingdom mean after subtracting effect of geodemographic composition. The class intervals represent equal numbers of cases. In other words three quarters of constituencies lie within one half a standard deviation of the national average

patterns of association between their respective residuals. In this way, we were able to establish whether nearby constituencies tended to have residuals that were closer to each other and of a similar sign than would have been anticipated on a random basis. The result of the analysis between the two sets of residuals consistently showed a positive correlation across each of the eight variables. In other words, constituencies which were very close to each other in location tended to show similar patterns of deviation from the national average even after having adjusted for the mix of MOSAIC neighbourhoods, the levels of deprivation in those neighbourhoods nationally, and the concentration effect of having high or low proportions of neighbourhoods with nationally high or low levels of disadvantage. From this correlation and from the standard deviation of the residuals from it, we could thereby establish the level of variance in each dataset that was explained by the effect of region, defining region not in terms of government standard region but in terms of moving blocks of ten contiguous parliamentary constituencies. The eight correlations are shown in table 6. By contrast with the contribution of MOSAIC and the multiplier/concentration effect, it is in respect of both unemployment and long-term sickness that local regional effects appear to prove most powerful. When this process was repeated just using the two nearest neighbours rather than the nine nearest, a still higher correlation emerged.

Unskilled occupation	0.408
Unemployed	0.565
Single parent	0.338
Off work through long-term illness	0.601
Professional/manager	0.451
Detached house	0.471
Two-car households	0.402
Persons with degree	0.444
Average of all categories	**0.460**

Table 6 Proxy variables : correlations between the residual of a constituency and the residuals of its nine nearest neighbours

8 Summary of analysis

The preceding analysis has identified the following:

- We have established for each of the eight proxies, the overall level of variance that existed at the level of 1991 Census respondent.
- We have been able to establish how much of this variance is lost by grouping respondents together on the basis of their geographical proximity, such that the constituency was the lowest level of reporting unit.
- We have established that less variance is lost if, instead, respondents are grouped together on the basis of the social similarity of the census enumeration districts (zones) in which they live, and based on categories assigned by a geodemographic system.
- We have established how much incremental variance might be retained, over and beyond that explained by grouping respondents together on the basis of neighbourhood similarity, if the characteristics of the broader community (in this case the Parliamentary Constituency) are known.
- We have established how much incremental variance might be retained, over and beyond the effects of neighbourhood and community, if we had known the broader characteristics of the region (measured as the nine closest constituencies).

Thus, for each proxy (as, for instance, is shown in table 7) we can separately isolate as a proportion of the total variability of each variable: (a) the share attributable to the demographic make up of the neighbourhood in which the person lives; (b) the share attributable to the broader community in which this neighbourhood is located; and (c) the share attributable to regional influences, regional being defined in a less arbitrary way than standard or TV region. The sum of this variance can be related to (d) the share that cannot be captured in a model which took no account of person or household as distinct from locational characteristics.

9 Conclusions

The chief conclusions that can be drawn from the findings described here are as follows:

	GB standard deviation		
Proxy for affluence / deprivation	**MOSAIC (type of postcode) %**	**Multiplier effect (constituency) %**	**Subregional (nearest nine constituency) %**
Unskilled occupation	98.18	1.55	0.27
Unemployed	79.44	18.30	2.27
Single parent	87.60	12.08	0.32
Off work through long-term illness	94.47	4.44	1.09
Professional/manager	87.48	11.85	0.67
Detached house	85.90	13.48	0.62
Two-car households	84.31	15.03	0.66
Persons with degree	91.59	7.67	0.73
Average of all categories	**88.62**	**10.55**	**0.83**

Table 7 Proxy variables: percentage of total geographic variance explained by the different effects

- Neighbourhoods are very much more uniform in respect of some target groups than they are in respect of others. In respect of housing, neighbourhoods are very uniform indeed. In respect of measures of rurality, they are also homogeneous. Measures of wealth and social exclusion vary strongly on a neighbourhood basis, more strongly often than social class and very much more strongly than would be expected on the basis of social class differences alone. Neighbourhoods are more uniform in respect of their family composition than in respect of the age of their residents. Better information and understanding of the behaviours whose summarisation to neighbourhood level loses most and least variance would be of considerable value both to the policy makers (for assessing the appropriateness of area-targeted as against person-targeted initiatives) and to modellers.
- The contention that locational context makes an incremental difference to life experiences over and beyond purely household level circumstances is strongly supported by this evidence. However, the level at which these contextual differences make a difference is seldom a single one. The relative importance of neighbourhood, social influences and regionality vary considerably.

- On the other hand what is consistently apparent is that the influence of the wider community tends to reinforce and exaggerate the effect of neighbourhood. In other words where, on the basis of neighbourhood effects one might believe a particular target group to be over represented, high concentrations of such neighbourhoods in close proximity lead to concentrations, even within a given type of neighbourhood, being significantly greater than in equivalent types of neighbourhood not surrounded by others with high propensities to contain that target group. Errors from the use of geodemographics in category multiplication models are systematic not random. This is particularly relevant to the modelling and calculation of demand since where differences between actual and modelled results are distributed randomly, then modelled and actual levels of demand would tend to converge as one aggregated small area with others in the locality. The fact that they are systematic results in any difference between actual and modelled demand being compounded as areal units are aggregated. There is a systematic tendency for the model to either under- or over-predict demand within all the constituent areas within a particular subregion or community
- For businesses and for government, the implication of the previous conclusion, at least in respect of measures of social welfare, is that category multiplication models are likely to be highly accurate in ranking small areas by relative level of demand, but less accurate in quantifying the exact level of demand.
- There is good evidence to suggest that the incremental predictive effect of social and regional levels is greatest in those behaviours whose distribution at a person level is influenced by overall availability at a local level. In the United Kingdom at least, unemployment and high (or low) house prices are good examples of target groups where, although neighbourhood contributes some of the variability, the circumstances of individuals are also affected by the overall level of competition for houses and jobs within wider labour market areas. The distribution of single parents, by contrast, is almost entirely driven by neighbourhood effects because its distribution in the wider labour market area is not subject to broader economic influences or restricted supply. Most consumer products, one would suppose, would behave more similarly to the distribution of single parents than to the distribution of high house prices.
- Where distributions of deprivation are significantly explained both by neighbourhood-level and coarser-level geographies, it is likely from a policy

point of view that their amelioration requires a twin approach. For instance, where high levels of unemployment are explained principally by geodemographic conditions, skills development is likely to be a more effective response than in areas with high levels of subregional residuals, where regional economic development is likely to be a more appropriate policy response.

- The mapping of the residuals after the geodemographic and social effects have been taken out of a distribution can provide a much clearer view of underlying regional and subregional problems, for instance that of sickness in the South Wales mining valleys and of unemployment on Merseyside. The isolation of these underlying patterns may be helpful in focussing on specific locational factors, for instance, the softness of the water supply in respect of certain diseases, the particularly dusty nature of South Wales coal seams, and indeed the social problems that require a concentrated and coordinated regional solution, such as unemployment on Merseyside.
- The variability of data at neighbourhood level and the reliability of the modelling at various levels suggests that the modelling of demand using tools such as MOSAIC can provide a very useful measure against which to benchmark operational performance in those operational areas where social and regional effects are likely to be low. More research is needed to establish which these operational areas are. Health would seem a better candidate for such analysis than crime or education.
- Regional variations, on the evidence of the eight variables studied so far, are much less significant than one might suppose on the evidence of the emphasis given to regional initiatives. A very large amount of regional variability is explained by the regional mix of geodemographic clusters. Both in public policy and in marketing, the independent effect of regionality is much exaggerated. Most regional differences are merely a result of their differences in demographic and geodemographic mix. Genuinely regional residuals arising from the modelling process in almost all cases apply to regions whose boundaries map very poorly with those of the United Kingdom government's standard regions or the TV regions by which commercial market research data are reported.
- The datedness of census data is seen by most policymakers to be much more of an impediment than need be the case. Evidence from this study shows that the distribution of target groups in 2001 is not materially less well predicted by a 1991-based geodemographic classification than equivalent target groups in 1991.

- This supports the contention that in circumstances where target groups are thinly distributed, when they are not available from administrative records and when they relate to events rather than conditions, much more use could be made of category multiplication modelling than has been made by government in its estimation of local demand.
- Given the robustness of these models and because of the tendency for social differences to average out so rapidly as contiguous census areas are grouped up into administrative or electoral units, it would seem that significant improvements in programme effectiveness could be achieved by selecting finer geographic areas for priority area identification even at the expense of having to use some smaller area estimates (as well as census statistics) rather than administrative datasets as statistical justification for their selection.

These and related topics in geodemographic analysis are the themes of ongoing research at CASA.

Acknowledgements

This research was funded by Economic and Social Research Council grant RES335250020 (E-Society Programme).

Chapter

Visualising spatial structure in urban data

Daryl Lloyd, Mordechai (Muki) Haklay, Mark Thurstain-Goodwin and Carolina Tobón

The proliferation of large yet finely granulated datasets has created the need to develop new and sophisticated methods of analysis in order to reveal the information locked within them. Visualisation plays a major role in this process, and enables the creation of information products that communicate the outcome of digital representation. However, the complexity of the underlying phenomena and their digital representation means that the final message may perplex the user. This is most apparent in the representation of data that describes complex urban agglomerations such as town centres. In this chapter, we provide examples of visualisation techniques that have been used at CASA's town centres project toward the goal of enhancing the project's methodology and making its outputs clear.

1 Introduction

In most of the world, town centres remain the economic and cultural heart of free-standing towns, or an established nexus of such activities within the built up area of larger city extents or conurbations (figure 1). In the United States, the locus of economic development has most usually long shifted from such 'downtown' sites to suburban locations (see Besussi and Chin, this volume). Town centre statistics are important for a wide range of users who need to make strategic and tactical decisions about the management of retail outlets, local transport policy, or local public-service planning. In the United Kingdom, most users have had to undertake their analyses in the absence of geographical detail. The shape, form and

Photograph courtesy of Jamie Quinn

Figure 1A British Telecom tower seen from Parliament Hill. The built up area of London encompasses over 100 town centres

Photograph courtesy of Jamie Quinn

Figure 1B The popular Covent Garden location

internal geography of the town centre has thus been of secondary importance to other attributes, such as the total number of shops present or what overall shop turnover might be.

There are two key reasons for this. The first, and perhaps most obvious, is that the spatial definition of town centres is inherently problematic because of the variety of possible interpretations of what a town centre actually is and what town centre activity necessarily entails (Thurstain-Goodwin and Unwin 2000). The second reason is that, traditionally, town centre data have been, on the whole, presented in largely aspatial form. For example, in the 1971 Census of Distribution (the last comprehensive source of U.K. town centre data), detailed information on town centres within local authorities was presented in tabular form—the standard method of presenting government statistics. The only geographical marker was the name of the centre itself. Commercial sources of town centre information also tend to be presented and analysed in data tables. The Hillier Parker multiple rank scores, long used to assess the relative strength of a town centre's retail economy, also relegated geography to nomenclature.

The analysis of data in this form is usually undertaken in a spreadsheet or in a database. Each individual record in the spreadsheet represents a town centre and is described by a number of separate attributes. To use a metaphor from the computer science literature, each town centre is treated as a discrete object that comprises a number of different attributes (name, number of multiple retail outlets, employment levels, and so on) that uniquely describe that object. Any analysis performed on these data assumes spatial consistency: that is, that data collection, processing and quality control are standard across an entire study area. The implication is that any differences in the data are attributable to material differences between town centres rather than to vagaries and inconsistencies in the data collection, processing or aggregation.

Yet when town centres are mapped, differences in their spatial definition become more apparent and suspicions may be raised that these differences are attributable as much to irregularities in spatial definition as to real differences on the ground. This is further complicated because the perception of a particular town centre will be different according to whether an individual is a planner, retailer or chartered surveyor—and whilst these differences are never apparent when using tabular data, they jump to the foreground when mapped. This is perhaps why town centre data that are disseminated in map form (such as the U.K. GOAD plans from Experian Plc, Nottingham, United Kingdom: see www.micromarketing-online.com for further details) are generally treated with more caution, and perhaps unfairly so, because the user is likely to become aware of the definitional differences that exist between town centres. Indeed, the map of the town centre form becomes the prevailing and defining characteristic of the town centre, with scant regard to the function of the entity.

This presented major problems for our research in creating a new statistical compendium of the United Kingdom's town centres for the Office of the Deputy Prime Minister (ODPM).[1] One of the key requirements of the research was to develop a method that created statistical aggregations that were spatially consistent. Thus, differences in any measured attribute of a town centre (such as retail turnover) would be clearly attributable to differences in the characteristics of the town centres themselves, rather than to the definitional process. In order to achieve this, the data were to be manipulated within a GIS. During the project, an analysis method was devised to manipulate large spatial datasets and to enable the creation of a robust statistical definition of town centres, and the aggregation of statistics for each one, using a consistent definitional method. This method has been developed and refined over a five-year period in an iterative process that took into account the requirements and needs of the diverse user population of the output statistics. As such, it may be considered as an example of user-centered design that involves professional users from start to finish (see Tobón and Haklay, this volume).

However, as well as being robust, output data were presented visually in a host of new ways both to explain them to the stakeholders and users and to encourage confidence in the methodology used. First, communicating the method that was used in the process of compiling and calculating the boundaries and the statistics was a vital part of the project. We needed to build the users' and stakeholders' trust in a complex method that was largely based on abstract concepts of spatial analysis, such as the representation of socio-economic data as continuous surfaces. Second, a visualisation method was vital for the communication of the final outputs of the project—the boundaries themselves. The expected user community was larger and more diverse than simply those who participated in the consultations throughout the project. Whilst visualisation was used to explain the method to the latter group, the same visual outputs formed the basis to the final reports which had wider availability.

In this chapter, we provide examples of some of the visualisation methods used during the project and describe how we used them to communicate the methodology and outputs. The following section discusses the project methodology and the move from point data to surfaces for purposes of data analysis. This is followed by a description of the critical contour concept—the idea that a certain threshold in the data surface can be used to delineate the spatial extent of town centres—and the use of a flooding metaphor to communicate it. The final visualisations discussed are those used in the town centres report to the government (Office of the Deputy Prime Minister 2002), which were intended for a wide audience. Here considerations of the ease of reading the reports and clear communication of the nature of town centre boundaries were the primary motivation for exploring appropriate data representations. We conclude the chapter with a discussion of the main difficulties that remain unresolved in our own research and relate these to issues of cartographic visualisation of socio-economic objects. Our particular concerns are boundaries that are in constant flux. By this, we mean that the town centre object is defined by its place, in an abstract sense, and name, rather than its spatial extent and boundary. Thus, the boundary is an attribute of the place, and may change over time.

2 Visualisation of the town centre representations

Visualisation through mapping has long been considered to be a fundamental geographic method (MacEachren 1995: 2). More recently, visualisation has been characterised according to the stages of scientific research in which it is used (DiBiase 1990). In this context, visualisation is used both as a private exploratory method at the initial stages of data investigation and also as a means to present and communicate results to a wider audience. MacEachren (1995) extended this view to define visualisation in terms of map use, specifically with regard to the level of interaction between the user and the map. Although some visualisations were produced for the private use of the research team during their exploration and investigation of the large multivariate dataset, those discussed here were principally devised in order to communicate complex concepts to a professional but non-expert-in-GIS audience. They are non-interactive in terms of the user's ability to manipulate or change the data representation. Nevertheless, the visualisations were a crucial ingredient in the understanding and communication of the problem being addressed as well as a means of explaining the technical intricacies of the modelling technique to the end users of the project statistics. This section provides a brief overview of the methodology[2] and provides the same explanations, ideas and visualisations that were utilised to communicate it.

Data structures and model outline

In order to allow spatial analysis with the town centres data, the data require georeferencing. Data from both the Annual Business Inquiry (ABI) and Valuation Office Agency (VOA) Floorspace databases are georeferenced at the unit postcode (UPC) level that is broadly equivalent to the ZIP Code +4 system in the United States. Each UPC represents on average about 15 households, but can vary from a single delivery point (a large user) up to one hundred such points. In addition to this, it contains markers to differentiate between residential and non-residential delivery points.

UPCs are untransformed points, and as such have no prescribed boundaries. A postcode is defined as the 'postman's walk' to deliver the mail to all of the addresses associated together. In Code-Point™, produced by Ordnance Survey, Great Britain, the point itself is positioned as the average of the national grid co-ordinates of all the delivery points contained within the postcode. This produces a fine-scale representation of the local geography. For example, figure 2 shows

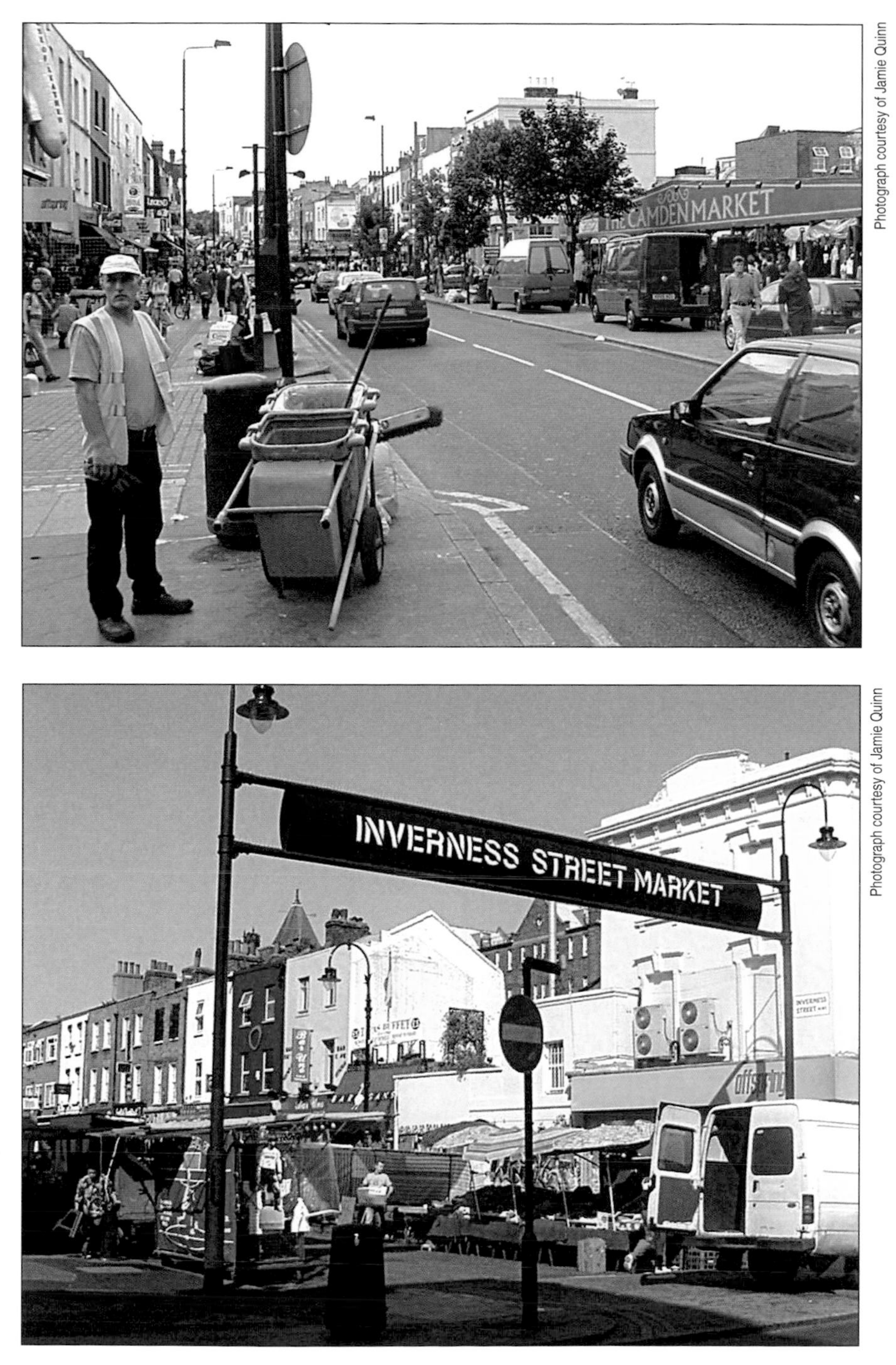

Photograph courtesy of Jamie Quinn

Photograph courtesy of Jamie Quinn

Figure 2 Camden Town centre, Inner London

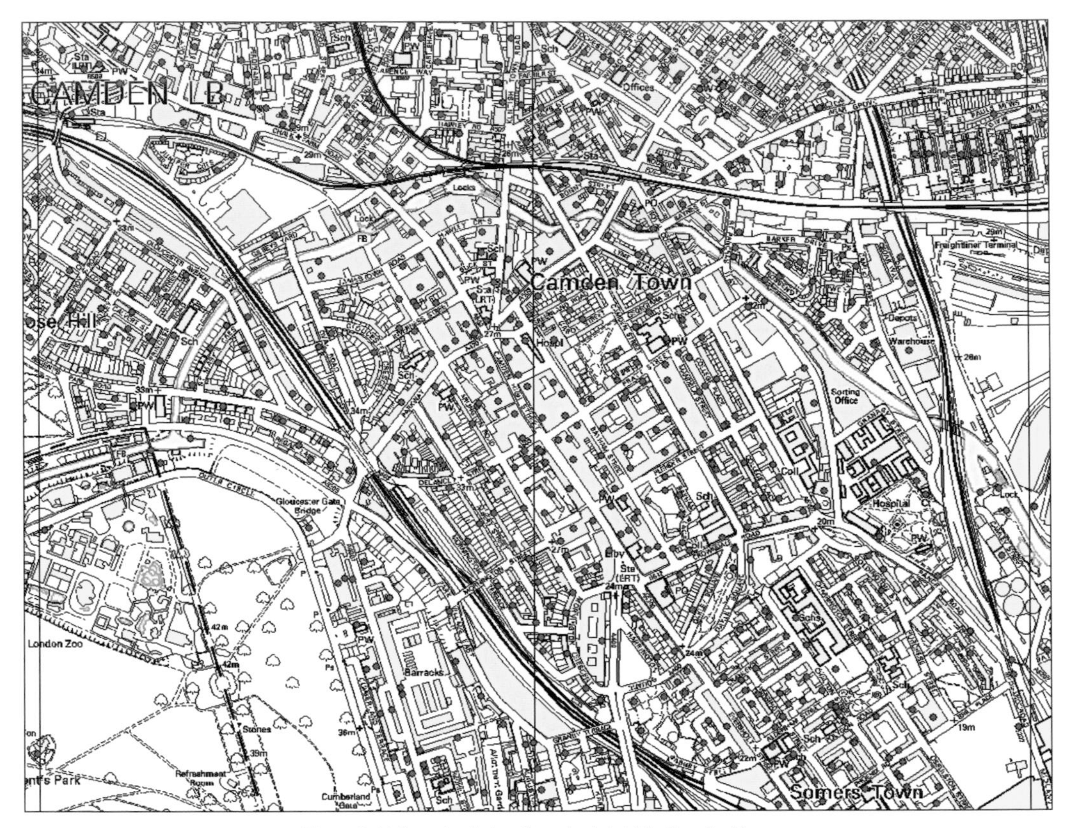

Figure 3 Unit postcode locations (red dots) in Camden Town

two typical pictures of a densely populated urban area in inner London—Camden Town—with the UPCs for the same area shown in figure 3.

Spatial units with fixed boundaries can suffer from the modifiable areal unit problem (MAUP), which has an impact upon statistics and data aggregation (Openshaw 1984). It is very rare that fixed tracts such as enumeration districts (EDs), wards or even postcodes reflect the true urban form, and, therefore, to attempt to carry out analysis using them could produce misleading results. The MAUP can be very restrictive on the type and validity of spatial work carried out, but by using UPCs to generate surfaces, some of these problems can be overcome. This is because the level of granularity is finer than most other socio-economic data sources, allowing more realistic representation of the urban form and structure (see also Thurstain-Goodwin, this volume; Schietzelt and Densham, this volume). The additional detail and greater precision added by using the UPC

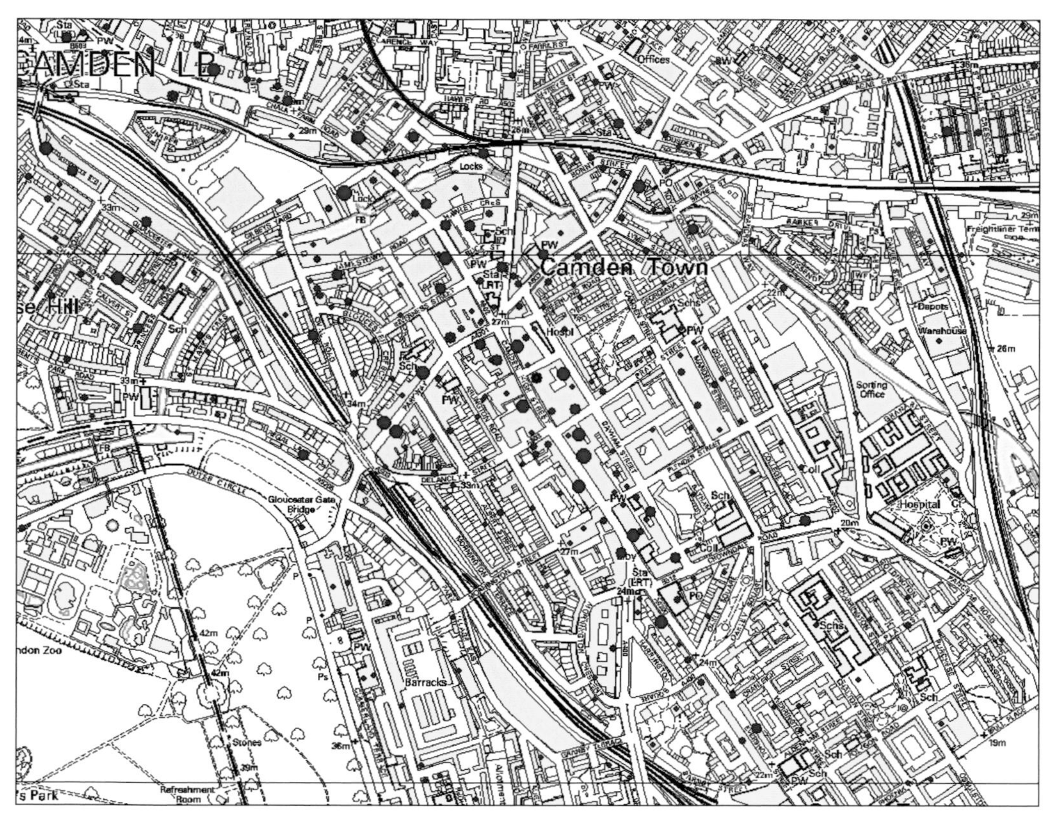

Figure 4 UPC locations graduated by number of employees in convenience retail in Camden Town

rather than the ED is also substantial, as the data require less aggregation to the representative unit.

The graphical representation of point data can be used to convey information about other data dimensions by plotting them as symbols that may vary in size, shape, colour, hue or saturation. If the data are multidimensional, this technique can not only improve the clarity of the graphical display but also portray information about the data in ways that may induce viewers to discover patterns or trends. For example, in figure 4 we demonstrate the use of graduated symbol mapping to represent the aggregate number of employees in the same area that was presented in figure 3. As can be seen, the use of thematic maps provides a way to visualise the data, but because of the large number of data points, it is difficult to identify a coherent pattern.

One step removed from the notion of data representing discrete objects is to treat them as continuous. Bracken and Martin (1989) have suggested that a field or surface approach provides a useful way of handling socio-economic data. Considering the size of the datasets, and that variations in data values between point observations are of interest, we have used surfaces as the main way to understand and analyse the constituent datasets of our town centres representations. The main rationale for this conjecture is that this permits more plausible and readily-comprehensible representations of geographical distributions (Bracken 1989). Fundamentally, there are two methods of producing a surface from point data. For sampled data (used, for instance, to create a digital terrain model), a spatial interpolator such as inverse-distance weighting or Kriging is used. Complete or almost complete population datasets, like the Annual Business Inquiry (ABI), require a density estimator, which spreads out the data from each point over the surrounding area. The resulting raster layer in a GIS represents the data surface.

The datasets used in the town centres project are made up of complete populations rather than sampled data, and thus kernel density estimation (KDE) was used to produce our surfaces. Details of the estimation procedure are explored elsewhere (for example, Diggle 1985; Brunsdon 1995) but we sketch the rudiments here. Whereas with sampled data a continuous data model needs to interpolate values in between points, a density estimator utilises the values at the known points to allocate intermediate values. A kernel of an applicable size is passed over the points, smoothing the data within it to allocate a proportion of the data values to each grid cell. The grid cells further from the original point receive progressively smaller proportions of the data. The size of the kernel used is the most critical factor in KDE. Planning Policy Guidance Note 6, the document that provides some of the motivation for this project, suggests that people are reluctant to walk more than 200 to 300 metres in a town centre (Department of the Environment 1996: paragraph 3.14), a figure that is similar to the average distance between UPCs. This range was therefore used as a basis for a series of tests carried out to find the best bandwidth and raster resolution to reflect the underlying distribution of the data. The optimal bandwidth was found to be 350 metres, and therefore the bandwidth used for the final model was 300 metres, a compromise between the two. As these data are spread between the known points, a density surface is produced, covering the entire study area, density referring to a given value, for instance total people employed per hectare.

Figure 5A is one such surface for the same data and area shown in figures 3 and 4. Once familiar with the concepts of surfaces, users can understand the data easily in spatial terms, and surfaces are significantly superior to the point format for studying the inter-point locations. In addition to this, there are other advantages that are not immediately apparent. For instance, it is easier to spot errors as anomalies stand out more clearly. This assists with error checking of the input data and its effects upon the output.

The concept of a surface can be an unfamiliar concept for many non-GIS users. Therefore, a number of experimental visualisations were created from the surface to introduce the concept, including presenting the data as contours. People with minimal, if any GIS or spatial data experience, would likely be unfamiliar with data in surface form, but would probably have experience with physical landform maps that use contours to represent the shape of the land. By putting the values of the 'Index of Town Centre Activity' (ITCA) surface in terms of contours, they were presented with a visualisation with which they have some familiarity and, thus, could understand. Figure 5B shows an example of a contour map output, and this can be compared to the surface in 5A. In terms of communicating the

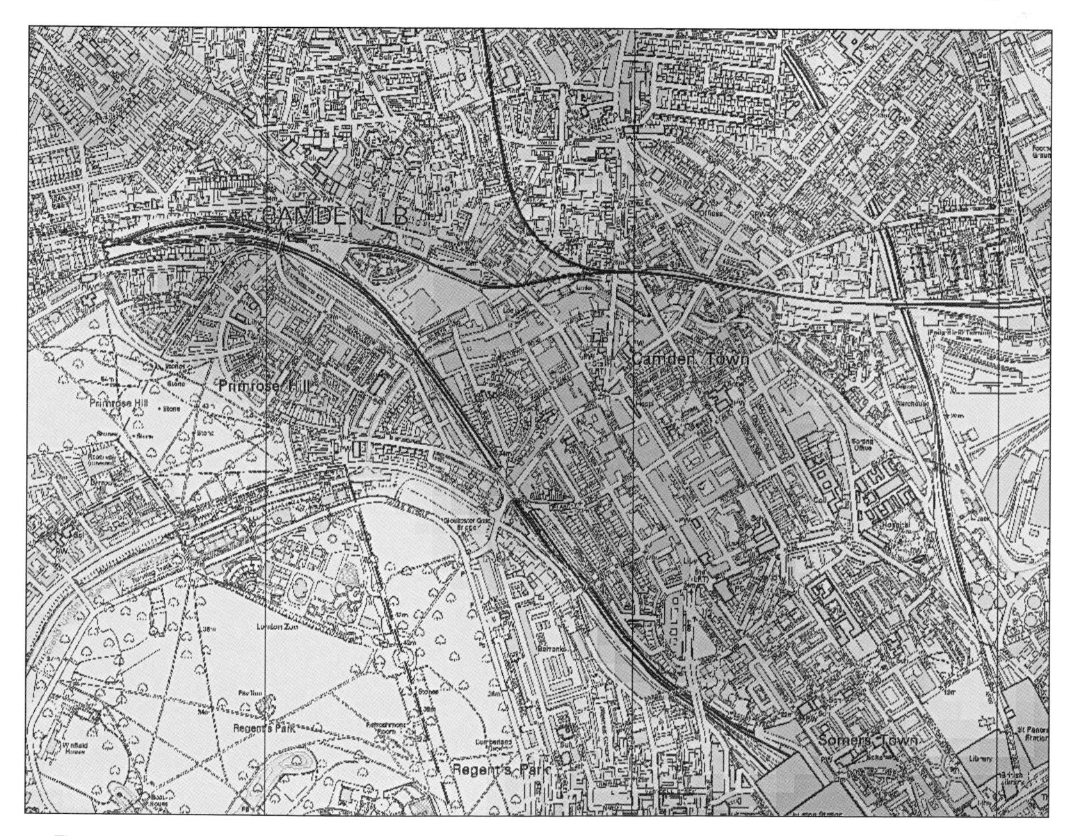

Figure 5A A data surface representing the ITCA (graded from little activity in light red through to more activity in darker reds.)

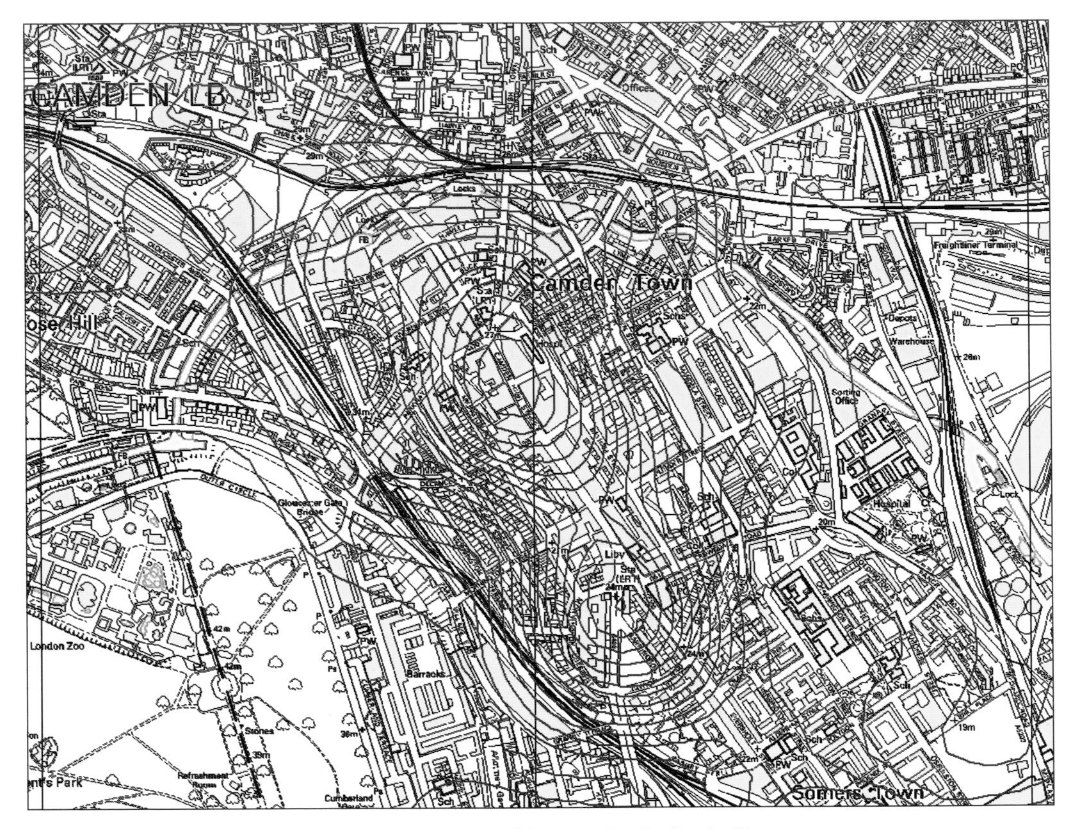

Figure 5B A contour map of the same data for Camden Town

data surfaces to users, either method could be used, the final choice being dependent on the audience and aim.

With the data in surface form there is greater flexibility in the use of spatial analysis as some of the worst excesses of the MAUP can be overcome (Martin 1989). The format also makes it easy to combine and contrast different datasets. In the CASA town centres project, a number of surfaces were used—three in total (Office of the Deputy Prime Minister 2002). These were:

a. Economy: a measure of local economic activity of a 'town centre nature' in terms of numbers of employees. This includes employment types that are commonly found in town centres, for instance, people employed in retail stores, offices of all types, restaurants and hotels. Each of these were used as a positive indicator, which contributed to each location's level of town centre characteristics. Other employment not commonly found in town centres, such as manufacturing and warehousing industries, were also used, but as negative indicators, lowering the level of town centre activity for the region.
b. Floorspace: a measure of the amount of floorspace within 'town centre type' buildings, such as shops and offices. Buildings used for primary industries were not used in constructing this indicator.
c. Diversity: a measure of the diversity of the employment base in the area. This is a count of the total number of different town centre type jobs that exist in the local area. Again, the employment that was included was made up by industries from the retail, office, leisure and services sectors. Factory and manufacturing jobs were not included.

As all three surfaces were built from a postcode point origin, and they all had the same spatial attributes in terms of resolution of grid cells and area covered, then they could be combined together using simple map algebra to produce a final output surface that reveals the ITCA. In this case, a simple linear additive method was used, giving each input surface equal weight. It would be possible to modify this to give each input a different weight, in order to account for the relative importance of each one, if desired.

The cell values on the ITCA surface indicate how strong a particular location is in terms of its town centre characteristics. The higher the value of the cell, the more characteristics it has that would make it be described as part of a town centre. To successfully identify the location and spatial extent of a town centre it is necessary to select a critical threshold deemed to define the town centre boundary. This critical value was selected in consultation with representatives of local authorities and using the research team's own knowledge of London. Areas that had ITCA values greater than this threshold were declared as being potential 'areas of town centre activity'. It also proved necessary to undertake a filtering process in order to impose a size threshold (areas of less than 4 hectares were too small to generate accurate output statistics, and the statistics that are produced were usually disclosive) and a diversity threshold to screen out areas that, whilst seeming to be town centres, were actually similar objects, such as business parks.

Using metaphors to communicate meaning

Whilst the actual selection of the critical contour was based on standard maps, visualisation based on a flooding metaphor has proved to be an effective tool for communicating this concept to the user community. A modern GIS, with the capabilities of draping a standard map over a 3-D surface and wrapping it, can be used for this purpose.

A useful analogy was to imagine the ITCA surface as a terrain model for London, with the critical threshold value being a pseudo-sea level value. This

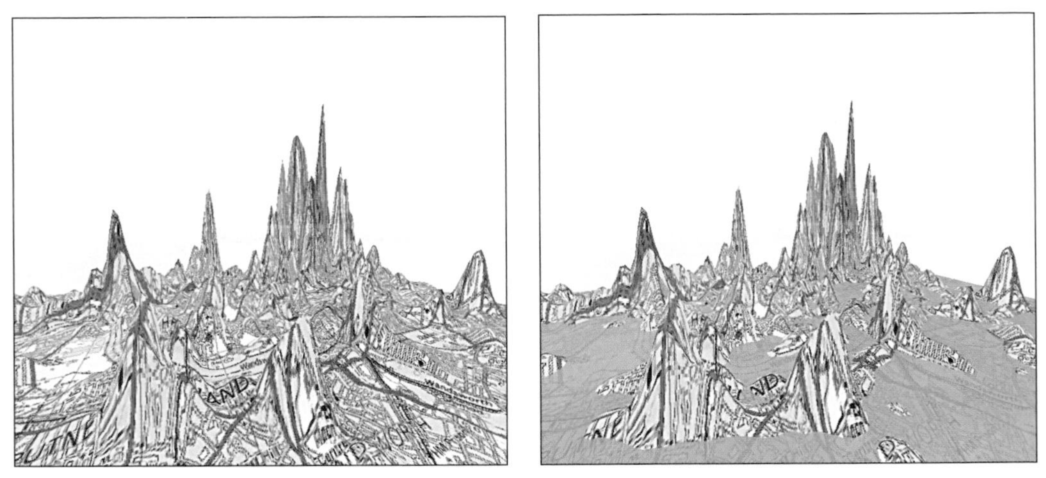

Figure 6A

Figure 6B

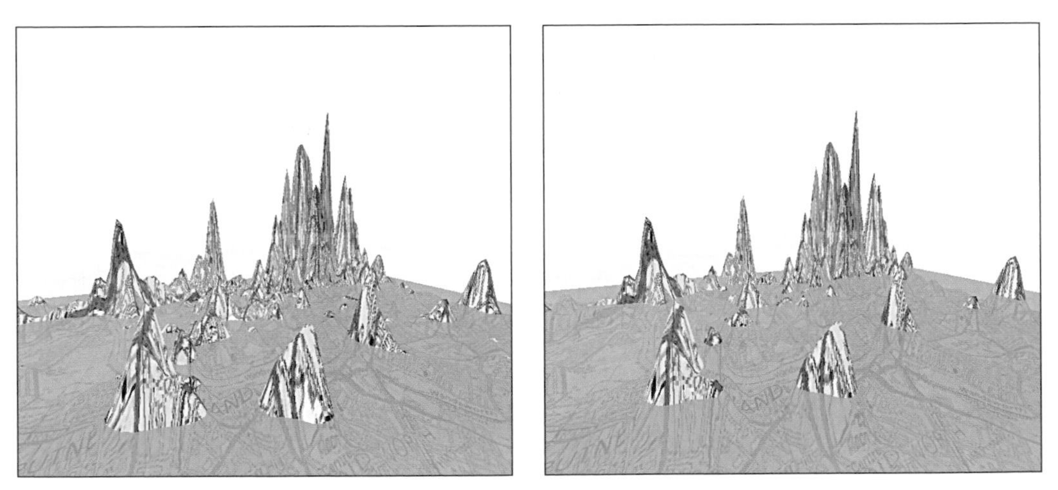

Figure 6C

Figure 6D

Figure 6A-D Flooding the index surface—looking from Putney towards Central London

threshold could be set at one of the contours shown in figure 5B. As the model is flooded with water, everywhere that is below the town centre threshold becomes submerged. The only parts of the land surface still left exposed are isolated islands—islands that represent the town centres (see figure 6). Landscape metaphors have been used to explore data such as population density and distribution (Wood et al 1999), as well as to investigate text documents of an online catalogue of references where surface elevation represents the amount of documents belonging to a particular topic (Fabrikant and Buttenfield 2001). In both cases, the metaphor is used to generate insights into the data by representing them in such a way that data values define the undulation of the terrain.

Communicating conceptual constructs to users

Thus far, the visualisations described in this chapter have been developed to communicate the technical process that was carried out to create the model of town centres. These visualisations have two crucial roles: first, they have helped to explain the project methodology to its stakeholders and end users; second, through this explanation, it is possible to build trust in the project's outcomes. This has been especially important in this project, as many stakeholders have voiced concerns about the methodology and its relevance to the definition of town centres. Their concerns have derived primarily from a lack of familiarity about the methods followed, and through our use of analogy and visualisation, the stakeholders were more able to understand and accept the methodology as being adequate and robust.

In addition to the explanatory visualisations used in presentations and in the reports, the project also included another type of visual product—local area maps on which the boundary of the town centre is marked. These maps, along with their associated statistics, are the major output of the project and whilst the data surfaces might be useful for sophisticated users, most of the end users are concerned to obtain robust statistics in an easy-to-access form. For each town centre, a report that contains a map with marked statistical boundaries is provided along side the relevant statistics presented in a tabular form. Whilst earlier in the project a printed paper version of the town centre report was considered, this was increasingly seen as inadequate for such a large-scale project. London, as with most U.K. cities, is made up of many smaller villages, towns and cities, all of which independently grew up before converging together to produce larger conurbations. Therefore, today's urban form of London contains over a hundred town centres, and producing a printed report with detailed coloured maps and accompanying statistical tables would have made the publication expensive for little reward—many readers would typically be interested in only a few particular town centres. The costs are exacerbated by the need to produce this compendium on a regular basis. Thus, it was decided to produce the reports in an electronic form. This decision enabled the creation of digital maps in which the boundaries are represented at four scales: the whole area of London, the borough level, and two local scales—one based on Ordnance Survey's (GB) 1:50 000 maps and the other based on their more detailed 1:10 000 series. The small-scale London and borough maps allow the user to drill down through them straight to the individual town centre maps. The report itself was formatted using HTML to make it compatible with Web browsers. Figure 7 shows a sample report for Camden Town. In some cases the different scales of maps were not all found to be suitable—for instance, in large town centres (such as Central London) the use of 1:10 000 maps did not produce appropriate images, and in the later stages the algorithm for map

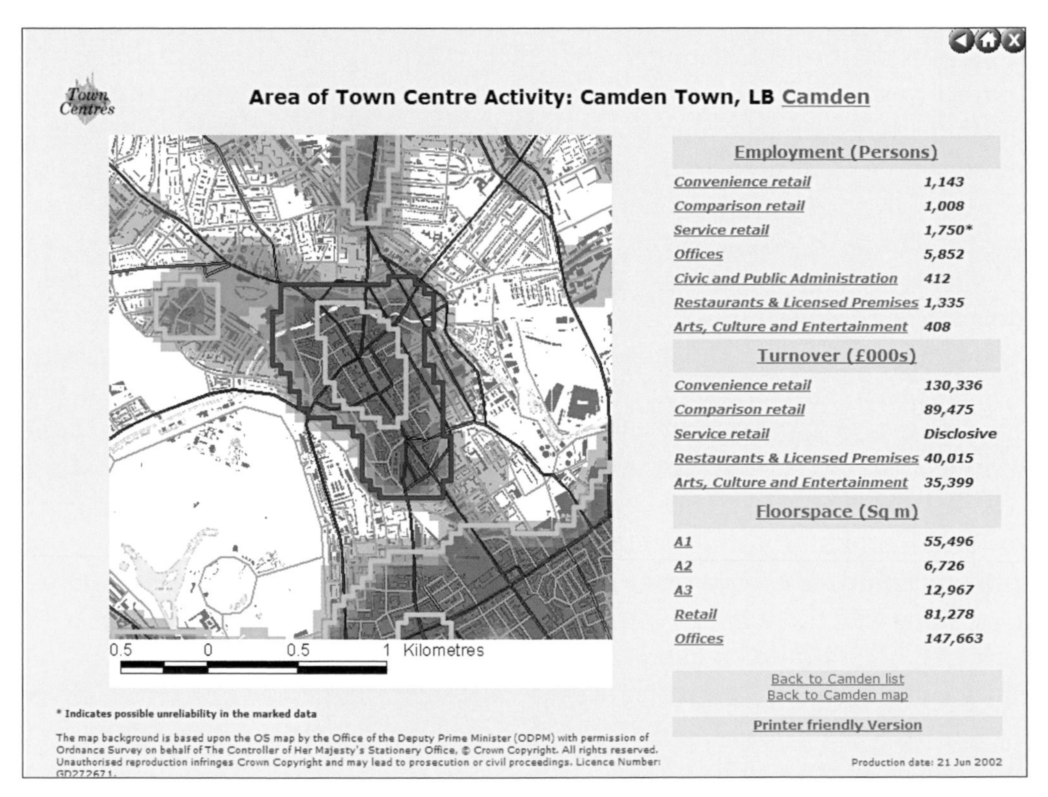

Employment (Persons)	
Convenience retail	1,143
Comparison retail	1,008
Service retail	1,750*
Offices	5,852
Civic and Public Administration	412
Restaurants & Licensed Premises	1,335
Arts, Culture and Entertainment	408
Turnover (£000s)	
Convenience retail	130,336
Comparison retail	89,475
Service retail	Disclosive
Restaurants & Licensed Premises	40,015
Arts, Culture and Entertainment	35,399
Floorspace (Sq m)	
A1	55,496
A2	6,726
A3	12,967
Retail	81,278
Offices	147,663

Figure 7 The Camden Town report from the CD-ROM

production was adapted to produce output using only the 1:50 000 maps for the larger centres.

Aside from the Ordnance Survey base maps, perhaps the most useful element in the reports are the town centre boundaries. This straightforward delineation needs to convey a complex message and boundary. Whilst a specific threshold value is used for the critical contour, which produces a sharp and clear boundary on the data surface, the boundary itself is an artificial statistical construct, and as such a fuzzy object in flux. The edge of a town centre, therefore, is usually not necessarily anything that could be identified on the streets. In some cases, it may fall along a real feature—for instance, a railway track, a large road, or a water body. In such instances, the real-life feature helps to divide the town centre type postcodes from the non-town centre type postcodes and can indicate a physical barrier to the extent of the town centre. In most cases, however, there is no defining point such as this. Thus whilst the boundaries are very clearly defined in statistical terms, most town centres have no distinct physical features that guide users seeking to delimit that area within which it feels like being in a town centre—viewed from the ground, most town centres appear to have no crisp and well-defined start or end.

In addition to this, once a map is produced, users have tended to regard the defining element of the town centre as being its shape, form and location. However, it has been important to explain to users that these are but some of the attributes, with the name being the core of its definition. Hence, the label 'Camden Town centre' defines a place, whilst its spatial extent and location is free to move and change from one time series to the next. This is important because a town centre is a fuzzy object and is also temporally unstable—that is, it can change in size according to its short-term economic performance for each release of data. If the shape and location became rigid defining characteristics, a growing, shrinking or moving town centre object would not be picked up, and only changes in statistics would be recorded. Furthermore, it is highly likely that many of the users of the statistics and the maps would hold their own concepts about the boundary, and that they would compare the maps to their own ideas when assessing the statistics. Therefore, it is important to communicate that the boundary of the town centre defines the area that was used to calculate the statistics at a snapshot in time and are subject to change with different releases.

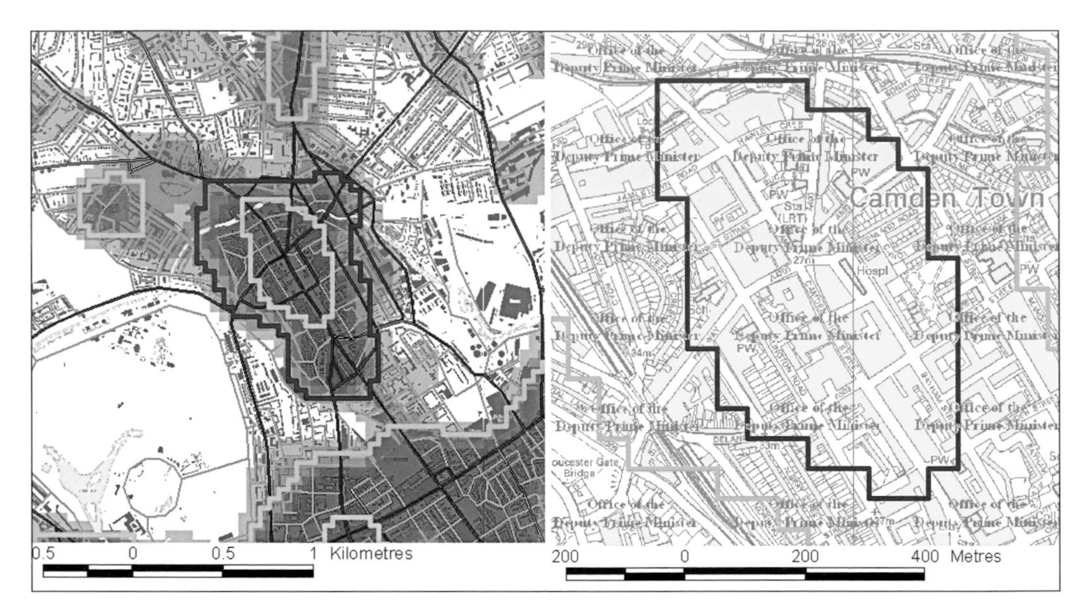

Figure 8 The Camden Town report from the CD-ROM. The town centre boundary of Camden Town is in blue, whilst the orange lines are nearby town centre and retail core boundaries. The ITCA is represented by the red colouring—the darker the colour, the more town centre characteristics the area shows

Representing a fuzzy object

Because of these issues, the use of crisp, rounded contour lines was considered unsuitable as users interpreted them as well-defined boundaries rather than pragmatic, generalised constructs. However, in an off-the-shelf GIS, the nature of contours is to interpolate across cells, inferring smoothly curving contours between ITCA values, like those commonly found on landform maps. As figure 8 illustrates, the solution that has been selected is the generation of a jagged-edge boundary that follows the shape of the surface grid cells. The method to create such a boundary is to produce a two-value raster (town centre and not) at an appropriate level of granularity from the ITCA surface with the threshold being at the same value as the critical contour. A GIS raster-to-vector routine was then used to produce a boundary line on the outer edge of the town centre, and the resulting jagged-edged boundary is marked by incremental right-angle steps. The raster-to-vector algorithm in the GIS applies a mid-point rule to cells in order to produce the boundary. This pushes the boundary towards the centre of edge cells, thus displacing it inwards by half a pixel. However, this is not considered as a major problem, as the spatial location of the boundary is not the most significant element on the map. In practice, the statistics are calculated on the basis of the smooth contour and not on the graphic representation. In terms of communicating the boundary, the general visual impact of the boundaries on the user is more important than attempting precise delineations. The loss of precision of the maps is a necessary trade-off in order to communicate the concepts in the fullest and simplest way.

In addition to the base maps and town centre boundaries, various experiments were undertaken in order to add more background information to the output. For instance we experimented with the addition of both the UPC locations and the data surface, but in both cases it made the reports more cluttered and harder to comprehend. Colour schemes also have an impact upon the visualisation, and again experimentation was required until a combination was found that was both aesthetically pleasing and communicative for users and stakeholders. The final maps are easy to read and provide a clear indication of the boundary including all fuzziness, thus achieving the necessary aims.

3 Conclusions

The set of visualisations described here demonstrates how a modern GIS is capable of producing effective and useful cartographic representations that convey clear and understandable messages to map users. Many of the socio-economic phenomena explored within the town centres project share common characteristics in terms of their ambiguity and fuzziness of definition. It is in an attempt to represent this complexity that most of the visualisations have been developed—a pragmatic attempt, in effect, to define and describe the inherently indefinable. However, there are still some problems that remain unresolved and despite our efforts to clarify them, these need to be placed on the agenda for future work.

The most important aspect that we have touched upon throughout the chapter is the role of the geographic boundary in the definition of a town centre object. When data were presented in tabular form, stakeholders did not perceive this to be a problem, for the simple reason that the definition was hidden from them. As long as the different users were aware that the table provided data about this or that town centre, they could interpret the values according to their personal understanding of the specific locale. However, once a map was presented alongside the values, there was a tendency to interpret the line on the map as a precise defining characteristic of the object. In practice, the lines were intended more as place labels than precise definitions of object extents. Moreover, any town centre is likely to change in form and function over time; for instance, it might expand or contract under the influence of local, regional or national socio-economic processes. Therefore, the spatial boundary might change too and with it the associated values of employment, turnover or floorspace. Thus, material to our modelling method is the notion that the geographic boundary is not a defining element but rather is a looser attribute of a town centre that has a given name. In effect, the geographical aspect (the boundary) is subordinated to the place name—an aspect that seems somewhat counterintuitive at first impression. All of this does not mean that the comparison of a single town centre from two time series should be meaningless or wrong. On the contrary, it can depict both the history and the change in structure of the town centre as the years go by. Such an interpretation of the boundaries of geographical objects is not intuitive to users, and we expect to develop visualisation methods that will clarify this point as research progresses.

The second valuable lesson from the project is the benefit of involvement of end users in the development of the statistics. All too often, statistical methods and visualisations are developed without continuous consultation with those whom will eventually use them. This might result in a misunderstanding of the method, distrust in the results, and the creation of maps that are not readily interpreted by many in the user community. Other chapters in this book stress the importance of accommodating and supporting user needs so to increase the likelihood of products from a project being adopted and used (Hudson-Smith et al, this volume; Tobón and Haklay, this volume). Within the town centre project, our ongoing interaction with professional end users at all levels—from planners at local authorities to retailers and policy makers—enabled us to understand their needs and to think carefully about the best ways to use cartographic and visualisation methods to communicate relevant information more effectively.

Finally, we note that the visualisation techniques that are available within modern GIS provide a basic toolset to create adequate cartographic output. It is wrong, however, to assume that sophisticated visualisation can be produced at the press of a button and without due consideration of its use, audience and content. Careful consideration and iterative development of appropriate representations remains the best strategy to ensure successful outcomes.

Endnotes

1 The ODPM (Office of the Deputy Prime Minister) is the U.K. government department with overall responsibility for planning guidelines.
2 For a detailed description of the town centres model, see Office of the Deputy Prime Minister (2002) and Thurstain-Goodwin and Unwin (2000).

Spatial modelling

Improvements in our ability to use GIS to understand what is spurring the development of applications invite us to consider *what if?* Nowhere is this interest more evident than in the use of GIS to develop scenarios concerning the sustainability of human activities—as illustrated here by Steve Evans and Philip Steadman. The sensitivity of model outputs to inputs, specifically with respect to the measurement of distance and path, is a theme that is explored in detail by Michael de Smith, while the sensitivity of model outputs to the assumptions made about the dynamics of change are investigated by Michael Batty and Naru Shiode. The interplay of measurement, optimisation and dynamics is further developed in the realm of location-allocation experiments using emergency vehicles in a chapter by Torsten Schietzelt and Paul Densham. In each of these contributions, the emphasis is on how theories are translated into models and how data and technology are thence used to structure particular model applications. In fact, what these illustrate is how the process of moving from model structure to implementation is affected and how algorithms are used to translate data and formal structures into applicable forms of spatial analysis and simulation.

Chapter

Interfacing land-use transport models with GIS: the Inverness model

Stephen Evans and J. Philip Steadman

This study demonstrates the interfacing of the TRANUS land-use transport model of Inverness, Scotland with a desktop GIS package developed as part of the European Union-funded PROPOLIS project. While the coupling of land-use transport models with GIS is nothing new, this system has been designed to automate the process, enabling the outputs from the land-use transport model to be rapidly viewed in a consistent manner in the GIS. The new software also enables the user to pass model data on to several new software packages developed by the PROPOLIS team. These packages calculate a set of indicators that measure the environmental, social and economic components of sustainability. These comprehensive assessment methodologies look at the long-term sustainable strategies modelled by the system in the context of a diverse range of processes and conditions. These include carbon dioxide emissions, energy efficiency in transport and land use, employment, regional economic competitiveness, biodiversity, and the incidence of the negative externalities of noise and pollution.

1 Introduction and an overview: PROPOLIS

More than three-quarters of the population of Western Europe live in cities. Their quality of life, their health and their safety are, to a considerable extent, affected by environmental quality, the provision of and access to services, and the safety of their cities of residence. These cities are continually changing and are the focus of a substantial share of society's problems such as degeneration, social exclusion, employment insecurity, environmental deterioration and traffic congestion. As a result of these and related problems the economic efficiency of cities can be impeded, for example, because of urban congestion and pollution. Methodologies are needed for anticipating and mitigating negative changes and for encouraging positive ones. The PROPOLIS (Planning and Research Of POlicies for Land use and transport for Increasing urban Sustainability) project addresses these issues by enabling the prediction of the impacts of urban transport and land-use policies. Integrated land use and transport models offer one method of simulating the spatial interaction and dynamics of a city or region and lie at the core of PROPOLIS.

The objective of PROPOLIS is to research, develop and test integrated land use and transport policies, tools and comprehensive assessment methodologies in order to define sustainable long-term urban strategies and to demonstrate their effects in European cities. PROPOLIS is a European Commission Fifth Framework Programme and University College London is one of the partners involved. The project is even more significant, since it compares and contrasts these sustainable long-term urban strategies in a range of settings right across the European Union. The other project partners include LT Consultants Ltd (Finland), Marcial Echenique & Partners (United Kingdom: part of WSP group), Universität Dortmund, Institut für Raumplanung, (Germany), Trasporti e Territorio srl (Italy), Marcial Echenique y Compañía S.A. (Spain) and STRATEC S.A. (Belgium) (source: PROPOLIS Web site; www.ltcon.fi/propolis 2002).

The work involves developing a set of indicators to measure the environmental, social and economic components of sustainability. Values for these indicators are calculated using enhanced urban land-use and transport models and new GIS and Internet-based modules. A decision support tool is used to evaluate the sets of indicator values in order to arrive at aggregate environmental, social and economic indices for the alternative policy options. To anticipate the long run land-use effects, a time-horizon of 20 years or more is used. The system is used to systematically test and analyse policy options in seven European cities to derive optimum combinations of different policy tools for land use, transport, investment, pricing and fiscal measures. The land-use and transport models have, in most cases, been developed through close contact with client-partners (often the city or regional authorities responsible for planning).

The land-use and transport models of the seven case study cities are the driving engines of the PROPOLIS system. These models were previously calibrated to correspond with the observed behaviour in the test cities. The case city models belong to three different, leading integrated urban land-use and transport model types: Helsinki, Naples, Bilbao and Vicenza (MEPLAN model, supplied by Marcial Echenique & Partners, United Kingdom); Dortmund (IRPUD model, supplied by Institut für Raumplanung, Germany); Inverness and Brussels (TRANUS model, supplied by Modelistica, Venezuela).

The innovative nature of the project is its integrated and comprehensive approach. GIS are used extensively at the core of the PROPOLIS project for visualising model outputs, calculating specific indicators, and for passing model outputs to other PROPOLIS software packages in a suitable format. Rapid and consistent interfacing between the land-use and transport models and the GIS is essential. Interfacing between such models and GIS has been achieved in the past, but in most cases was performed by converting, tweaking and transferring the necessary files from the land-use and transport model to the GIS by hand (work usually carried out by a GIS specialist). Not only is this slow and inefficient if it is carried out more than two or three times for a model, but it also can be subject to user errors. The approach developed by CASA is far more automated and consistent.

The nature of the models used makes it possible to move from simple transport and land-use variables towards important and more relevant environmental, social and economic indicators. These new indicators will address, among other things, biodiversity, regional economy and competitiveness, employment, the need for new construction, and the justice of distribution of the negative/positive economic effects of policies. The policy-testing process has been designed to study different pricing measures and their combinations with other policy options. This is expected to produce innovative and effective results that are also reflected in reduced carbon dioxide emissions and greater efficiency in energy use. An important goal of the project is to study the traffic generated by different land uses and thus estimate the total energy use and amount of emissions implied by a given land-use pattern.

The desired outcome for PROPOLIS as a whole is the production of general recommendations for all European cities based on the findings from the policy testing, together with city-specific demonstrations of their effects. It is hoped that the work might produce innovative policy recommendations, since the system is able to reveal interactions and multiplier effects by monitoring changes in the representation of the system. We will focus on the experience that CASA has had integrating the TRANUS city model of Inverness and the Highland Region of Scotland, United Kingdom, with a GIS interface and present some of the results from this particular case city.

2 The model software: TRANUS

TRANUS is an integrated land-use and transport model, developed by Modelistica (Caracas, Venezuela) that can be applied at an urban or regional scale. The software fulfils dual functions: first, the simulation of the probable effects of applying particular land-use and transport policies and projects; and second, the evaluation of these affects from social, economic, financial and energy points of view. TRANUS has its roots in spatial interaction and discrete choice theories and creates dynamic simulations with feedback loops between the land-use and transport modules (de la Barra 2001). The advantages of integrating models of land use and transportation are well known and have been documented extensively in the transportation science literature (de la Barra 1989). However, there are very few such integrated modelling packages in practice. For the transport planner, integration of land use and transport provides a means of making medium and long-term demand estimates, which are impossible with transport-only models in which demand is a prespecified input. For the land-use planner, integrated modelling, whether at an urban or regional scale, makes it possible to assess the impacts of transport policies upon a full range of activities at a particular location. But it is in consistent land-use and transport planning that the TRANUS system realises its full potential.

There are a number of different software modules that have to be run in order to process a TRANUS model and three are key. First, the land-use module deals with the location and interaction of activities and the representation of the real estate market. Second, the transport module deals with trip generation and the subsequent loading of the transport network. The third is an evaluation module. For a detailed description of the functions and modules behind TRANUS the reader is referred to de la Barra (1989) and de la Barra (2001). From a GIS perspective,

Image courtesy of Frank Brown

Figure 1 Inverness and the Moray Firth bridge

Figure 2 One of the more rural parts of the Inverness model

whether we are dealing with inputs to feed a TRANUS model or outputs from a TRANUS model, the important thing to remember is that data are either related to zones (polygons) or to links on the transport network (lines/routes). Some node data (points) also exist in the form of transport network connectors, and there are more aggregated inputs and outputs that are related to the model as a whole.

3 The case study region: Inverness and the Scottish Highlands

Inverness is the capital of the Highland region of Scotland (population approximately 65 000 in 2001), and its largest community. Inverness developed originally as a port and market centre, and owes its prominence to its strategic position within the topography and transport system of the Highlands, being located where the Great Glen and the Moray Firth converge. Today it is well served by road, rail and air (with its own airport) as well as by sea, and acts as the gateway to the Highlands for international tourism. Strategic road improvements carried out over recent decades, in particular the A9 bridge across the Moray Firth that connects Inverness to the north (figure 1), have made the city the most accessible service place for the Highlands as a whole. The surrounding region comprises farmland, moorland and mountain landscape of great scenic beauty (figure 2). Several small towns—Dingwall, Invergordon, Tain and Dornoch—are strung out along the coast to the north of Inverness, while others—Nairn, Forres and Elgin—follow the coast to the east. The deep-water harbour of the Cromarty Firth provides facilities for the maintenance and construction of North Sea oilrigs at Invergordon and Nigg Bay.

The region as a whole has its problems that arise from the rugged terrain, the extreme dispersal of a rural population, an inadequate communications infrastructure, and the seasonal nature of the tourism business. There is a constant threat of losing jobs and retail services to the larger Scottish cities to the south, Aberdeen, Glasgow and Edinburgh. In the 1980s a substantial proportion of local retail trade was lost, both to these competing centres and to the mail-order trade. Despite this, the region has gained population in recent years (by approximately 20 per cent between 1971 and 1991), and there has been sustained growth in tourist spending. Because of its relative accessibility, Inverness attracts shoppers and visitors from the entire region, some of whom make shopping trips by car over long distances on a weekly or monthly basis. The population of the Highland region is expected to grow from 204 000 in 1991 to nearly 220 000 by 2011.

Over a similar period the population of Inverness itself will rise from 64 700 to more than 72 000.

The challenge for the planners on the local governing Highland Council is to channel this growth to sustain the vitality both of Inverness itself and its remote satellite communities. Growth must be accommodated without compromising the beauty of Inverness and the region, the principal asset in attracting tourists. The Council hopes to attract new information and high-technology businesses and retain them with high-quality services and housing. There are particular challenges in shifting trips from car to public transport, given the extreme low density of the population, but new light rail services linking settlements in linear patterns along the coast seem to offer one potential solution.

4 Model and scenario design

Although the population of Inverness and the Highlands is low, it would be too time-consuming and practically impossible to attempt to keep track of the movements, choices and decisions of every individual in the region. There is, then, the need for data about individuals and detailed geographic locations to be aggregated. Hence, individuals in the Inverness model have been grouped according to common social and household characteristics. Individual buildings and the location of firms and houses have all been aggregated into larger discrete polygons or zones in order to keep the spatial representations in the model at a manageable level (figure 3).

The model has been built primarily so that the Highland Council can predict the impact of their Structure Plan, a detailed environmental, economic, community and infrastructure master plan, including maps, of the region over the next 20 years. The model has some 162 zones and 830 transport links. The land-use model distinguishes nine economic sectors: agriculture, fishing and forestry; energy and water supply; manufacturing; construction; distribution, hotels and restaurants; transport and communications; banking, finance and insurance; public administration, education and health; and other services. Eight types of urban land use are distinguished. The population is divided into five household types according to family composition (single households, couples, single parent family, couple with children, retired).

A special feature of the land-use model is that the market in floor-space is represented explicitly in terms of a competition for different types of building. Domestic buildings are classified into detached houses, semi-detached houses,

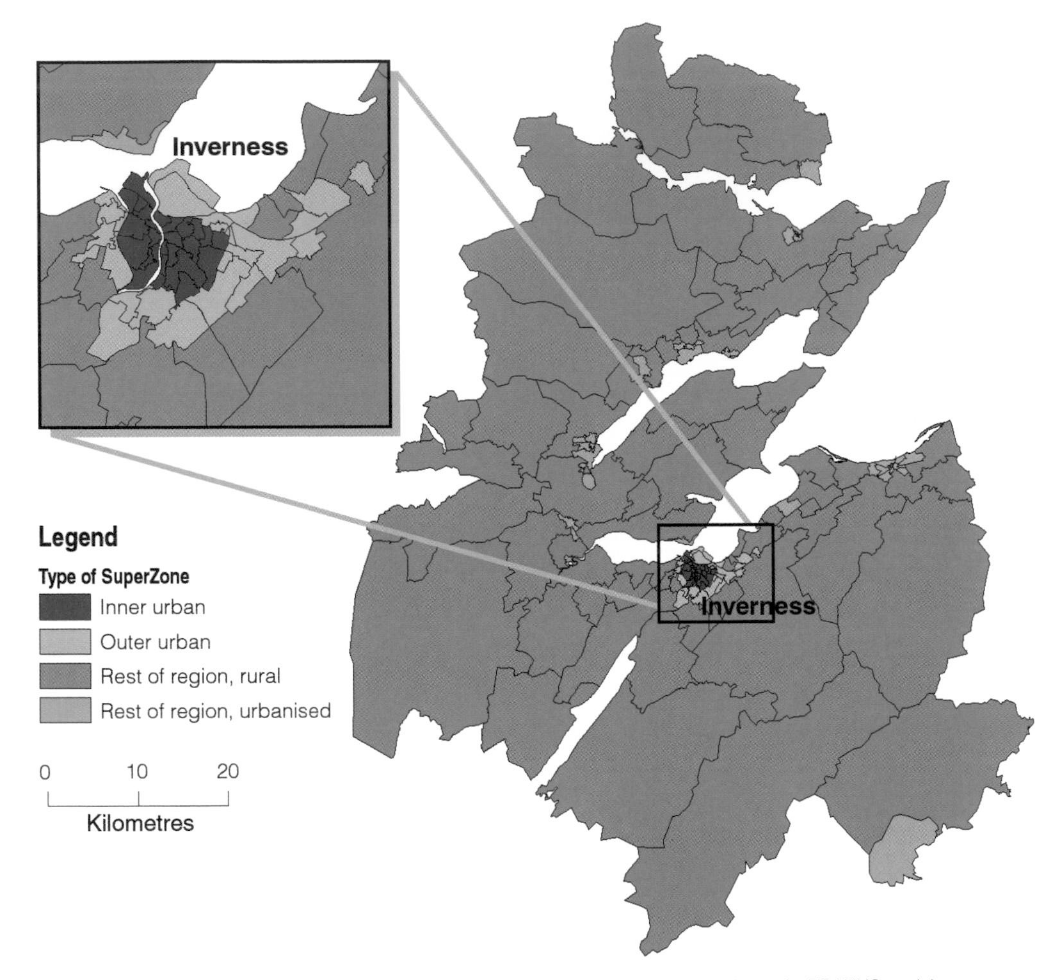

Figure 3 Inverness and the Highlands in Scotland showing the zones that make up the TRANUS model

terraced (row) houses and flats. Non-domestic buildings are classified as 'framed' structures or 'sheds'. Certain commercial and retail uses are then allowed to compete for some domestic building types (as well as for framed and shed buildings). Figure 4 illustrates the structure of the land-use and floor space submodel. Here it is simplified into three columns: activities, floor space and land. The lines connecting activities to floor space types and to land represent the demand relationships that were assumed to take place. For instance, heavy manufacturing industry may choose between framed structures or sheds, while commercial activities and light industry may also occupy terraced houses. Households may choose between only terraces, flats, semi-detached and detached homes. Sheds and framed buildings

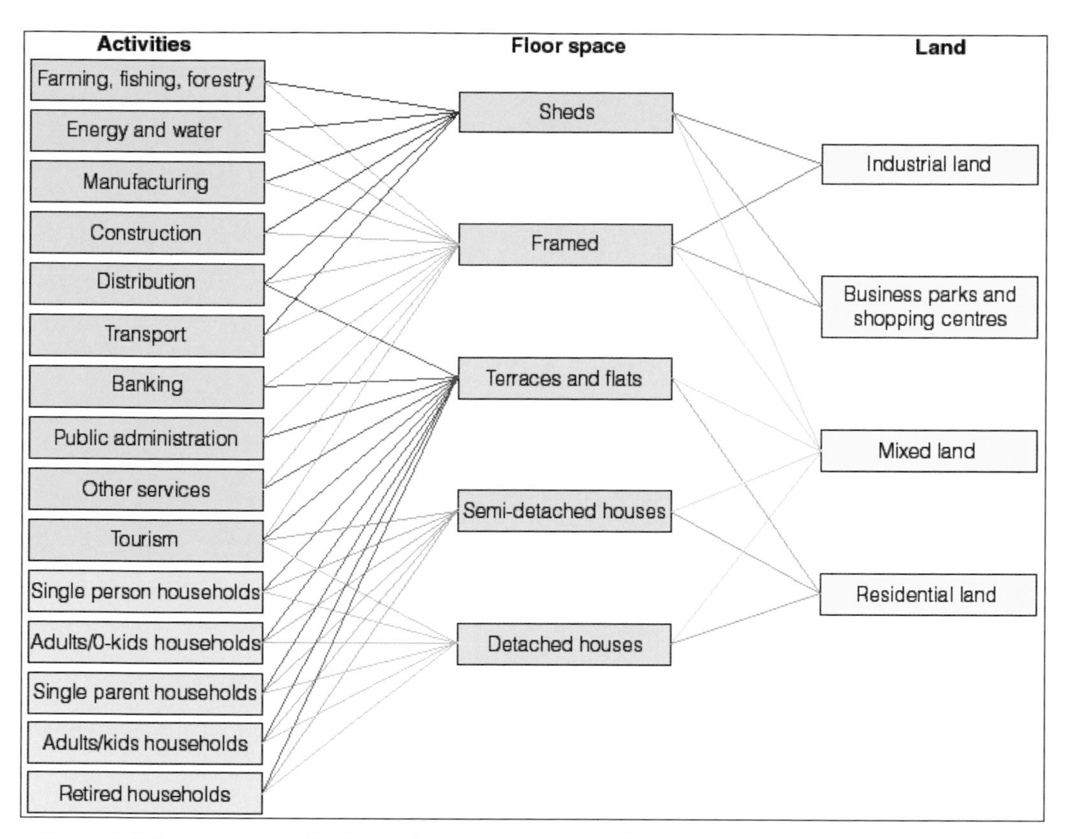

Figure 4 Activities, floor space and land categories in the Inverness model. The diagonal lines show which activities may occupy which types of floor space, and which types of floor space may occupy which types of land

cannot be built on residential land, while detached houses can only be built on residential land.

The transport model divides transport demand into eight types of journey according to purpose: five different types of work-induced trip (made by households of different types); travel to and from places of education; shopping trips; and trips to obtain other services. Road transport modes include walking, cycling, car with single occupant, car with multiple occupants, urban buses, rural buses, and minibuses. The rail network is separately represented. A bus-based park-and-ride scheme is being introduced into some of the future scenarios. The model was developed using multiple spatial and non-spatial data sources, and base data were derived from the 1991 Census of Population. Many of these data were loaded into the model software by hand, often sourced from the existing Highland Council GIS. This was a slow and complex process that took several months. Once the data were loaded, the model was run for a single projection for 1996, a horizon for which data already existed. The model was then calibrated until it produced model outputs that matched the existing 1996 data. This procedure continued until the model was fully calibrated. Future scenarios were then modelled with some confidence that they should produce realistic outputs. A series of five scenarios was created for five-year increments to the scenario horizon in the year 2021.

The scenarios developed for the Highland Council (by consultants Rickaby Thompson Associates Ltd, United Kingdom and Modelistica, Venezuela) are defined within a three-level structure. A **context** level deals with broader political, economic, demographic and environmental circumstances; a **policy** level deals with individual land-use and transport policies; and a **simulation** level defines scenarios as combinations of policies within contexts. Three contexts are envisaged: business as usual, Highland renaissance and natural change. Each of these entails different assumptions about future economic growth, increases in population and households, changes in land and transport costs, and external environmental policies and priorities. Business as usual means things will proceed just as they are. Natural change is a worst-case economic, low-growth scenario. The Highland renaissance scenario envisages a future where strong emphasis is placed on environmental priorities, but where these are compatible with, and indeed contribute to, economic growth.

At the simulation level, six options are compared: intensification of centres, development corridor, new settlements, small towns dispersal, rural dispersal, and current policies. 'Intensification of centres' confines new growth to existing developed areas. 'Development corridor' distributes new development in a wide arc following the coast, from Nairn in the east to Tain in the north. 'New settlements' introduces new small towns or villages along this same arc. 'Small town dispersal' directs all growth towards several selected existing centres. 'Rural dispersal' distributes new housing at low densities across the region and envisages that the occupants will be relatively self-sufficient and make major use of telecommunications. 'Current policies' continues the Highland Council's present directions and initiatives. Most of these scenarios embody transport policies that will greatly improve bus and rail services, discourage car use where feasible, and encourage walking and cycling at the local scale. Because not all six scenarios at the simulation level are compatible with all three contexts, it was decided to reduce the total number of permutations under consideration from 18 to 10. This way, those scenario and policy combinations that would probably never exist side-by-side in the real world are avoided.

5 Interfacing TRANUS with the GIS

One of the key requirements to achieve the aims set out in the PROPOLIS project is rapid interfacing between the land-use transport models and GIS. This enables the user to quickly visualize the results from the model and to pass these results on to other software packages in a common format. This ensures that the policy testing process can be carried out rapidly while providing consistent post-processing of the data and a standard presentation of model outputs.

CASA has developed a TRANUS GIS Module to enable the automated transfer of files from TRANUS to a fully customised GIS interface. The module comprises several interlinked tools that enable the data structuring and data visualisation of the TRANUS outputs. This way the user can visualise and analyse results from the TRANUS model using a map interface in tabular format and as graphs. The aim is that a user with little experience using a GIS can obtain basic but consistent results from TRANUS within a short space of time.

The front end TRANUS GIS Module is built on ESRI (Redlands, California) software, specifically ArcObjects™ technology from the GIS package ArcGIS® 8.1. Microsoft VisualBasic® has been used to customise the interfaces and to develop

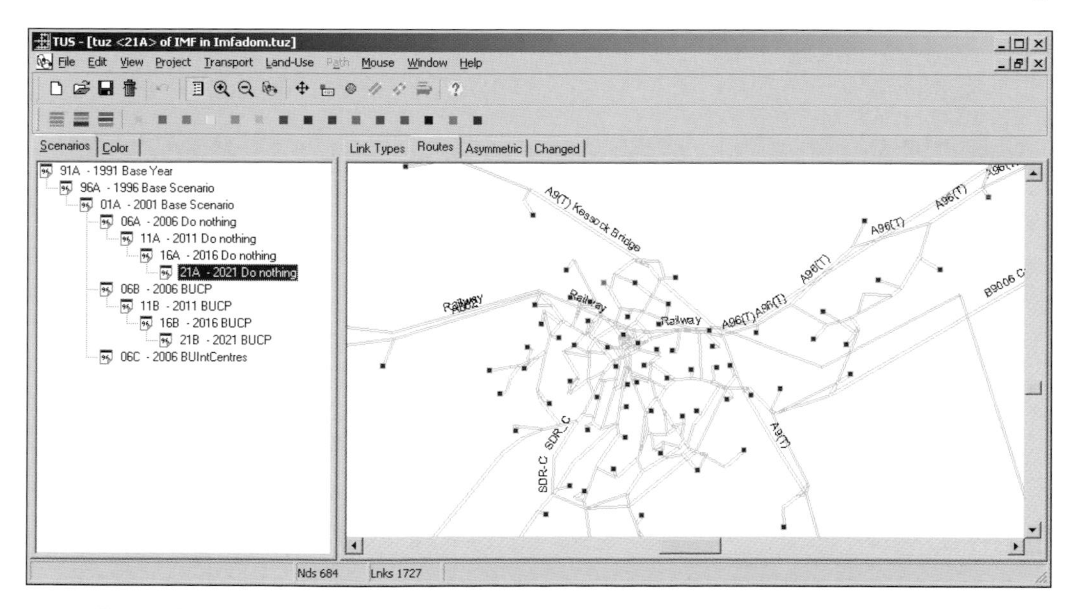

Figure 5 TRANUS model of Inverness and the Inner Moray Firth as represented in the TRANUS software. Here we have zoomed in on Inverness town centre. Transport links are shown as straight lines, while zones are represented by points (blue squares). The scenario tree can be seen on the left hand side

necessary additional software modules. The system has two main interfaces—a data management interface and a GIS/analysis interface.

The data management interface is driven from the ESRI data management package ArcCatalog™, which manages the structuring and translation of files from TRANUS into the GIS package. This interface also handles all the relationships between data entities from the TRANUS model and automatically creates a certain amount of metadata to enable the efficient management of the large number of files.

The GIS/analysis interface is designed to automatically create a map once each data transfer has been run. This allows the user to quickly check and compare the model outputs. The interface can be used to analyse data cartographically and to identify spatial implications from the TRANUS outputs. Setting up the model and preparing the individual year, scenario and policy combinations are managed within the TRANUS software (figure 5).

Within the TRANUS software, binary files generated by each model run can either be output to the GIS via the data management interface described below, or they can be read back into the TRANUS user shell in order to generate another run of the scenario for the next five-year increment. This process can be continued until the user has reached the particular five-year increment that they would like to view in the GIS interface. Our data management interface can then be used to couple the model output to the GIS and also to ensure that it is correctly named,

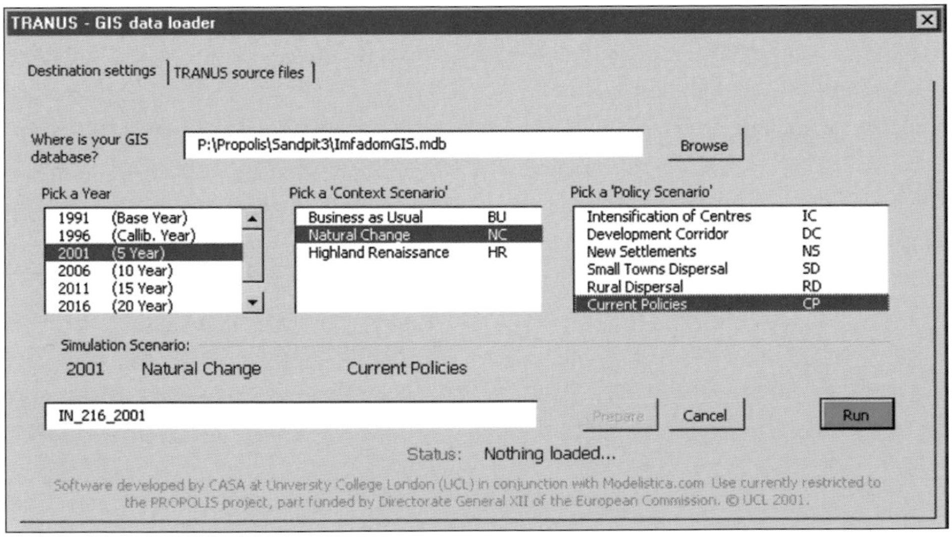

Figure 6 TRANUS GIS data loader

based upon the combination of year, scenario and policy that has been used in TRANUS, as is shown in figure 6.

A series of automated procedures grab the relevant TRANUS output files, reformat them where necessary and generate metadata where necessary. Finally the files are coupled to the relevant spatial data that have already been stored within the GIS. Depending on the size and complexity of the model and the speed of the processor, this whole process can take between two and five minutes. Once the data have been correctly structured, a customised version of ArcMap™ is launched with the data from the model preloaded. Within the software, we have created a customised menu for dealing with TRANUS files. This TRANUS menu offers a range of options for displaying the zones and transport network according to a range of outputs derived from the model.

The menu (shown in figure 7) allows the user to dynamically change the map interface to show zones as a choropleth map using any of the TRANUS outputs (for example, see figure 8). Likewise the transport network can be displayed based

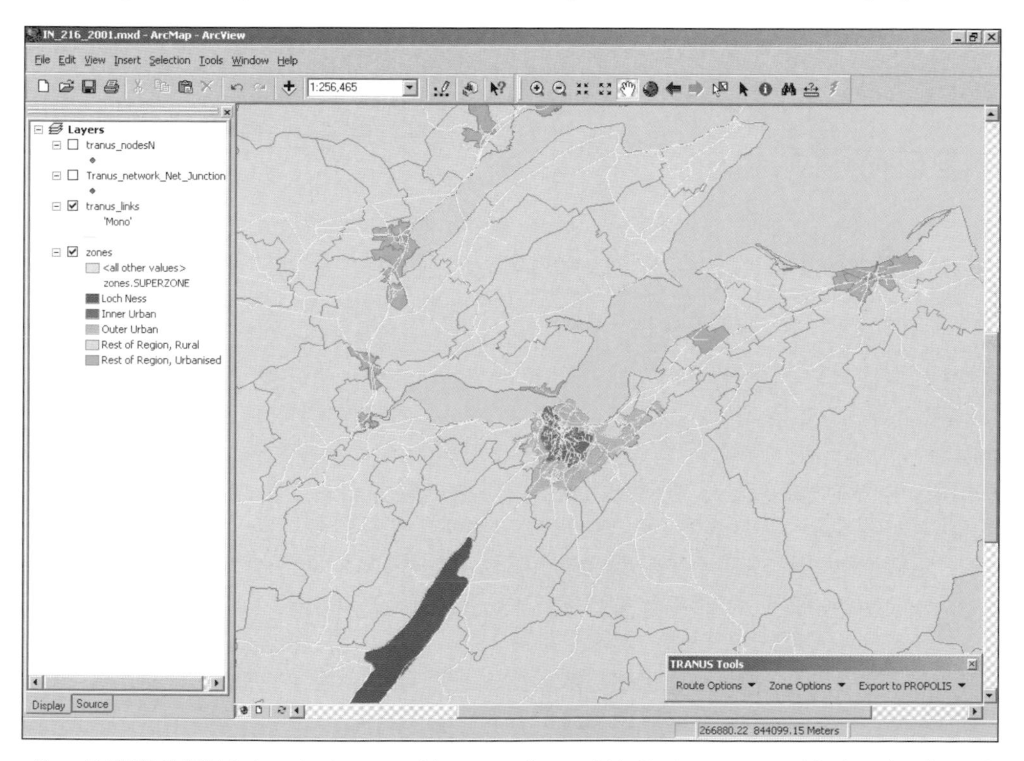

Figure 7 TRANUS GIS interface showing some of the menu options available. The transport network is shown in yellow and Inverness is at the centre of the map

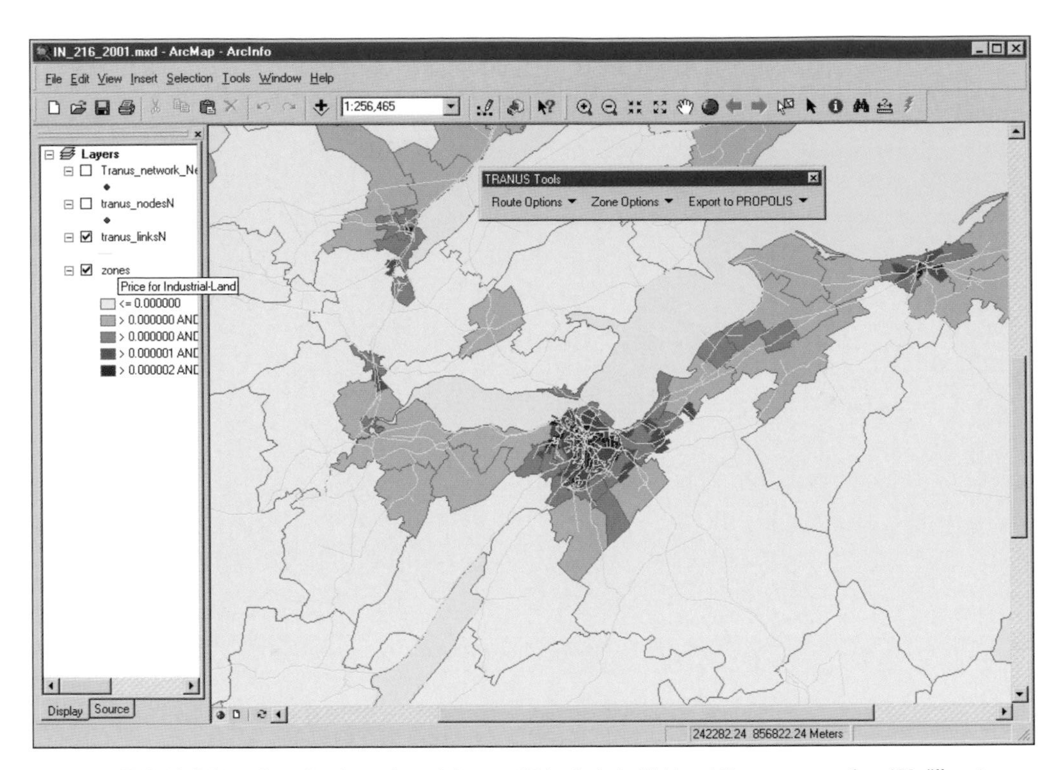

Figure 8 TRANUS GIS interface showing a choropleth map of Price for Industrial Land. There are more than 150 different zone-based outputs that can be displayed from the TRANUS tools menu based on the TRANUS outputs

on a number of TRANUS outputs. It can be shown with the different road types and railway lines displayed in a traditional cartographic style. Alternatively the routes can be displayed using proportional symbols according to the volume of vehicles that the route is designed to cope with. More importantly, the routes can be displayed according to predicted average traffic speed or by the volume of vehicles that TRANUS predicts will travel along each link according to the chosen year, scenario and policy combination (figures 9, 10, 11 and 12).

Our original aim was to create a system that would allow us to display all types of transport modes, by the predicted number of vehicles, on the map interface using dynamic segmentation. This was attempted by converting the lines and nodes that make up the transport network in the GIS into a route-system. Although our initial experiments were reasonably successful, it became apparent that we would need to reformat a large number of TRANUS output files for every model run in order to achieve this. In addition, a simple request, such as displaying average traffic speed on each link, proved to be extremely computationally

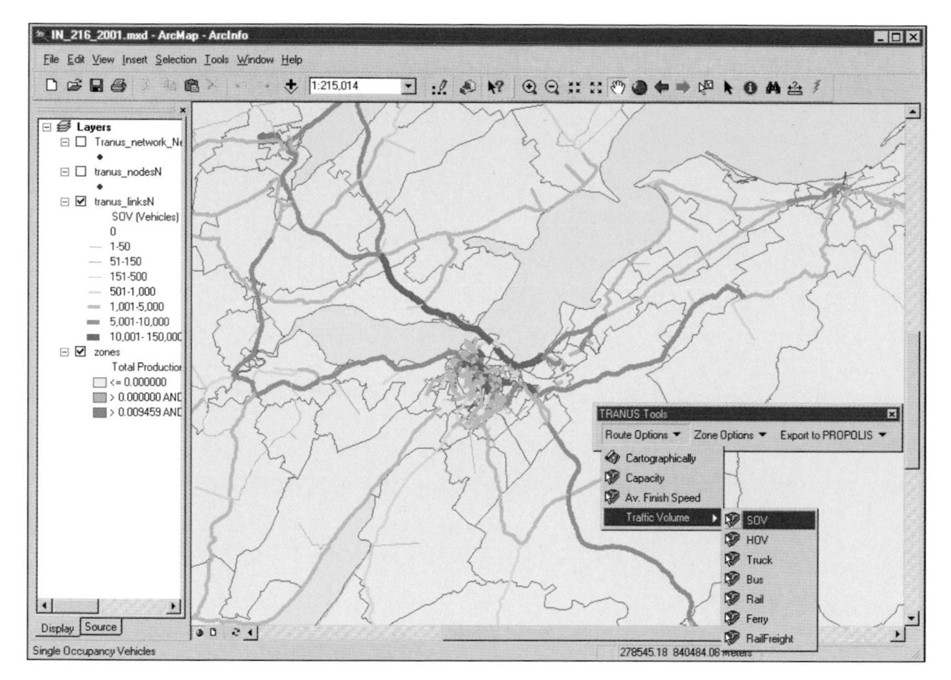

Figure 9 TRANUS GIS interface showing the predicted loading of the transport network with single-occupancy vehicles

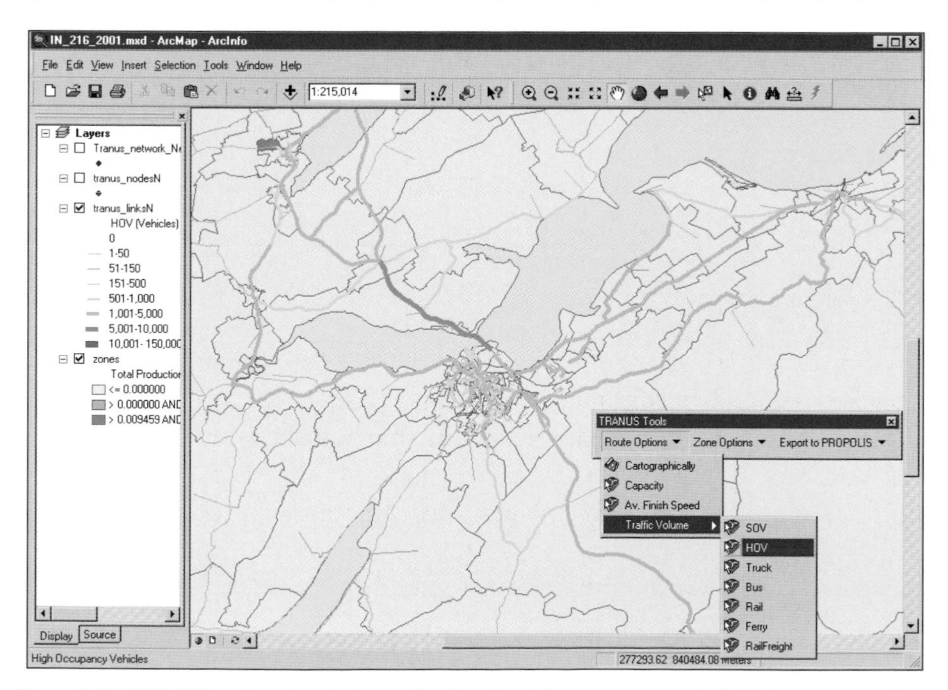

Figure 10 TRANUS GIS interface showing the predicted loading of the transport network with high-occupancy vehicles

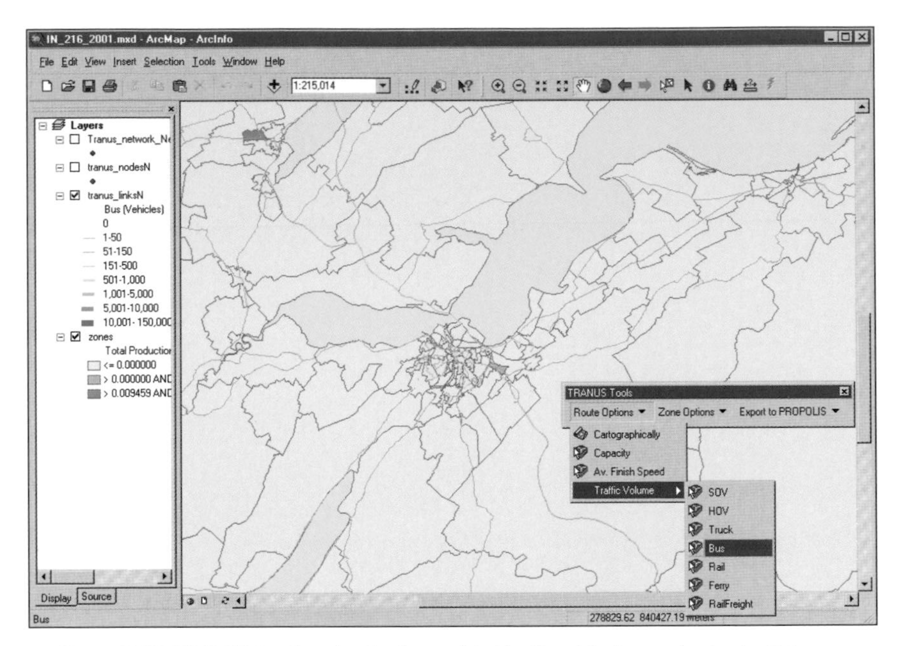

Figure 11 TRANUS GIS interface showing the predicted loading of the transport network with buses

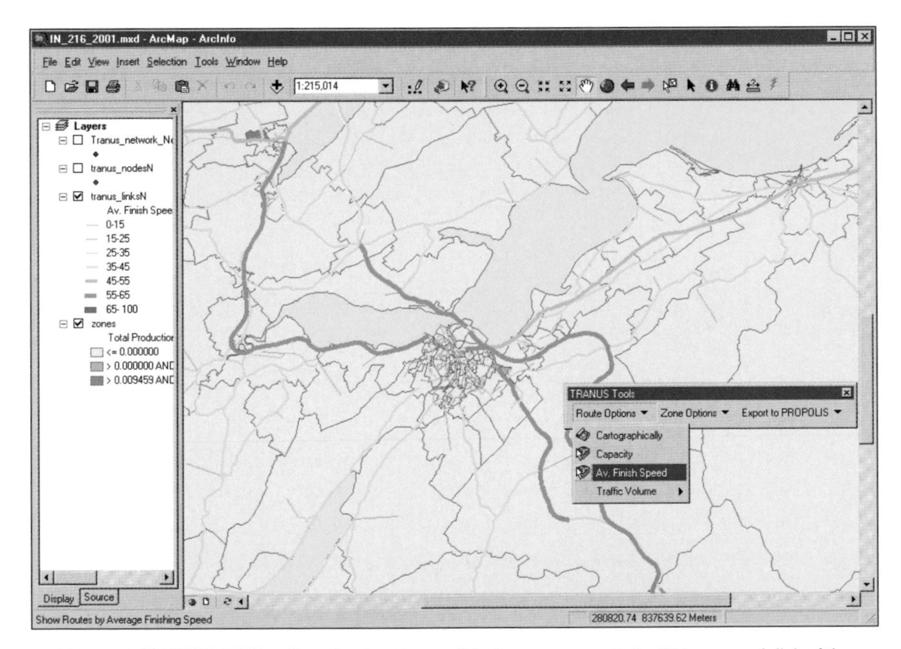

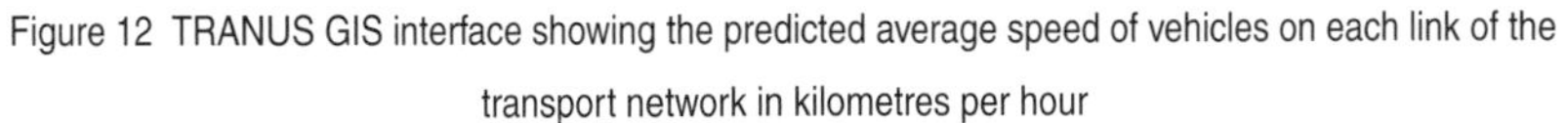
Figure 12 TRANUS GIS interface showing the predicted average speed of vehicles on each link of the transport network in kilometres per hour

expensive since the raw TRANUS files included data for every single combination of user, mode and reason for each journey. These had to be filtered before the data could be displayed. In the end it proved simpler and quicker to precalculate the required outputs and to avoid having to maintain the network topology, especially since the primary aim of the GIS interface is to provide a rapid visual check of the model outputs.

Once the model outputs have been checked using the GIS, they can then be passed on to the other software packages that have been developed for the PROPOLIS project. Automating the data flows between different modules has been one of the key objectives, since it is clear that this will result in a more efficient overall tool for policy testing. To make this process as smooth and consistent as possible the PROPOLIS consortium has developed a common data format. This enables the different modelling groups to view and compare outputs from different city models that have been generated by completely different modelling software packages. Interoperability is a key issue within projects that rely on data being moved between partners. It is also essential if the results from the different

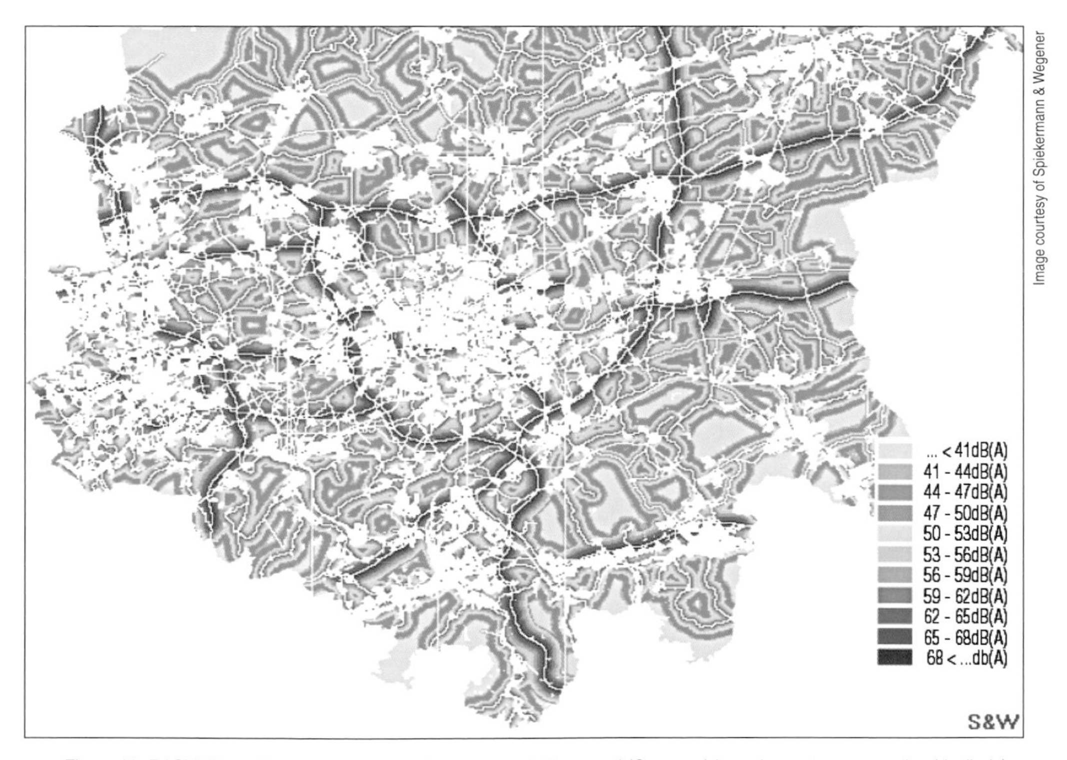

Figure 13 RASTER module showing quality of open space in Dortmund (Germany) based upon transport noise (decibels).

models and the different cities are to be compared. Within the TRANUS tools menu developed by CASA, there is a third drop down selection that triggers a set of programs that convert the TRANUS data into this common data format.

Several PROPOLIS software packages utilise the common data format output. The most important, from a GIS perspective, is the RASTER module (figure 13) developed initially for a preceding project called SPARTACUS (Lautso 2003), by the Institute of Spatial Planning (IRPUD) at the University of Dortmund (irpud.raumplanung.uni-dortmund.de/irpud/index_e.htm)

TRANUS, like many of the other city-modelling software packages, is not directly capable of capturing important aspects of urban sustainability because the zone-based spatial resolution is too coarse to represent smaller scale features such as individual houses, individual shops, or even an office on one specific floor of a building. This problem has been addressed in the past by using vector-based disaggregation (as described, for example, in Johnston and de la Barra 2000). However, this approach usually results in a huge number of polygons and can be a relatively slow and computationally inefficient process. The Dortmund RASTER module avoids this by using both spatial and non-spatial data from the city model and converting them into a cell-based (raster) GIS format. This allows much faster processing, although this is, of course, dependent upon the chosen cell size. The RASTER module maintains the zonal organisation of the aggregate land-use and transport models and adds a disaggregate raster-based representation of space for calculating local environmental and social impacts of policies. The existing methodology can provide indicators of land coverage and exposure of different socio-economic population groups to particulate matter, nitrogen dioxide and noise, based upon the model outputs. Each of these indicators requires disaggregate treatment of space provided in the module. These indicators are important, since one aim is to identify policy packages that are likely to achieve the following goals without compromising economic efficiency and social sustainability:

- reducing greenhouse gases from urban land-use and transport system by 20 per cent
- reducing energy use of the urban transport systems by 20 per cent
- reducing traffic accidents by 15 per cent

These policy packages are also likely to reduce urban pollution and congestion while at the same time ensuring accessibility and mobility. If PROPOLIS can achieve this, then it will clearly demonstrate the potential of integrated land-use and transport planning as well as the effects, on a strategic level, of emphasising collective and other sustainable transport forms.

6 Conclusions

The tight coupling of the outputs from a land-use and transport model (such as TRANUS) to a GIS is possible and enables rapid visualisations and post-processing of a model that has already been created and calibrated. This improved interfacing and interoperability between TRANUS and a GIS is essential if such modelling techniques are going to be used to compare a reasonable number of policy and scenario combinations. By contrast, the process that is still extremely inefficient in generating the actual land-use and transport model of each city is loading the data into the model and calibrating it. Since much of the data used to feed these models are derived from existing GIS databases, the next logical step is to develop an automated technique for loading the necessary data from a GIS into the land-use and transport model. Zone-based data, stored as polygons, would hold important information, such as population, building density, land use, and other demographic information, while line-based data would hold important information about the transport network.

Once correctly formatted, this data could be automatically loaded into the land-use and transport model. This would enable many more cities and regions to be modelled with these techniques since the necessary data are far more likely to already exist in a GIS than in a format ready to load into the land-use and transport model. However, with increasing developments in the capabilities of GIS to handle complicated transport models and utility networks, the question is whether it would be more beneficial to build the land-use and transport model into the GIS itself, and, therefore, avoid the need to transfer data in the first place. Although GIS seems to thrive in areas where it bears the brunt of routine digital tasks, a question mark hangs over the real demand for such an integrated land-use and transport GIS (Batty 2002) or even software that allows the tight coupling of a separate software model with a GIS.

Acknowledgements

PROPOLIS is a research project supported by the European Commission under the key action 'The City of Tomorrow and Cultural Heritage'. Support for this work from the European Commission, CASA, The Highland Council, and ESRI is gratefully acknowledged.

[Spatial modelling]

Chapter

GIS, distance, paths and anisotropy

Michael J. de Smith

This chapter examines the adequacy of existing distance measures available to the spatial analyst and their relation to path definition. Initially we discuss familiar measures based on Euclidean geometry and a number of extensions to these. We then discuss approaches to the determination of feasible and optimal paths in regions of variable topography and cost metrics. Finally, the development of powerful new computational procedures is described in some detail.

1 Standard distance measures

Providing capabilities to calculate and extract distance measures is a fundamental part of any GIS, and indeed, all spatial data tools. Most software packages offer at least three categories of distance measure: (1) coordinate-based distance measures in the plane or on a sphere (which may be adjusted to a more close approximation of the earth's shape); (2) network-based measures determined by summing the stored length or related measure of a single link or multiple set of links; and (3) polyline-based measures, determined by summing the length of stored straight line segments (sequences of closely spaced plane coordinates) representing a given feature, such as a boundary. Sets of these various measures provide the basic components of most forms of spatial analysis. But what assumptions underlie such measures and to what extent do the instantly available numerical values disguise and possibly distort the complexity of underlying real-world spatial patterns?

The first and third of these distance measures utilise the simple and familiar formulas for distance in uniform 2-D and 3-D Euclidean space and on the surface of a perfect sphere or spheroid. These formulas make two fundamental assumptions: first, that the distance to be measured is calculated along the shortest physical path in the selected space—this is defined to be the shortest straight line between the selected points (Euclidean straight lines or great circle arcs on the sphere); and second, that the space is completely uniform—there are no variations in terms of direction or location. These assumptions facilitate the use of expressions that only require knowledge of the coordinates of the initial and final locations, and, thereby, avoid the difficult question of how you actually get from A to B.

In most cases measured terrestrial distances are reduced to a common base, such as the WGS84 reference ellipsoid, as shown in figure 1. Modern electronic distance measurement (EDM) devices, which are based on light waves or microwaves, provide very precise figures for distances between measurement stations. But these measurements are subject to a number of well-defined errors, principally the result of atmospheric refraction, which in turn is dependent on the temperature, pressure and humidity along the interstation path (these affects are minor below 15 kilometres in normal conditions). In general, it is impractical to measure conditions along the path itself, and measurements at each end are taken to be sufficient. The curved wave path distance is then adjusted (reduced) to take account of this path curvature and the relative heights of the station points. The resulting slope distance can then be reduced once more to the reference ellipsoid or to the modern equivalent of mean sea level (geoidal/sea level) to provide ellipsoidal distance. Note that the reference ellipsoid may be above or below the geoidal surface, and neither distance corresponds to distance across the landscape. EDM measurement is rarely used for distances of 100 km or more, and increasingly DGPS measurements of location followed by direct coordinate-based calculations avoid the requirement for the wave path adjustments discussed above.

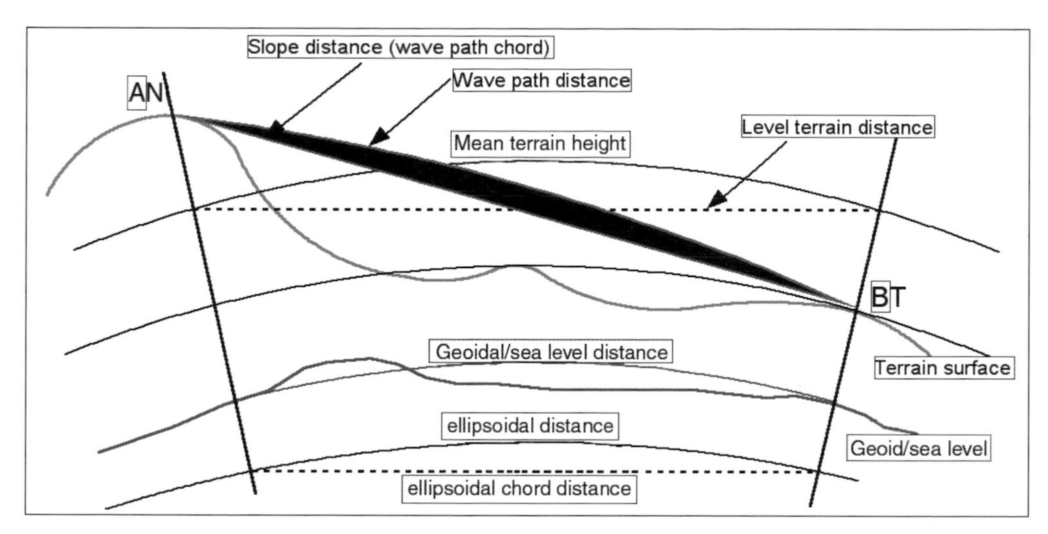

Figure 1 Reduction of measured distances

Whilst coordinate-based formulation is very convenient, it generally means that a GIS will not provide true distances between locations across the physical surface. It is possible for such systems to provide slope distance, and potentially surface distance along a selected transect or profile, but such measures are not provided as standard in most packages. For problems that are not constrained to lie on networks, it is possible to derive distances from the surface representation within the GIS. For example, this can be achieved using a regular or irregular lattice representation along whose edges a set of links may be accumulated to form a (shortest) path between the origin and destination points. For the subset of problems concerned with existing networks, such questions could be restricted to locations that lie on these networks (or at network nodes) where path link lengths have been previously obtained. Both of these approaches immediately raise a series of additional questions. How accurate is the representation being used and does it distort the resulting path? Which path through the network or surface representation should be used (for example, shortest length, least effort, most scenic)? How are intermediate start and end points (that is, points not explicitly represented in the GIS dataset) to be handled? Are selected paths symmetric (in terms of physical distance, time, cost, etc.)? What constraints on the path are being applied (for example, in terms of gradient, curvature, restriction to the existing physical surface, etc.)? Is the measure returned understandable and meaningful in terms of the requirements? How are multipoint/multipath problems to be handled?

A careful consideration of almost any spatial problem will highlight the inadequacy of the assumption of isotropic space. Lack of uniformity is at the heart of spatial studies, as was argued convincingly by geographers from the late 1950s and throughout the 1960s, culminating in David Harvey's seminal work *Explanation in Geography* (1969). This observation forces us to ask the question: what measures of separation are appropriate? In many instances the standard measures of Euclidean and spherical geometry suffice, especially over small areas. They also work well over larger areas that are reasonably isotropic, such as paths through air or space, although even here optical and, ultimately, gravitational distortions come into play. The measures are less useful if paths are restricted to a two-dimensional surface, and even less useful if functional distances are to be considered. For a cyclist, horse-drawn cart, truck or train the total amount of climb and maximum gradient of a route may be just as important as planar or even surface distance; for a commuter the primary considerations are likely to be those of cost and time rather than distance; for construction engineers the main concerns are usage costs, operating costs, environmental costs and construction costs (related strongly to vertical and horizontal gradient considerations along the path), and these are rarely linear functions of coordinate or even true surface distance.

In the real, anisotropic world the measures of distance we use should reflect the spatial (and in some cases temporal) variation we experience and the purposes for which we require the measurements. Such measures should be defined in terms of path and function and not merely upon an abstract computation derived from the values of the coordinates at the start and end points. We need, therefore, more general formulations of distance measures in accordance with these principles and seek their inclusion in GIS and spatial tool sets.

2 Alternative coordinate-based formulations

Most GIS software uses the standard Euclidean formula when computing distances between plane coordinate pairs and for calculating 'local' distances, as in the case of the buffer operator. However, for applications that require network distance measures between locations, Euclidean distance is generally inadequate, and computing large numbers of (shortest) network distances is often impractical or inappropriate.

In order to obtain more useful measures of interlocation distance, many authors have suggested that the standard Euclidean distance formula be amended in various ways. The most frequently used modifications are based on replacing the use of the square and square root operations, that is, replacing the powers of 2 and 1/2 with other values, typically p and $1/p$. We shall refer to such modifications as the family of p-metrics. The familiar diagram below plots 'circles' using various values for p. The lines show the position of a point at a fixed distance from the centre using the simple p^{th} power/p^{th} root model.

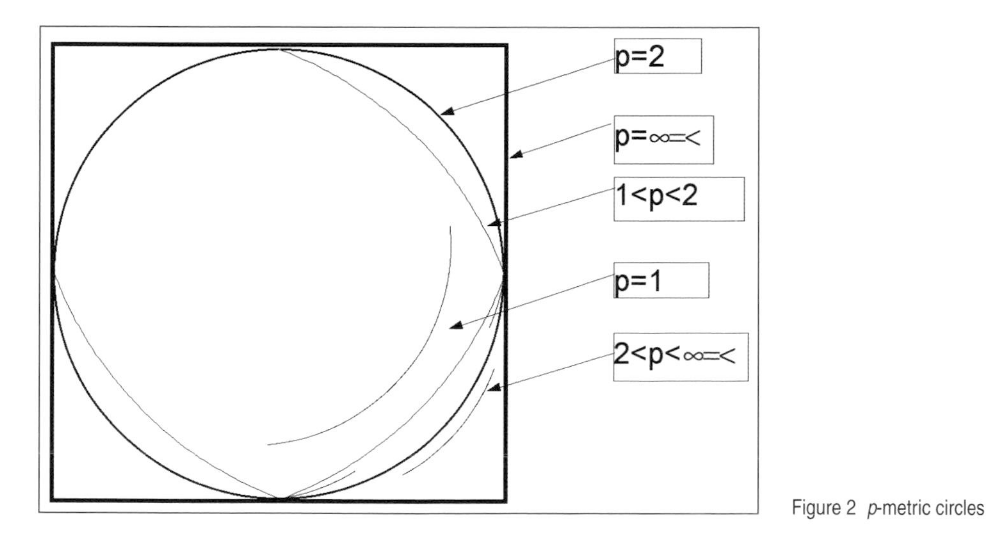

Figure 2 *p*-metric circles

In geographical literature, the case p=1 is variously called the Manhattan, taxicab, rectilinear or city-block metric. Values for p of approximately 1.7 have been found by many studies to approximate road distances within and between towns, and such estimates may be used in place of direct-line distance (where p=2) in problems such as locational analysis and the modelling of travel patterns (see, for example, Love, Morris and Wesolowsky 1988). An alternative is to use a simple route factor (typically around 1.3) as a multiplier and to continue using Euclidean distances, but inflated by this route factor[1]. A recent study by Peeters and Thomas (1997) analysed the impact of alternative values of p in simulated location-allocation problems and concluded that optimal locations are not substantially changed by the choice of p when $1.2 < p < 2.2$. Significant divergence does occur with $p < 1$ and $p > 2.4$. Also, note that if $p < 1$ the triangle inequality no longer holds; that is, it is possible for the distance (or time) from A to B plus the distance from B to C to be less than the distance (or time) from A to C. This characteristic is quite common in transport infrastructures, where one-way systems, timetabled routes and traffic congestion alter the expected paths significantly. Numerous variations on the simple p-metric model have been studied, including using a mixture of powers for the x- and y-directions (p and q), adding weighting factors (a and b) for the two directions, including a different power, s, for the root, etc. Most of these variations provide modest improvements in estimation of transport route distances at the expense of increased complexity in determination of the parameters.

One of the most striking features of the p-metric diagram is that for every value of p except 2, the 'circles' show rotational bias maximised at 45 degrees. Similar bias exists if other variants of the p-metric family are plotted. This highlights the fact that a distance value obtained by measurement and application of any of these modified formulas will change if the reference frame is rotated. If this seems a bit obscure, imagine measuring the distance between two points randomly located on a rectangular street network oriented north-south/east-west. In general, this will involve finding the length of a route in the north-south direction and adding the length of a route in the east-west direction. If, however, we could ignore the physical network and rotate and translate the pattern of streets about one of the points until the two points both lay along one road (a rotated north-south route), the length measured would be shorter. Although the p-metric family provides real value in analytic studies, this lack of rotational invariance is undesirable in the context of distance measurement. Clearly it would be simple for suppliers to include such distance expressions in their GIS packages, but the estimation of the parameters and the interpretation of their results would require considerable care.

3 Incremental distance and paths

Earlier we noted that the standard Euclidean formula is very effective over short distances and is rotationally invariant. These characteristics provide a sound basis on which we can construct a more general framework for distance calculations and are preferable to the p-metric family. If we denote the distance between two points A and B by the symbol s, we can denote the Euclidean distance between two points very much closer together by *delta-s* or *ds*. In the case of the perfectly uniform plane with constant costs, s is simply the sum (or integral) of lots of segments of length ds along the straight line between the two points A and B, for example, along the *surface* transect from A to B in our first diagram. This observation is obvious, but it highlights the fact that distance measurement in continuous space assumes that you can define at least one continuous path between the start and end points.

Suppose now that costs (in a general sense) are not constant in all directions and locations. For example, suppose that we have to take into account the cost of land at various locations, or the density and direction of traffic, or the variable density and pattern of routes across the region—how do we compute meaningful distances in such cases? And why not explicitly include the physical variations in the land surface, required, for example, if we are seeking to construct a new road or rail link, a sewerage network, or to make a least-effort or least-cost journey? The computation of a meaningful measure of distance s now appears to be far less obvious. If the costs (or generalised costs) of crossing a very small region can be determined by a single composite measure, say F, and values can be determined for all points of interest, then cost distance can be computed by the sum (or integral) of Fds. This sum must take place along a path, and in some (very few) cases it is possible to work out in advance what the shape of the path must be in order to ensure that the result is a least cost path. This minimisation problem in its most general form is extremely complex and may have no feasible solutions depending upon the constraints imposed. Of course, the Fds model is not the only way of defining cost variations, but it does embrace a substantial group of realistic problems.

To illustrate this approach further consider the two diagrams in figure 3. The first shows paths for a constant cost uniform plane, emanating in various directions from the point (2,5). All are straight lines, as expected. The second diagram is very different. The sample region and initial point are the same but costs (for example, land costs or traffic intensities) are much higher as one proceeds up the diagram. In fact, in this case costs F increase as a linear function of the form $F=10y+1$, so at the point [2,0] the cost function is $F=1$ whilst at [2,10] it is $F=101$. In this case paths diverge markedly from the straight line routes. For nearby locations, even if heading into the 'expensive' region, fairly direct routes make sense, but there comes a point where the cost of a direct route is so high that a massive diversion is worth considering. This is a familiar pattern to anyone travelling by car in a modern city!

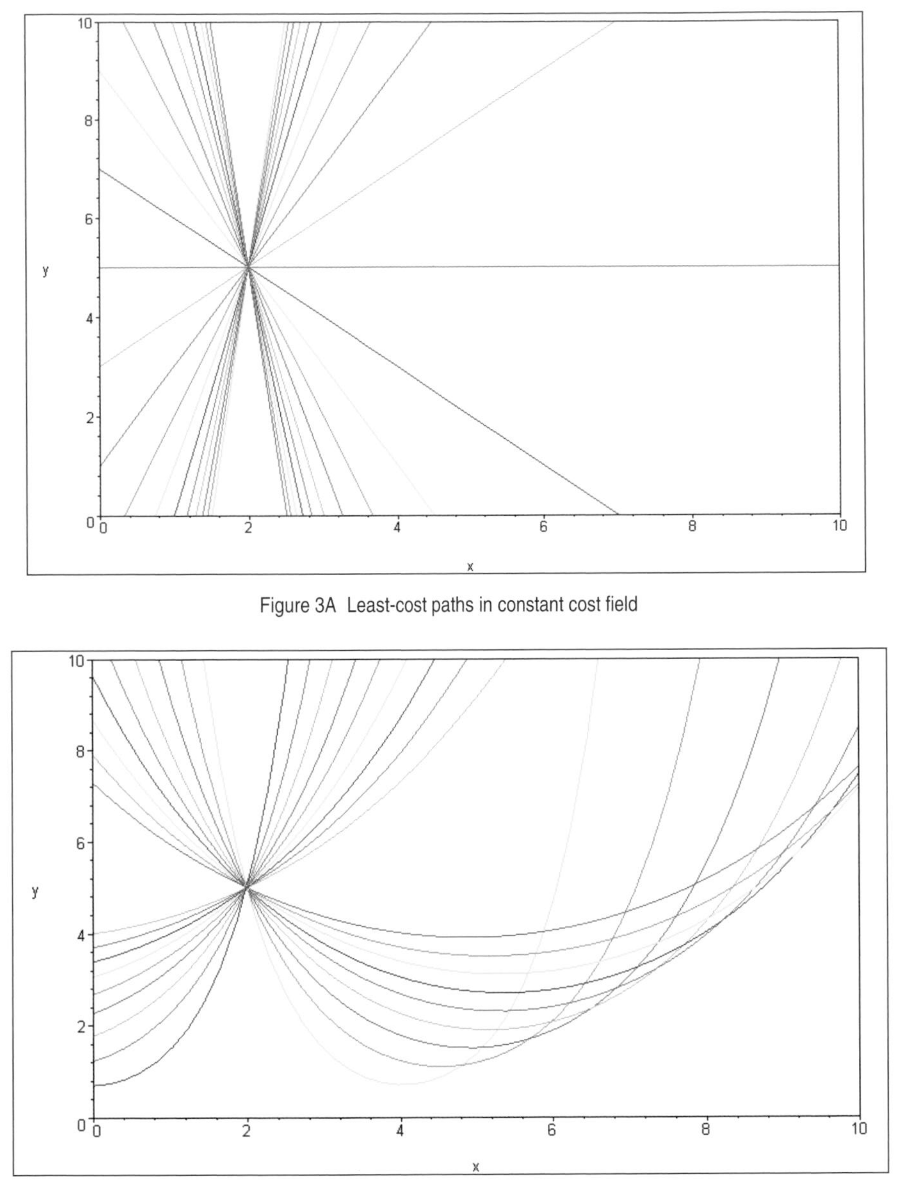

Figure 3A Least-cost paths in constant cost field

Figure 3B Least-cost path in linearly variable cost field

Solution paths such as those shown in figure 3 can be calculated analytically only for the simplest of cost functions (linear functions of one variable or simple radially symmetric functions), with no variation in landform or any other constraints. However, analytic solutions and numerical approximations do provide guidance on the expected form of solutions, rules for local optimal path determination, and templates against which alternative approaches can be evaluated.

4 Path finding

Identifying optimum paths can be very difficult. Current GIS programs may incorporate options to compute least-cost paths (which are not network constrained), typically using a method known as Accumulated Cost Surfaces (ACS—see, for example, Douglas 1994). This is a powerful technique and one that clearly works in many cases, but it requires considerable care in both creating the cost surface and interpretating the resulting paths found. For example, the physical surface is not an explicit variable, since the model operates on a lattice (or raster) cost model defined over a uniform plane. If slope, for example, is an important variable, it must be handled as a component of cost. Curvature constraints pose greater problems, as would a requirement, for example, to ensure that a waste-water pipeline path only flowed downhill. Even assuming a generalised cost surface and accumulated surface can be defined satisfactorily, and for a substantial number of problems this may be the case, it still leaves the question of optimum path determination. Normally this is taken to be the path of steepest cost descent from the source to the destination, but this can be shown to be incorrect in some cases; and what if this path is not unique or is unrealistic because of gradient or curvature considerations? Furthermore, Goodchild (1977), Rowe and Ross (1990), and others have shown that the use of a lattice structure generates path distortions in both direction and length that remain no matter how fine the lattice.

The problem of finding a least-cost path over a flat surface and the extended problem of the physical surface no longer being flat are closely related (de Smith 1981; Rowe and Ross 1990). If the cost function and the physical surface can be described in terms of well-behaved mathematical functions, the problem reduces to one of finding the shortest path across some other, composite, surface. The details are not important here, but the result does allow us to start by solving problems in the plane with variable costs, knowing that similar methods are likely to provide guidelines for solutions when the surface is no longer flat.

Problems of this kind occur in many disciplines. The field of robotics[2] provides some of the most useful pointers to tackling more complex path problems using computationally intense but fast (possibly real-time) search procedures. In the example in figures 4A and 4B, the challenge is to navigate the robot (a computer

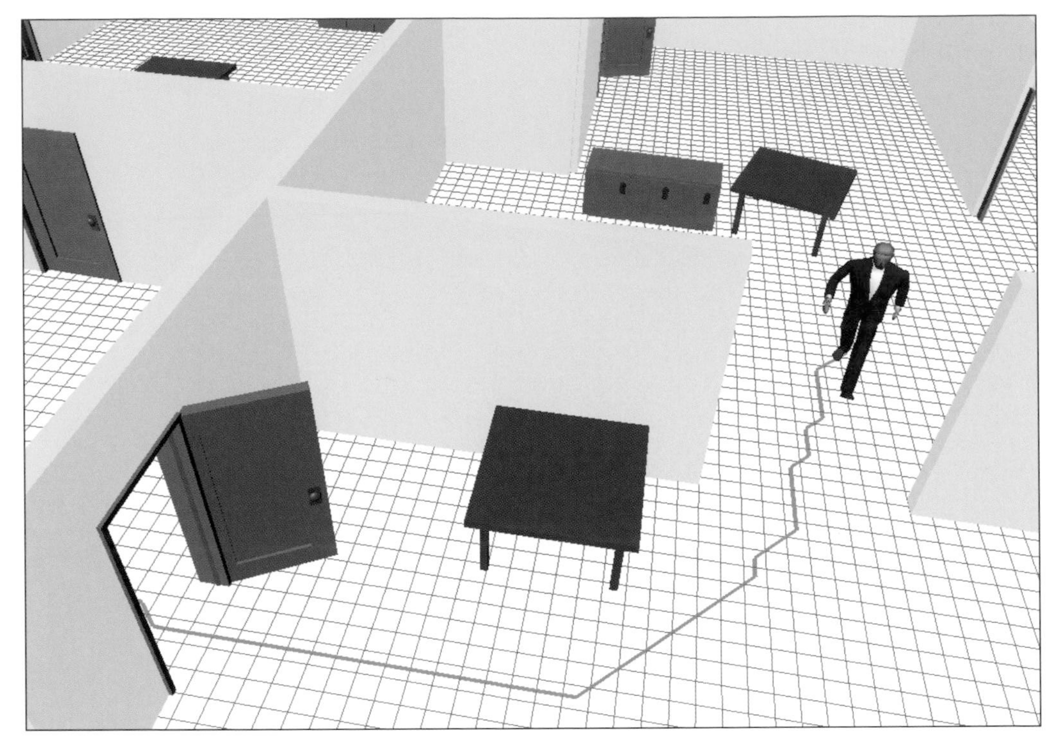

Figure 4A Pathfinding in a room

animation of a person in this case) to a target point (the green dot) without colliding with any of the obstacles in the rooms of an office.

Collision avoidance requires allowing for a buffer zone using some measure of the proximity of the robot to the obstacles and doorways. In this example the movement path is assumed to consist of a circle whose centre can pass through any point of the floor represented by a pixel map (that is, a square lattice). The path shown can be computed using a deterministic or heuristic shortest-path algorithm over the lattice. This is a form of 'path-planning' process, prior to movement of the robot along the preselected path. It normally requires prior knowledge of all the objects in the scene and a procedure for searching for a feasible and possibly optimal path through the scene.

An alternative solution procedure would be to divide up the 2-D floor space into polygons or triangles, representing the boundaries of the overall floor area, the target point and the (buffered) various obstacles, and then to treat this mesh as a network, with feasible and optimal paths computed across the network edges. This can be likened to seeking out acceptable routes for a city bypass, with buffer

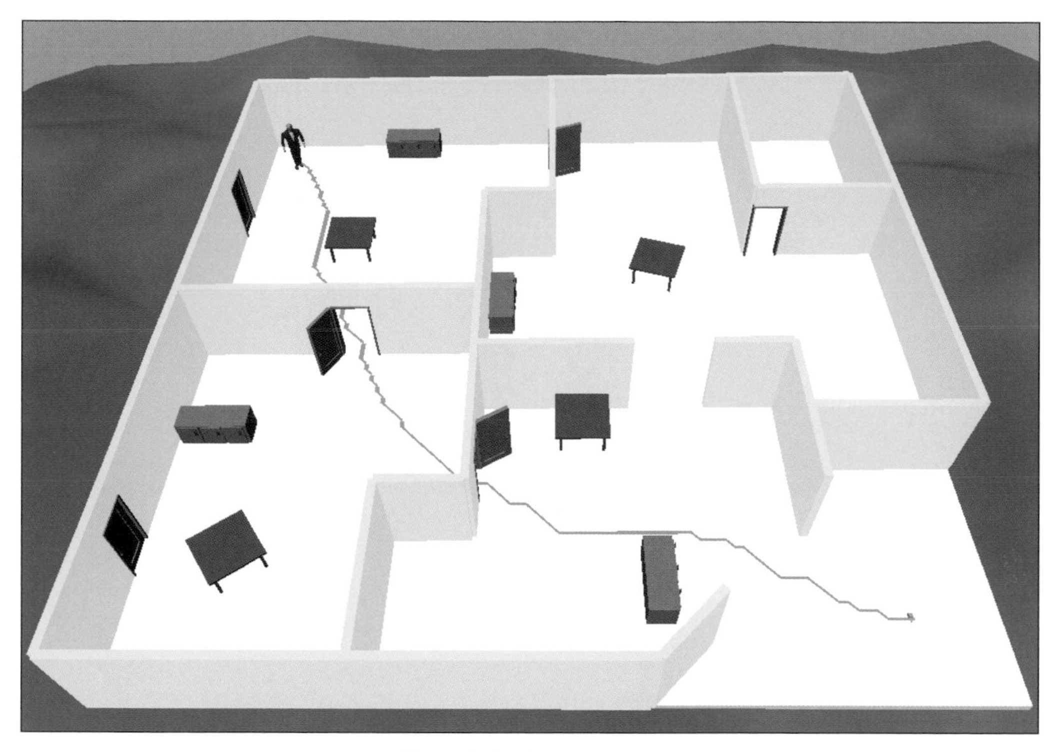

Figure 4B Pathfinding in a room

constraints relating to environmental considerations such as noise and air pollution. The path length (cost) is simply the incremental step length times the cost or effort of completing each of these steps.

As problems of this kind become more complex and larger, for example covering multiple paths, including more complex obstacles, allowing for three dimensions, applying constraints such as limits on path curvature or gradient, and including variable costs, so many solution procedures run into difficulties. These include representational and computational complexity, solution time, and solution failure (for example, becoming stuck in a subregion). Search procedures that are deterministic and systematic appear to suffer from such problems to a greater extent than procedures that incorporate some element of random search. In the next example, we apply a simple search procedure that incorporates a limited statistical component to locate feasible routes satisfying gradient constraints.

5 Pathfinding in East Creech region, Dorset, England

Paths that involve motion (for example, of robots, vehicles or people) normally incorporate gradient and sometimes curvature constraints (vertical and horizontal). In the example illustrated in the two diagrams and photograph in figures 5A and 5B we have applied a simple search algorithm to the problem of locating a path across a ridge structure in a small part of Dorset, England. A transect line, shown in green, identifies the straight-line route between two selected points and the red lines show the existing roads. Two gradient rules were applied, one based on approximately 1:10 maximum (rural road with motor traffic) and the other based on 1:5 maximum (steep, but usable by walkers and riders). The search procedure adopted was as follows:

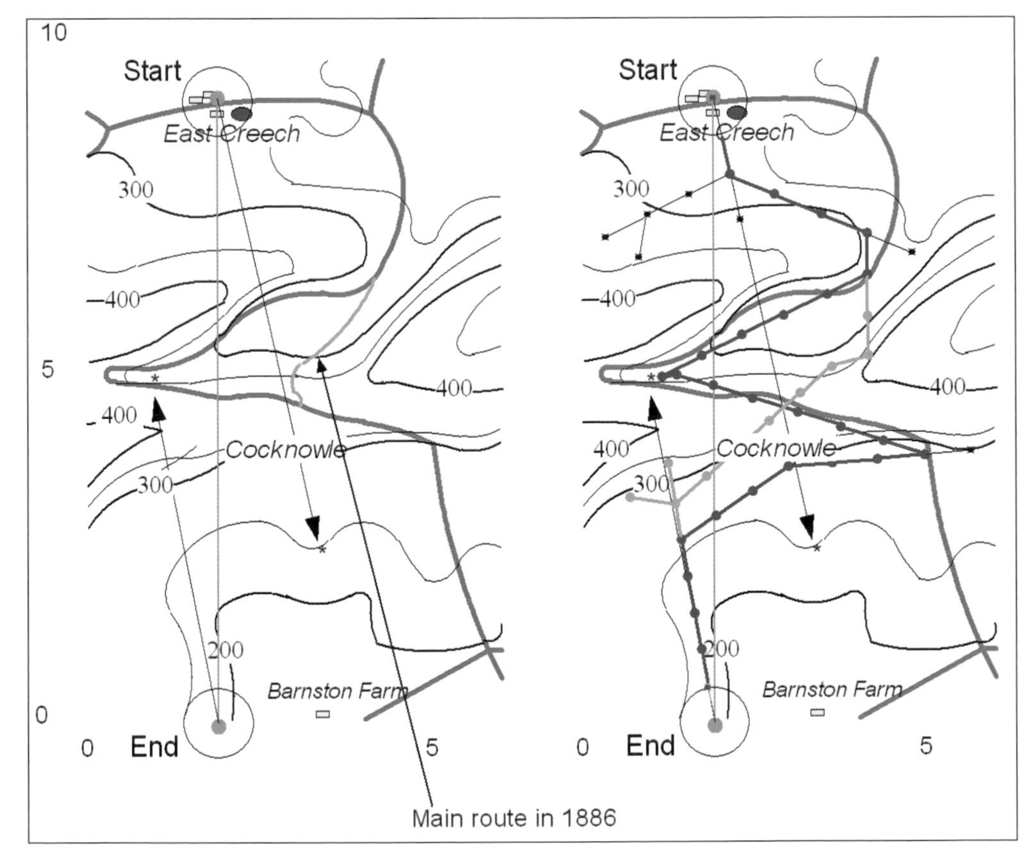

Figure 5A Gradient constrained search model

Figure 5B Original routes near East Creech, Dorset

(i) Define the step length, D, to be used in the search process.
(ii) Select a random point in the sample region.
(iii) Extend a short straight line segment, length D, from the start point towards the random point.
(iv) Check that this line does not (a) cross the edge of the sample region or any other obstacle/no-go area; (b) satisfies the gradient constraint; and (c) is not closer than D to the target or another path originating from the target.
(v) If the gradient exceeds the constraint, select a new random point in the sample region (optionally use the subregion defined by the current point and the region 'north' or 'south' of this point) and continue. If this does not result in a feasible path repeat the process with a new random point, and continue. If this leads to a dead end, step back along the path and try again from the preceding point.
(vi) Alternating between the end point and start point, repeat (ii) through (v) and iterate until the paths can be joined or the solution fails.

The main arrows on the maps show the initial (accepted) search directions selected for the start and end points, and the blue line on the right-hand map indicates a solution path located by the more strictly constrained search. The grey path on the same map shows the less-constrained route, based on the same initial search pattern.

It is interesting to compare the second path with the historic map from 1886. The current road follows the line of an older cart route winding away to the west. The thick grey line on the first diagram shows the location of the main route in 1886 (not shown on recent OS maps, but clearly visible from a current photograph, shown in figure 5B). This latter route was more direct and corresponds quite closely to the grey route found by the less-constrained search method.

Many improvements and variations on this kind of search procedure are possible (and necessary), but it does provide a much simplified illustration of how such models can be applied.

6 The VORTAL algorithm

The simplified procedure described previously does not seek an optimal path in terms of length or cost, nor does it address more complex problems in which there are obstacles, multiple sources and destinations, and varying costs. For such problems more powerful search procedures are required. The computational methodology we describe in this section we call VORTAL—Variational Optimisation of Random Trees ALgorithm. It is designed to solve complex real-world pathfinding problems in two-dimensional free space (sometimes called the 'geometric domain'). For the purposes of the present discussion, we describe the basic algorithm and its application to a simple unweighted three-point pathfinding problem that includes obstacles. If there are no obstacles and the bounding triangle has no angles of 120 degrees or more, this problem yields a single optimal solution with an intermediate point (known as a Steiner point) linked to the three vertices. These three paths meet at 120 degrees, and the solution point can be determined by an iterative procedure. The no-obstacles case provides one test of the procedure's optimality.

Definitions

The procedure requires as input a number of variables including the location of the initial three points and the obstacle(s) if any, the initial step size to be used for the random trees search phase, the step size to be used for the variational phase, and (optionally) the cost per unit distance for the simplest cost model.

Search phase

The algorithm grows random (self-avoiding) trees from each point. As the trees grow, the branches are checked to ensure that they lie within the sample space, do not cross any obstacles, or come closer to any obstacles or other branches than permitted. When two of the trees are close enough to join, their growth is stopped and a third path is generated between the point at which the first pair meet and the tree emanating from the third point. We now (hopefully) have a feasible but suboptimal path.

Optimisation phase

The optimisation phase involves a combination of tension considerations and variational optimisation. The tension process is equivalent to pulling on each path in turn, as if it were a string, until it is 'tight'—a process that involves removing intermediate points until no further path reduction is achieved. The simplified path has far fewer steps, is closer to the optimal path, but still consists of intermediate vertices from the original feasible path that may not lie on the true optimal path. The variational phase of optimisation involves migrating each point of this limited subset towards locally optimal positions by moving each one by the variational step size in any one of eight directions until the total path cost is no longer significantly improved. The variational and tension optimisation procedures are iterated to ensure that local optima are eliminated as far as possible. Checks are made at each step to ensure the constraints (proximity and retention within the solution space in this case) are maintained.

This process is illustrated in figure 6A. The random tree growth process is biased in order to encourage growth towards the target, and the black line shows the initial feasible path. The three coloured lines then show the result of the optimisation phase.

On a modern desktop PC this algorithm runs in a second or two on the problem described, unless the configuration of obstacles is such that a solution is impossible or near impossible. Reduction of the step size in the search phase does not tend to improve the likelihood of finding feasible solutions and increases the search time—reducing the proximity size does not tend to have such effects. Obviously performance of the algorithm will be affected by the spacing of obstacles, and solutions will not be found if step sizes are chosen which are large in relation to

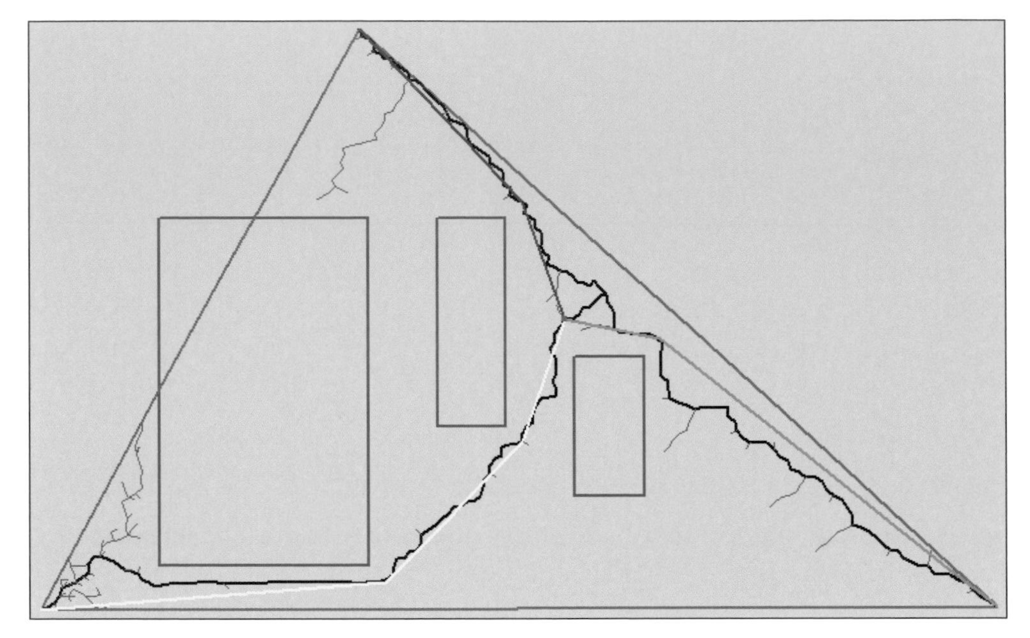

Figure 6A VORTAL procedure applied to a three-point, three-obstacle problem

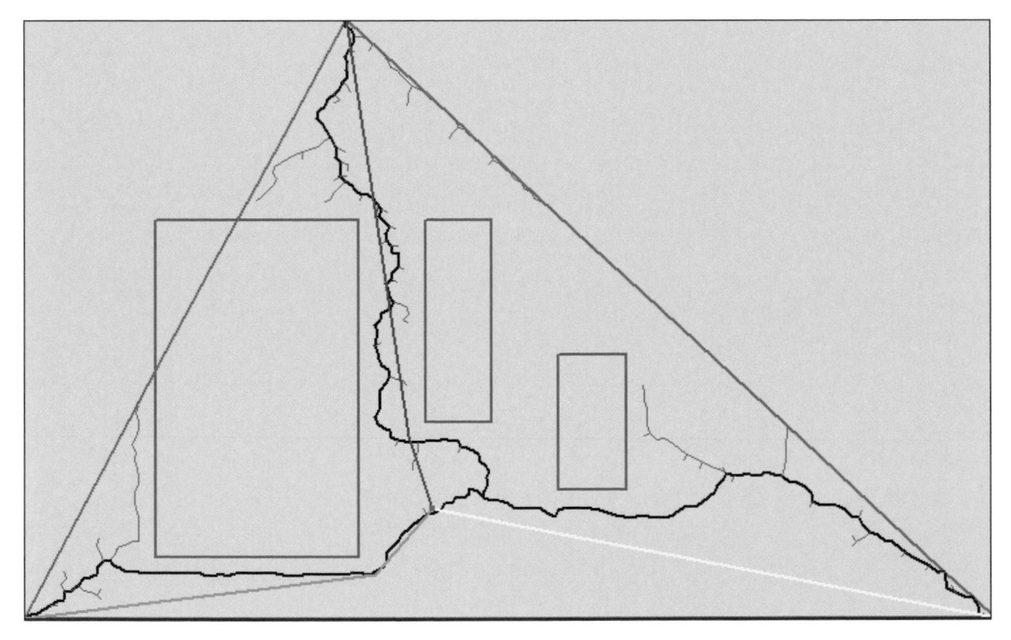

Figure 6B VORTAL procedure, alternative solution topology

object separation. Multiple runs of the process yield alternative feasible topologies, whose length and cost/benefits may then be compared. Figure 6B shows an alternative solution topology with a total cost 5 per cent less than that of figure 6A.

Because the VORTAL algorithm involves a tree search process from each vertex, additional constraints such as gradient can be readily incorporated in the manner described earlier. Curvature constraints could be included within the initial search phase or within an intermediate smoothing phase prior to optimisation. Cost and physical landscape variation can then be handled within the optimisation phase based on the procedures described, and these enhancements are being implemented at present. Note that this process does not require the calculation of accumulated cost surfaces nor the use of analogies to the paths that light waves might be expected to take.

7 Conclusion

A range of computationally based exploratory methods is becoming available for pathfinding and optimisation. In addition to the approaches outlined in this chapter, there are methods based on the use of agents (Batty, this volume; Batty, Jiang and Thurstain-Goodwin 1998; Mitasova and Mitas 1998), cellular automata (Torrens, this volume; Batty 2000), and developments of the ACS method discussed earlier. In such cases, the exploratory phase seeks feasible paths and the optimisation phase(s) seek to improve such paths to ensure optimal or near-optimal behaviour. The length and cost of these paths is computed incrementally which ensures that the measure is both valid and unambiguous. Hybridisation of such techniques with well-established graph theoretic methods is an essential development, particularly where new infrastructure projects must connect to existing facilities. At present, research in this field remains at an early stage and sophisticated toolsets for the spatial analyst are not available. The approaches described, however, do offer the prospect of such essential facilities becoming available in the near term.

These new methods also raise the possibility of addressing a wide range of applications in addition to those of optimal pathfinding. Initial application areas include the solution of facility location problems, providing novel approaches to a variety of urban and transport planning problems, and augmentation of spatial decision support systems. However, such methods can be extended into less-familiar areas of spatial modelling and analysis; for example, within simulation modelling, to assist in the reconstruction of historic maps from their modern versions based on data relating to historic land use and settlements, known transport facilities and trading patterns; and within the earth sciences, in connection with the analysis of fluvial processes and in landform modelling. In all these diverse areas, computational methods benefit from their interactive and visual nature, providing an experimental framework that can be subjected to extensive testing and adjustment, in addition to delivering practical solutions to complex problems in an approachable, visually appealing and educative manner.

Endnotes

1 The ratio of the mean distance between random points as measured by the Manhattan metric to that measured by the Euclidean metric is 1.27, thus a value of 1.3 equates to a street pattern similar to but slightly more circuitous than rectilinear.

2 See for example, www.kuffner.org/james/anim.

Chapter

Population growth dynamics in cities, countries and communication systems

Michael Batty and Narushige Shiode

This chapter discusses the aggregate dynamics of population systems that maintain a consistent regularity through time in the sizes of their elements. City size distributions represent the classic example, with the size of cities being inversely proportional to their rank. This is enshrined in Zipf's (1949) rank-size rule that provides one of the strongest and most exact relationships throughout the sciences. We first review fundamentals and then suggest that a wide class of random multiplicative processes provide good models for systems that grow to produce the kind of rank-size regularities that emerge for cities, countries and communications systems. We then focus on the internal dynamics that take place between time periods marked by these regularities showing that dramatic changes can take place, even though such systems appear to be rather stable at an aggregate level.

1 Social physics and Zipf's Law

Most of the models and methods presented in this book deal with spatial systems ranging from entire cities to local neighbourhoods where analysis takes place in census tracts down to the geometry of buildings and streets. In this chapter, we depart from this focus on the meso and the micro, scaling up to cities and regions in nation states and the global space economy. Although we lose the rich detail that characterises more disaggregate spatial analysis, we gain the advantage of dealing with systems where events are aggregated to the point where trends and discontinuities in temporal and spatial patterns can be clearly detected. It is in this domain that some of the strongest regularities in the way spatial systems are organised become apparent, enabling us to build simple but effective models of their dynamics.

The earliest formal theories of how cities and regions were structured emerged from what came to be called 'social physics' (Stewart 1950). Social physics applied classical mechanics, fashioned around the concepts of force, gravitation and potential, to the way social systems were organised, usually in space. Initial applications tended to see city systems in equilibrium, because the signatures of such systems were usually based on regularities such as distance decay and diffusion, which appeared to follow well-defined power laws. However, in the last decade with the advent of complexity theory, this traditional analysis has been enriched with a concern for how robust and long-lasting regularities observed in such systems come to sustain themselves over many years in the face of quite volatile and effervescent dynamics. Here we begin by describing typical regularities and then show how stability in such patterns is consistent with rapid and sustained growth and decline, illustrating our argument with examples from city systems, population growth at the world level, and the penetration of new communications devices in nation states.

The best known and strongest regularity in the social sciences, some say even the sciences en masse, relates to the size and frequency with which different populations are distributed, in space as cities and through the economy in various forms of wealth. For such systems, it is widely agreed that there are a large number of small events and a small number of large events, with the relationship between being continuous and regular. For cities, this means that there are a large number of small places and a small number of large ones. Casual observation supports this regularity as we illustrate in figure 1, but it was Zipf (1949) who first popularised and formalised the relationship that he enshrined in the term the

Figure 1 Scaling laws: few large cities, many small towns. Prague to rural Dorset (Shaftesbury, United Kingdom)

'rank-size law' or 'rank-size rule'. For the most part, we will try to avoid formal algebra here, but it is still necessary to state the law mathematically since much of our subsequent analysis depends on being clear about this meaning.

If we rank all cities by size from the largest to the smallest with the largest P_1 associated with rank 1, the next largest P_2 with rank 2, P_3 with rank 3, and in general P_r with rank r, then the relationship between city size and rank is that population is inversely proportional to its rank. The formal relation can be written as $P_r = P_1/r^{\alpha}$ where α is a scaling constant. It was Zipf who first showed that $\alpha \approx 1$ not only for populations but also for word counts from novels and speeches, income distributions, and so on, thus establishing an even more curious characteristics of the law. Thus from the largest population P_1 which is often called the primate city, the second largest population is half the first $P_2 = P_1/2$, the third is a third of the first $P_3 = P_1/3$, and so on down the hierarchy with more and more smaller towns appearing, clustered ever nearer to one other in terms of size.

At first sight, this law seems to defy common sense. How, one might ask, can objects as complicated and complex as cities are ordered in such a simple manner? Krugman (1996) sums it up extremely well when he says: 'We are unused to seeing regularities this exact in economics—it is so exact that I find it spooky. The picture gets even spookier when you find out that the relationship is not something new—indeed the rank-size rule seems to have applied to U.S. cities at least since 1890!' (page 40). If this is a universal law, then we need to explain how such regularities persist at the macro level when we know that at the micro level all manner of changes are taking place. Moreover, cities have changed a lot during this period although in our analysis, we have kept the aerial definitions the same. The urban United States in 1890 was a very different place from 1990 with the focus changing from east to west and north to south, the rust belt declining economically in the face of a growing sunbelt, and the drift from farms to the cities reversing itself during this period. In this chapter, we first review some novel explanations, but from our own data, we suggest that we are but at a beginning in terms of seeking good explanations. In this sense, we illustrate how important it is to explain space in terms of time, statics in terms of dynamics.

To make progress, we must trace a little history and reformulate the law. Although it is impossible to say who first observed these types of distribution, it was Vilfredo Pareto in the late nineteenth century who first formalised a similar relationship concerning the distribution of income. Pareto said that if you count the number of individuals who have an income greater than a certain level, then this number would decline inversely with the level of income. In other words, as

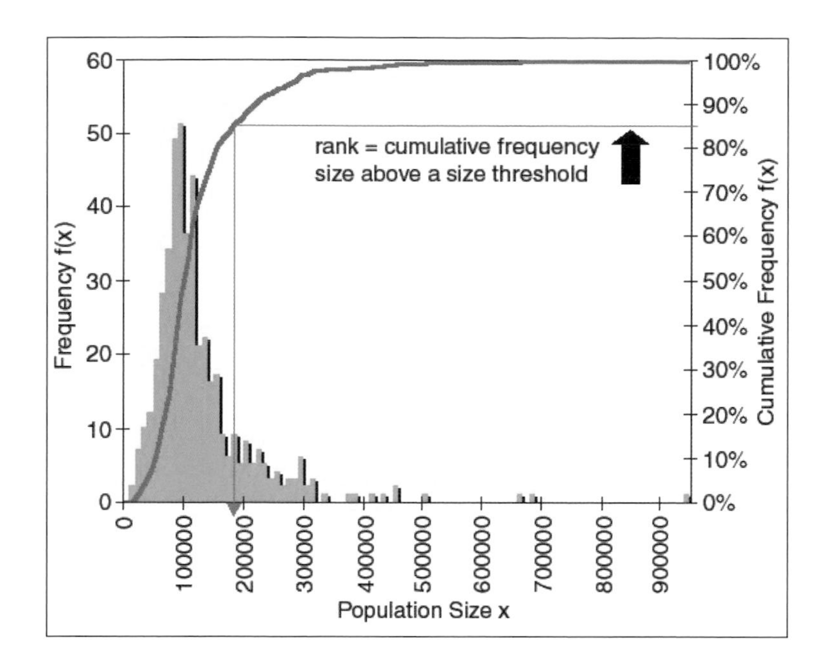

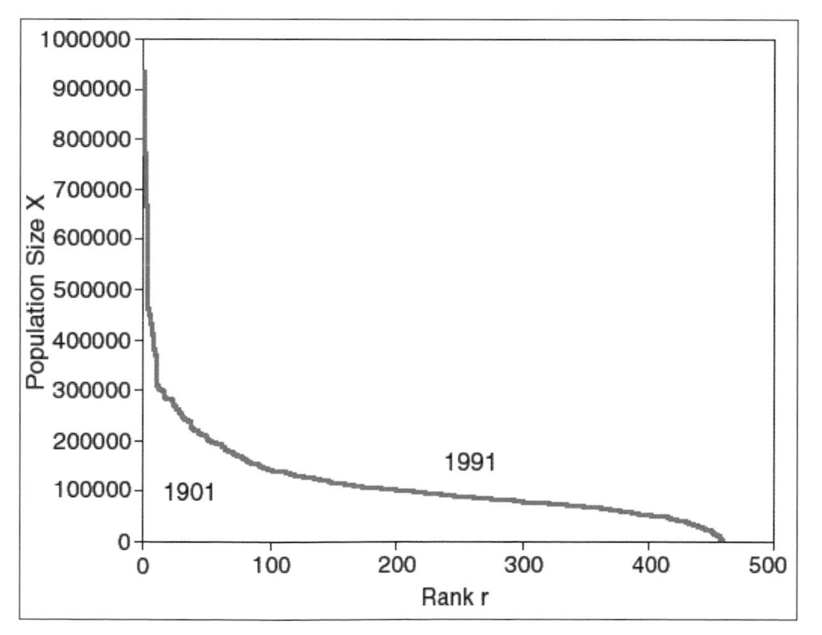

Figure 2 Frequency, cumulative frequency and rank size based on the British population in 1991

income gets larger, the number of persons who have that level of income gets smaller. In fact, this is none other than the rank-size distribution with the number of persons with an income greater than a certain amount the rank and the income level itself the size. It is worth noting that frequency and rank are consistently related. For every rank-size relation $P_r = P_1/r^{-\alpha}$, we can transform this into its frequency defined in terms of population as $f(P_i)$ and cumulative frequency $F(P_i)$ where i is now the unranked observation. The details are presented in the useful tutorial by Adamic (1999, Web access 2/9/2002).

We show these frequencies for the 1991 population of British urban areas in figure 2, where the link from frequency to cumulative frequency to rank-size is implied graphically. It is possible of course to fit power laws to any of the three relations, but the rank-size is more convenient. Frequencies do not need to be counted and binned. We will come back to figure 2 because the processes that we introduce will not, in general, yield distributions that are true Zipf laws. This might be anticipated in figure 2A; it shows that the frequency is a highly skewed normal distribution—a lognormal—rather than an inverse power law although in its right-hand tail, it might be thought of as a power law. This has been a source of considerable controversy, which we will return to later.

We conclude this brief introduction with a little more history. There is a massive research literature on Zipf's law. Wentian Li's reference archive (Li 1999, Web access 2/9/2002) only sketches the main contributions and he cites 220. In the physics literature of the last 10 years there are probably at least five hundred articles on power laws that reference Zipf, and for over 50 years there has been a steady stream of articles in regional science and urban economics. The earliest formal proposal to apply the model to cities was by Auerback in 1913, but Lotka in 1924, Goodrich in 1925, and Singer in 1936 speculated on city size distributions, all preceding Zipf's (1949) seminal text. Rashevsky and Lewis Fry Richardson, both writing in the 1940s, also made prescient contributions to the debate. Harry Richardson (1973) reviews the early work with an emphasis on how rank-size is generated using hierarchical models consistent with central place theory, while Carroll (1982) provides the most recent comprehensive review.

2 Explanations: multiplicative random processes generating order and growth

There are several models built around spatial economic theory that generate city size distributions consistent with Zipf's law, but most of these generate cities in terms of static spatial structure. Insofar as they are dynamic, they presume a spatial equilibrium. We have already noted models that generate size distributions consistent with notions about the hierarchy of central places. To these we must add various macro-economic analogues, often incorporating trade and growth theory of which one of the most recent is Krugman's (1996) and adaptations thereof (Brakman et al 1999). We do not propose to review these explanations further as we prefer a simpler and more parsimonious class of model. This model is based on random growth which, as we will show, is one generating considerable excitement at present, and the one which in terms of the systems we are interested in here seems the most applicable.

The most basic explanation of Zipf's law is so obvious that it is often overlooked. Imagine a set of objects all of which have the same size. Imagine that these objects can grow or decline by doubling or halving. For each object, flip a coin to determine if it grows or declines. Keep doing this for a few iterations and you will see that the chance that an object grows bigger and bigger is lower and lower because of the possibility each time that the object can decline rather than grow. This is a simple branching process that essentially says that the number of branches leading to large sizes relative to the total gradually decreases as the process continues. If you want to justify this for yourself, take say 10 objects and work through the process. This is quite labourious, but after a dozen or so trials, it becomes clear that the objects are beginning to order themselves into a small number of large ones and a large number of small ones. This process is even clearer if objects disappear when they fall below their lowest size with new objects at this smallest size replacing them. A particularly nice illustration of this process is given by Gabaix (1999) where he shows this even more clearly when the average growth of the system is constrained to a constant each time the objects are grown. In short, what this process is suggesting is that if things start small, the chances of them all growing big is very low because at each stage on their way to bigness, there is an even chance that they will get smaller.

Many plausible models are based on this kind of multiplicative growth where the notion of exponential growth arises from the process. The standard model in economics, which was developed by Gibrat in 1931 for the growth of firms, is essentially one of proportionate but random growth (for a history and detailed explanation, see Sutton 1997). What the model assumes is that growth is comprised of an average rate λ which applies to all the objects in the population—cities, firms, incomes, whatever—and a deviation from this average ε_{it}, which is a normally distributed random variate applicable to each object i at time t. The model can be stated very simply as $P_{it} = \lambda P_{it-1} + \varepsilon_{it} P_{it-1}$ where P is the population of object i at times t and t-1. The model simplifies to $P_{it} = (\lambda + \varepsilon_{it}) P_{it-1}$ where if we start with the population at time $t = 0$ as P_{i0}, and apply this equation, we generate a growth series that we can write as $P_{it} = (\lambda + \varepsilon_{it}) (\lambda + \varepsilon_{it-1}) (\lambda + \varepsilon_{it-2}) \ldots (\lambda + \varepsilon_{i1}) P_{i0}$. If we examine the logarithm of $(\lambda + \varepsilon_{it})$, assuming that ε_{it} is small relative to λ, then it is clear that we can simplify this as ε_{it} (using Taylor's expansion). We can thus write the series in log form as $\log P_{it} = \log P_{i0} + \varepsilon_{it} + \varepsilon_{it-1} + \varepsilon_{it-2} \ldots + \varepsilon_{i1}$.

This is none other than the expression that leads to a normally distributed set of objects, in this case a normal distribution of the logarithms of the populations or, when transformed back, a lognormally distributed set of populations. We have already illustrated what a lognormal distribution looks like in figure 2A, which shows the frequency of urban areas in Great Britain by size. This distribution is not a power law, hence our earlier caveat indicating that such distributions are likely to be lognormal, although one might be forgiven for assuming that its long tail could be approximated by such a law. Moreover, it arises as a result of a geometric process where growth is compounded in contrast to a normal distribution that is the result of additive growth. Gibrat called this the Law of Proportional Effect that is the result of a process where the statistics of the growth rates—the mean and the variance—are constant for all sizes of event. Its approximation using a power law assumes that the smaller events are disregarded and that the power law is fitted to the long tail, sometimes called the heavy or fat tail that occurs in the skew of the distribution to the right. It is not surprising that researchers have disregarded the lognormal distribution, for fitting Zipf's law just to the long tail of the distribution has been most successful. For example, Krugman (1996) took 130 metropolitan areas listed in the 1993 Statistical Abstract of the United States and demonstrated that nearly all the variance in the ranked city size distribution could be explained by $P_r = P_1 r^{-\alpha}$, where the value of the parameter was $\alpha = 1.003$, uncannily near the mythical value of unity!

During the last 10 years there has been an explosion of models built around Gibrat's law that show the links between proportionate random growth, random walks, and Brownian Motion in generating power laws for a whole range of physical and social phenomena (Stanley et al 1996). The solution to generate a power law from a multiplicative process is really rather simple: introduce a process that essentially cuts off the thin tail and leaves the distribution simply described by the fat tail, which then follows a power law or some related exponential. We have already anticipated this to an extent in our informal discussion earlier where we suggested that if an object got below a certain size, it disappeared and was replaced by a new object at the minimum threshold size. This is akin to not letting the object get below a certain size and it is this mechanism that, when added to the Gibrat model, leads to distributions that follow a power law. The most effective demonstration of this is provided by Levy and Solomon (1996a, 1996b) who initially presented the model as one that mirrored an economic market. They justified the lower bound for problems of this kind by arguing that income or wealth could not fall below a given limit due to subsidies in the economy. It is harder to say the same about cities, but Blank and Solomon (2001) suggest that there may be service thresholds below which there are no economies of scale, and cities would not exist. In fact what they also suggest is an extension to their model that embraces birth and death processes not unlike ideas originally developed by Simon (1955).

There are many developments at present with this style of model. One of the most intriguing issues is that mathematical analysis is throwing up all kinds of results that enable parameter values to be approximated from plausible assumptions about how such systems work. Gabaix (1999), for example, shows how this kind of model is consistent with various economic growth models, while Sornnette and Cont (1997) argue that this is but one of a very large number of multiplicative processes with repulsion that are related to random walks, Levy flights, diffusion and Brownian Motion. Distributions such as those shown in figure 2 suggest all kinds of ad hoc approximations. Laherrere and Sornette (1998) discuss a class of stretched exponentials, showing how these fit various ranked distributions from cities to earthquake magnitudes. Malacarne et al (2001) demonstrate that the Zipf law modified by Mandelbrot (1966) as $P_r = P_1(c + r)^{-\alpha}$, which they call the *q*-exponential, fits various data from cities in Brazil and the United States particularly well. Recently an even more intriguing suggestion has been made by Reed (2001). He shows that if a birth process is assumed for Gibrat's law, and if it is assumed that events are distributed exponentially according to their age, then in the steady state, the distribution becomes what he calls 'double Pareto'; this means that the smaller events follow a different power law from the larger events with the thin tail being simulated by a positive power, the fat tail by a negative power.

Our short review of where this field now stands barely does justice to the wealth of ideas currently being generated from both physics and, to a lesser extent, economics. Many of these ideas are being unpacked even further to link to new dynamic models based on conventional economic and physical theory that suggests how such power laws emerge. These models are much more explicit and causally complex than Gibrat's law, which we might take almost as the default or null hypothesis in this domain. These new models relate strongly to fractals, complexity and chaos, where various mechanisms are postulated as to how power laws emerge from simple assumptions. In what follows, we show that despite the stability implied by these distributions from time period to time period, it is critical to explore the underlying dynamics of change. To this end and with the premise that the kind of growth that occurs is random and proportionate, we show how this is consistent with very volatile dynamics where cities and other types of events move up and down the hierarchy quite dramatically while still preserving the underlying regularity that is Zipf's law.

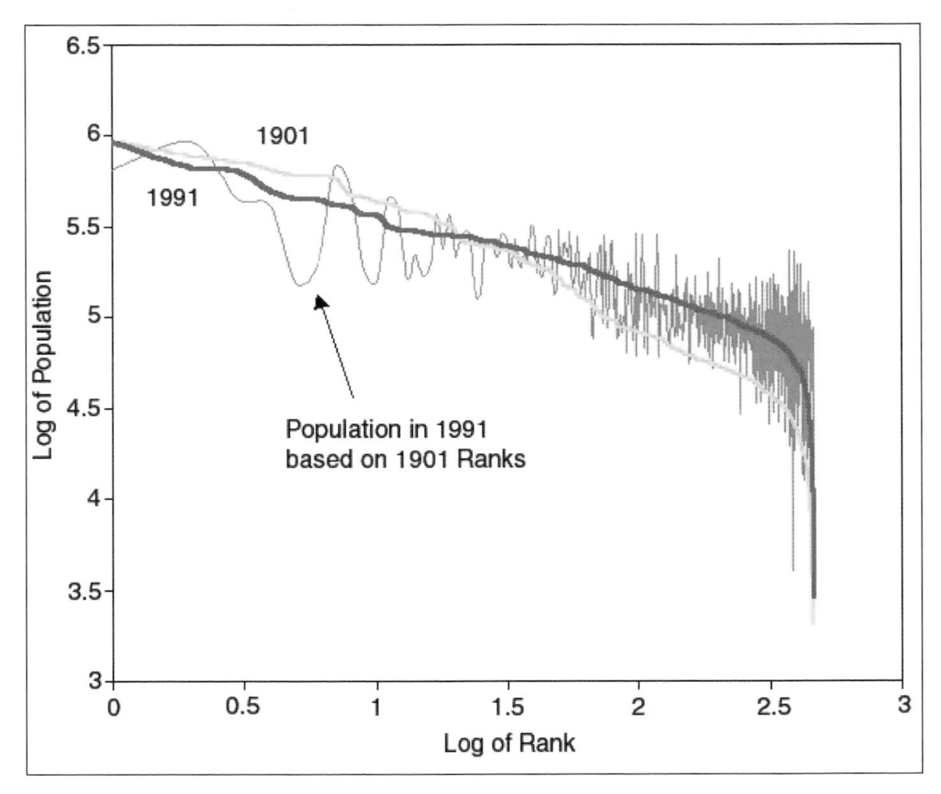

Figure 3 Rank-size relations for 1901 and 1991 with changes in rank

3 Searching for transitions in evolving systems

When we examine changes over time in the rank-size relationship for many different kinds of systems, we see a remarkable regularity in form. To anticipate our analysis, we show the log-log population rank-size relationship for 459 administrative units in Great Britain in figure 3, where it is very clear that the form is quite similar. Yet this similarity over long time periods, indeed over decades and centuries even, disguises quite turbulent dynamics at the more micro level. Our causal knowledge of how places have changed over such extended time periods tells us that there is a lot more going on than meets the eye with respect to the kind of stability implied by the relations in figure 3. We also show in this figure the changes in rank that have occurred over a century between 1901 and 1991, thus revealing substantial change.

To get a handle on such change, there are a variety of statistics that we can compute. If we normalise the populations for size, in the case of populations at times $t = 1901$ and $t+1 = 1991$, we can define the overall percentage shift in the population as $\Sigma_r \mid (P_{rt} / \Sigma_r P_{rt}) - (P_{rt+1} / \Sigma_r P_{rt+1}) \mid$. During this period, there is a large shift of some 46 per cent. This is in direct contrast to an average shift of 85 places in rank using the formula $\{\Sigma_i | r_{it} - r_{it+1} |\} / N$ where r_{it} and r_{it+1} are the ranks of place i at times $t = 1901$ and $t+1 = 1991$. This average shift in rank is some 17 per cent of the overall number of ranks (459), and this is considerably less than implied by the absolute shift in populations (Batty 2001c). In short, rank is a considerably more stable measure than absolute population from time period to time period, thus masking the microdynamics of settlement change. In the analysis that follows, we do not look any further at changes in the ranks of specific places but concentrate on trends and discontinuities that these kinds of statistics reveal through time.

We use a measure of difference between ranks due to Havlin (1995), not unlike the absolute average difference just stated but which has rather more tractable statistical properties associated with a measure of distance. This was originally used and extended by Vilenksy (1996) for the comparison of texts based on word frequencies, where it was found that books by the same author had rather more in common with each other in these terms than books by different authors. For any two times j and k, the distance is defined as $R_{jk} = \{\Sigma_i (r_{ij} - r_{ik})^2 / N\}^{1/2}$. This gives the average shift in rank from time j to k. We refer to this as the Havlin Matrix R which is clearly symmetric with respect to time, though we are mainly interested in the data that relate to differences between forward time periods where $k > j$. From this very rich matrix of differences, we can develop various temporal analyses. Specifically, we are interested in the trend through time, in discontinuities in these trends which are associated both with dramatic changes in the rank of different places, and the possibility that places fall or gain in rank only to reverse themselves back towards their previous positions in time.

There are many ways in which we can measure different changes between ranks between different time periods. In the various datasets that we explore in this chapter, we look at the cumulative change in ranks based on the forward series R_{tt+1}, R_{tt+2} . . . R_{tt+n} as well as the backward series R_{t+nt}, R_{t+nt+1} . . . $R_{t+nt+n-1}$. The plots that we make use a particular year as the numeraire in computing these forwards and backward series. We also look at the incremental forward series where the ranks change from time period to time period. This involves a comparison of R_{tt+1}, R_{t+1t+2} . . . $R_{t+n-1t+n}$. We can of course examine many different orders of difference in the Havlin Matrix but we restrict ourselves to looking at first-order change based on $\Delta_{jk} = R_{jk} - R_{jk+1}$, $k > j$. We would usually expect to find that these differences were positive unless there were reversals in the general trend. From this difference matrix, we can compute the overall drift as $\Delta = \Sigma_{jk} \Delta_{jk} / \Sigma_{jk}|\Delta_{jk}|$ where the range of the summation is defined according to the indices above. We have now introduced sufficient measures, and at this point we demonstrate the analysis on three very different datasets, each implying rather different kinds of dynamics.

4 Cities, countries and communications

The dataset we have for urban populations is based on one constructed around British municipalities from the decennial censuses of population between 1901 and 1991. These data have been standardised to the 459 municipalities that existed in England, Scotland and Wales in 1901. Strictly speaking, it might be argued that these municipalities are not cities, but we prefer to use these labels as this reflects an exhaustive subdivision of the space economy that is not complicated by changing definitions of what a city is. The dynamics implied in this series are rather conservative. The British population has not grown all that much over the last 100 years relative to other growing systems: in 1901 it was around 37 million growing to 54 million by 1991. We would expect the dynamics implied by this growth to be fairly smooth from our knowledge of the British spatial system where most of the big cities were already established by the beginning of the last century. This is in quiet contrast to what has happened in other parts of the world.

In figure 4, we plot populations for these 459 municipalities by their rank, and this shows remarkable stability in the aggregate form of this relationship. There are some crossovers, and the profile flattens slightly through time implying that the population is decentralising. In figure 5, we show the Havlin plots which reveal consistent changes in rank but without any real surprises: no discontinuities and no obvious switches back to rank orders that have occurred earlier. We have also examined the first-order change through time based on examining the series $R_{1901,1911}, R_{1911,1921} \ldots$. This reveals substantial changes in rank order in the mid-twentieth century during the period of economic depression and war, although these are difficult to interpret without a detailed examination of the disaggregate patterns in different places.

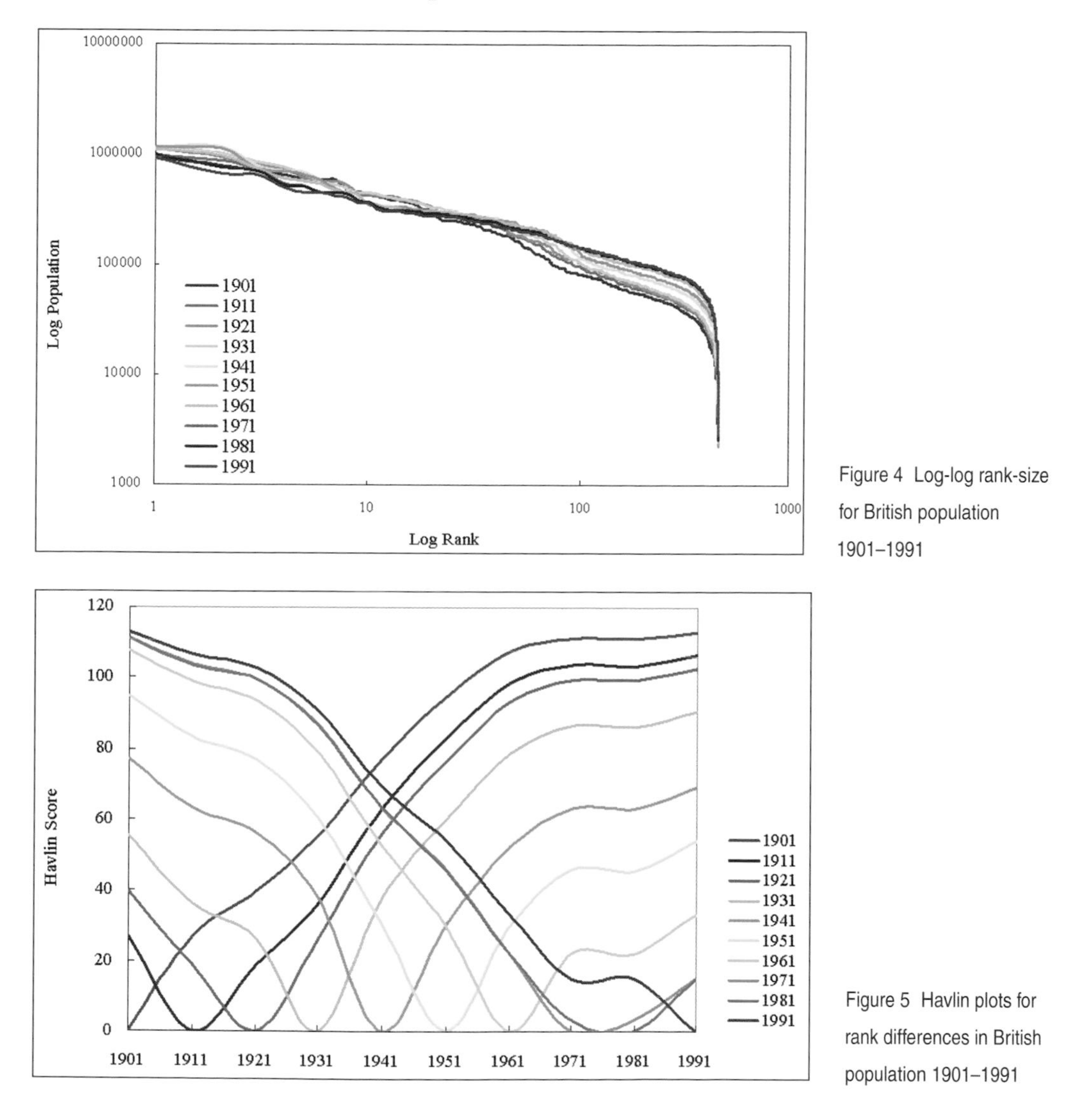

Figure 4 Log-log rank-size for British population 1901–1991

Figure 5 Havlin plots for rank differences in British population 1901–1991

Our second example is based on global populations in 210 countries. We have data on populations in these countries annually from 1980 to 2000 taken for ITU (International Telecommunication Union) published in 2001. These data are even smoother in a disaggregate way than the national British data as can be seen in figure 6, where the log-log rank-size plots show very little major change during this period.

The Havlin plots in figure 7 are quite regular with no surprising discontinuities. What this means is that the world-population system has hardly reordered itself at all over the last 20 years in terms of rank-size. The first-order change series $R_{1980, 1981}$, $R_{1981, 1982}$, . . . shows little change, even less than the British data, and this implies that as we aggregate, we lose details and we smooth variations. In fact, this is borne out very clearly if we compute the drift parameter Δ which is equal to unity, meaning that there are no reversals in overall rankings from time period to time period, although there are individual switches as the Havlin matrix reveals. For the British population system, the drift parameter is almost unity at 0.995.

Our last example demonstrates a very different kind of dynamics for it deals with the recent rapid diffusion of mobile communications devices by country. The data are taken from the same sources as the global-population series and again show change from 1980 to the year 2000. We show log-log plots of the size and rank of mobile devices by country in figure 8. These are considerably more skewed and have a much bigger thin tail than in the other two examples. In terms of growth dynamics, this would appear to be largely due to the fact that the system is growing rapidly and is still somewhat immature. This is clearly marked by the Havlin plots shown in figure 9, which display considerable discontinuity in the early to mid-1990s when mobile phones were first penetrating many countries. In fact, the first-order change based on the series $R_{1980, 1981}$, $R_{1981, 1982}$ . . . is fairly smooth, but the changes in rank are very substantial with almost half the number of countries having changed rank by the year 2000. The drift parameter too reflects this turbulence with a value of 0.674, implying that 33 per cent of the system has moved back and forth between previous rank orders during the 20-year period.

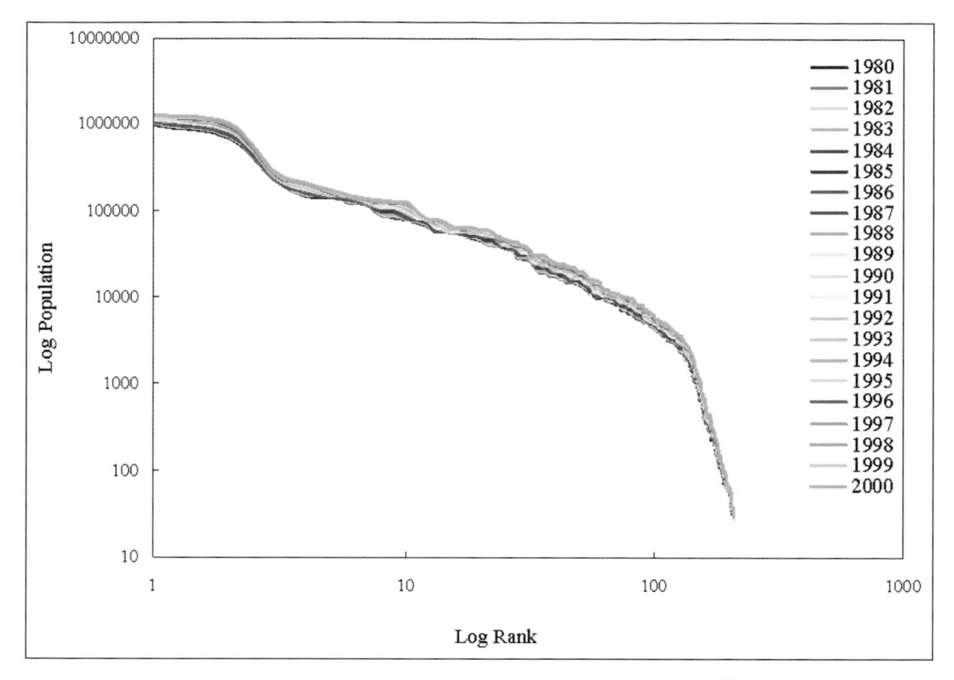

Figure 6 Log-log rank size for global country populations 1980–2000

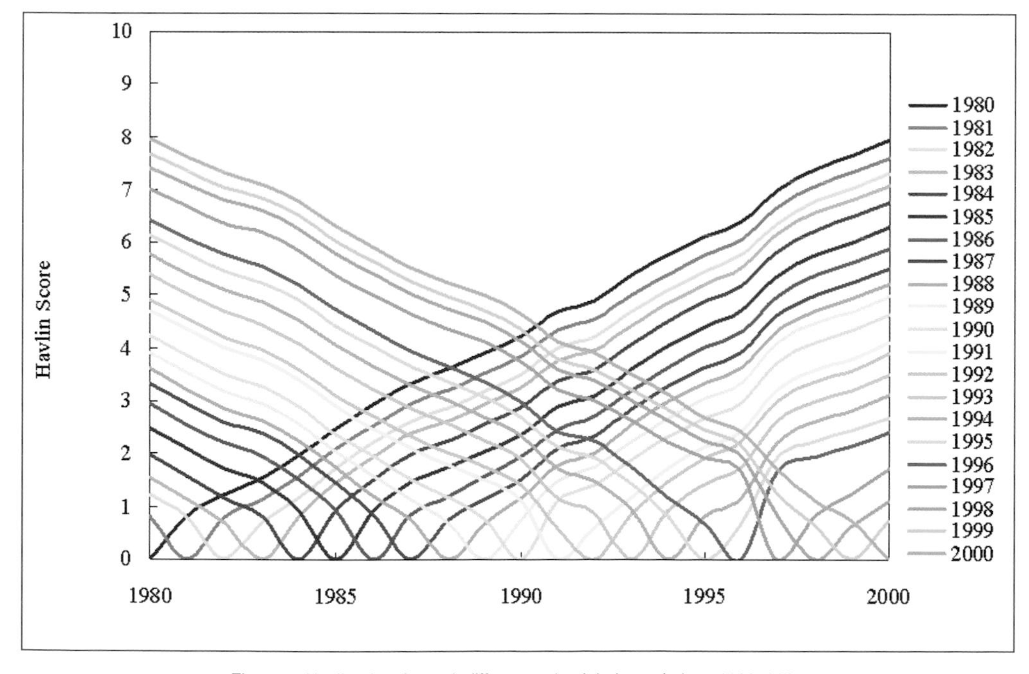

Figure 7 Havlin plots for rank differences in global populations 1980–2000

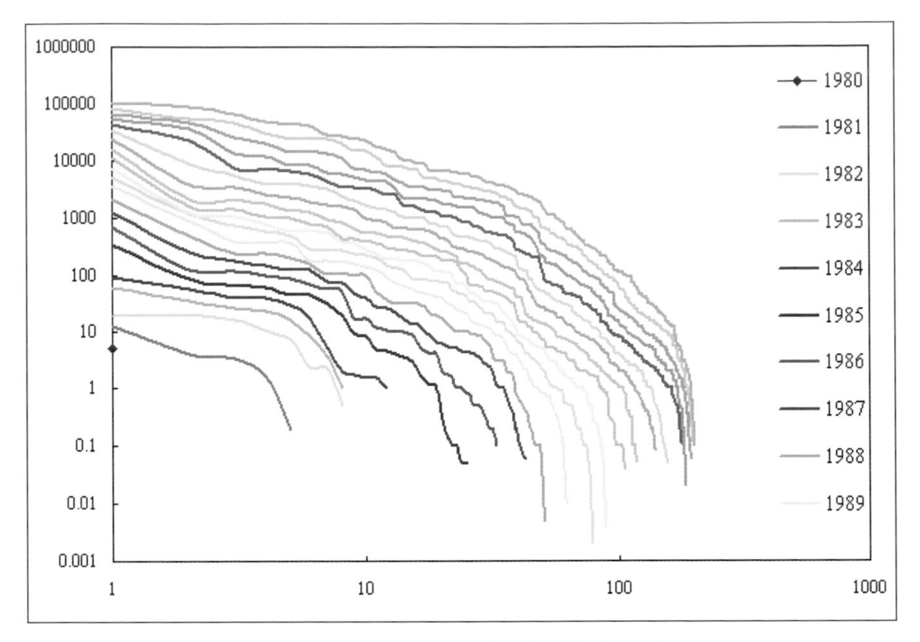

Figure 8 Log-log rank-size for penetration of mobile devices by country

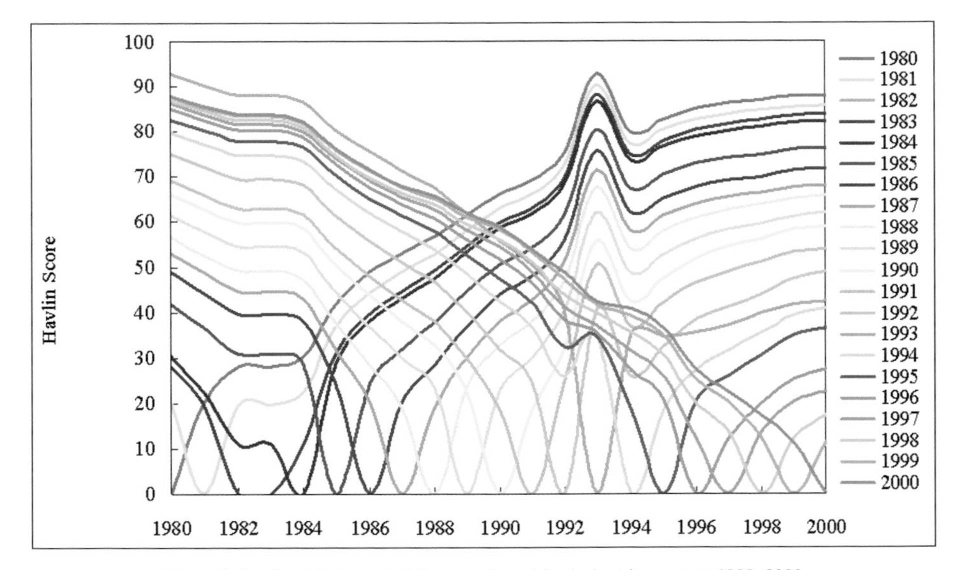

Figure 9 Havlin plots for rank differences in mobile devices by country 1980–2000

5 Next steps

This chapter has shown how aggregate spatial patterns marked by strong statistical regularities, implying stability at least superficially, are often simply masks for a deeper underlying dynamics. The rank-size rule is perhaps the best known of such regularities but to date, although there have been a flurry of explanations based on random growth theory, there has been little empirical exploration of what their dynamics means. We have not actually fitted any relationships here, for the various Zipf plots reveal very strong regularities in any case, and it is easy enough to guess that power laws will give very high correlations with the observed data. However, what we have shown is that by comparing different relationships through time using the Havlin matrix, we are able to unpack the dynamics of change in more detail, illustrating that seeming regularity can be consistent with turbulence at a more disaggregate level.

There is much work still to do with the data that we have explored here. In particular we need to consider the extent to which we need to truncate the distributions by getting rid of the thin tails so that we can fit the best power laws. We also need to assess the extent to which such systems give Zipf parameters close to unity, and ways in which we might develop random growth models that reflect the kinds of systems that these rank-size profiles characterise. We also need to explore the extent to which we can say that the Havlin statistics reveal that similarities through time reveal changes belonging to one system or another. Just as it is possible to compare two different authors and works by the same authors using these statistics, we need to pose and answer the question as to whether systems that change through time are nearer to other systems at the same time or to themselves at earlier time periods. In this manner, we would hope to extend this kind of analysis into ways of classifying spatial systems into different types of structural dynamics.

Chapter

17

Location-allocation in GIS

Torsten H. Schietzelt and Paul J. Densham

Location-allocation analysis was first developed long before GIS was first proposed, and, in fact, predates the invention of the computer. Formal approaches date back to the first quarter of the last century, and the fundamental advances in the field originate in operations research of the 1950s and 60s. Over the last decade or so, however, location-allocation has become intrinsically linked with GIS-based analysis and modelling. Modern GIS software includes much of the core functionality needed for interactive locational decision making. This chapter discusses the range of modelling, solution strategies and applications and describes a current location-allocation research project that focuses on emergency medical services.

1 Location-allocation applications

Location-allocation analysis can be used to determine optimal locations for multiple facilities with respect to the spatial distribution of demand for a given service and, frequently but not necessarily, the spatial variation of costs and risks associated with candidate locations. It has a broad range of application areas in the public and private sectors. The following examples are a representative cross-section of this domain without any claim to completeness:

- emergency services
- other public services
- retail and other commercial services
- industrial production and warehousing
- transport planning
- gas, water, electricity and telecommunications
- IT network design
- electoral districting and campaigning
- air traffic control
- humanitarian relief, and
- military defence and homeland security.

An infamous historic example of location-allocation decision making is Governor Elbridge Gerry's (1744-1814) redesign of the electoral district map of Massachusetts (1811-12). He modified the boundaries with the objective of increasing the number of constituencies that included a safe majority of supporters (figure 1). Spatial compactness of constituencies was not regarded as a mandatory constraint at that time. One consequence was a particularly odd, reptile-shaped district and that caused Gerry's name to be given to this approach: gerrymandering. However, the same optimisation method, when applied with fair and transparent objectives, is both legitimate and valuable (see, for example, Barkan, Densham and Rushton 2001).

A less controversial early example was Alfred Weber's attempt to develop a normative model for determining best production sites with respect to raw material sources and market locations (figure 2; Friedrich 1929). He used the technique of constructing isodapanes (lines of equal accessibility), today a basic data manipulation tool in GIS. Weber himself had to cope with mechanical drawing tools and no significant progress was made until the early 1960s when various researchers came up with general solution methods for the facility location problem. One landmark achievement was Cooper's (1963) extension of Weber's single

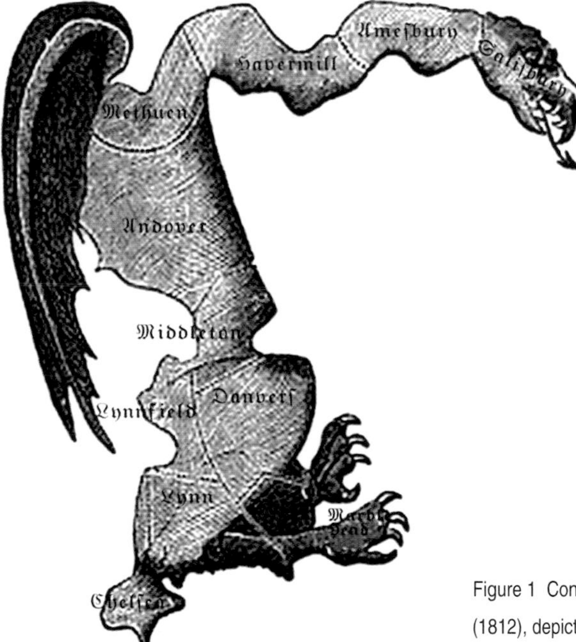

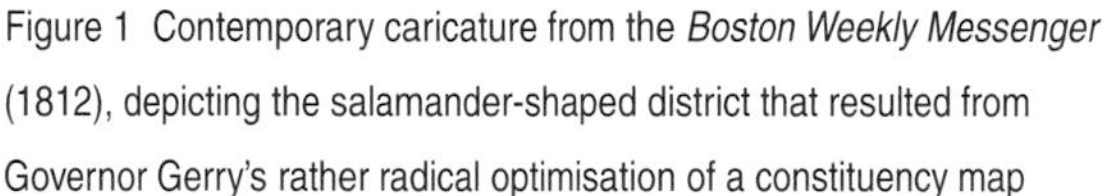
Figure 1 Contemporary caricature from the *Boston Weekly Messenger* (1812), depicting the salamander-shaped district that resulted from Governor Gerry's rather radical optimisation of a constituency map

facility model to the multifacility location-allocation problem, which then became known as the *p*-median problem. Another breakthrough was the transference of the problem formulation from continuous space to a network representation (Maranzana 1964). Many real-world problems are constrained by transport networks and Maranzana's work opened up a vast new area of applications.

The *p*-median problem belongs to a class of optimisation problems that is computationally intractable. No exact algorithm is known to guarantee the optimal solution within a computing time that is a polynomial function of the problem size (Garey and Johnson 1979). Subsequent study, therefore, chiefly concentrated on heuristic solution methods, that is, methods that employ a strategy of educated guesswork or trial-and-error. Despite significant progress, this remains a key research issue at present. As heuristics became more efficient, computers more powerful, and data more readily available, larger and increasingly complex location-allocation problems could be addressed. In the mid-1970s, research branched into four main new directions (Ghosh and Rushton 1987): first, researchers began to seek more realistic representations of the geography of demand, accessibility, costs and risks; second, the decision-making behaviour of individuals and organisations came under consideration, especially in cases

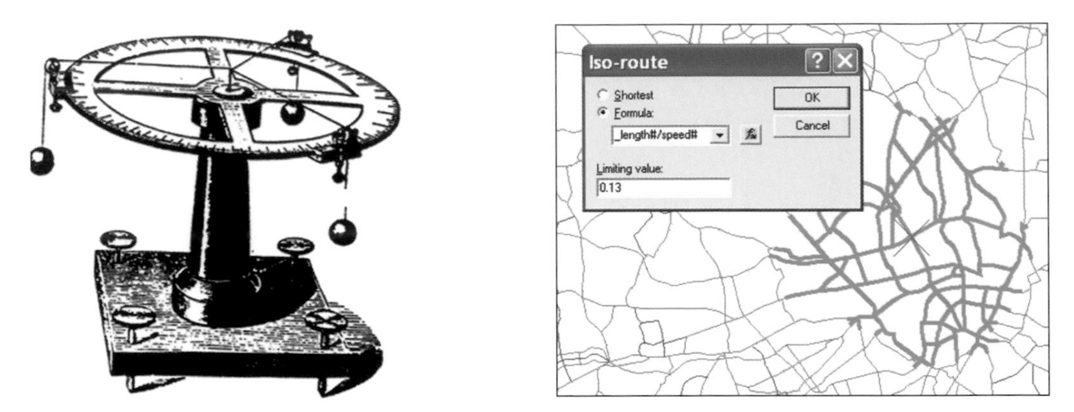

Figure 2 The Varignon Frame used by Weber to construct isodapanes and iso-routes in GIS. An iso-route is defined as a section of a network within a specified link distance (or other measure of accessibility)

with multiple or fuzzy objectives; third, consumer behaviour was included in model formulations; and, finally, data uncertainty and associated issues became important.

Location-allocation problems often represent complex spatial decision problems. The complexity manifests itself through multiple and conflicting objectives and an uncertain or ill-structured concept of the properties of an optimal solution (Densham 1991). Spatial decision support systems (SDSS) are specifically designed to address this kind of problem by guiding the user, or groups of users, interactively through the decision-making process. SDSS are not primarily a specialist type of software. They are better understood in terms of a conceptual framework that integrates key computer-based components to support spatial decision making: a data repository, a user interface and data manipulation, analysis and visualisation tools. A discussion of these concepts is provided in Densham (1994).

Within the SDSS framework, modern GIS are destined to form the central component. They include a wide range of cartographic visualisation, data manipulation and generic spatial analysis tools, thus making complex spatial structures and relationships accessible. Standard GIS functionality comprises, for example, distance measuring, route finding (shortest-path analysis), and the creation of iso-routes (on the basis of complex accessibility functions), 3-D surfaces, contours, grids, and Thiessen polygons. GIS have the capability to interact with external databases storing non-spatial data. Some, such as Cadcorp's (Stevanage, United Kingdom) Mapmodeller or ESRI's (Redlands, California) ARC/INFO®, support topological network structures with directional constraints on links and conditional connectivity at nodes. Specialised functionality can be added to GIS using either proprietary scripting languages (as in ArcView®, Arc/INFO or MapInfo®) or industry standard programming languages (as in Cadcorp or Geomedia™). Most GIS vendors also offer ActiveX controls that allow programmers to embed GIS functionality within bespoke applications.

Figure 3 A good ambulance location is critical to effective service delivery

2 A categorisation framework for location-allocation problems

Location allocation analysis simultaneously involves the location of facilities and the allocation of demand to each individual facility on the basis of some measure of accessibility. Depending on the practical context, location-allocation problems can be distinguished and categorised on one or more axes with ranges that include (with examples):

- Static to dynamic facility locations
  - Static: locating fixed facilities such as hospitals
  - Dynamic: locating and re-locating mobile facilities, such as ambulances, according to dynamic changes in the state of the system

- A single objective that captures the problem to multiple, sometimes fuzzy, objectives
  - Single objective: locating radio transmitters in such a way that achieves full coverage of a service area with a minimum number of units
  - Multiple objectives: locating ambulances in such a way that system-wide performance and equality of service provision are optimised in a balanced way
- Uncapacitated to capacitated facilities
  - Uncapacitated: the siren location problem—within the covered area, a siren can serve unlimited demand
  - Capacitated: ambulance base stations, having a finite service capacity
- The provision of a unique service to a diverse set of services
  - Unique service: an ambulance system that employs only one type of vehicle
  - Diverse set of services: an ambulance system that employs a diverse fleet of vehicles

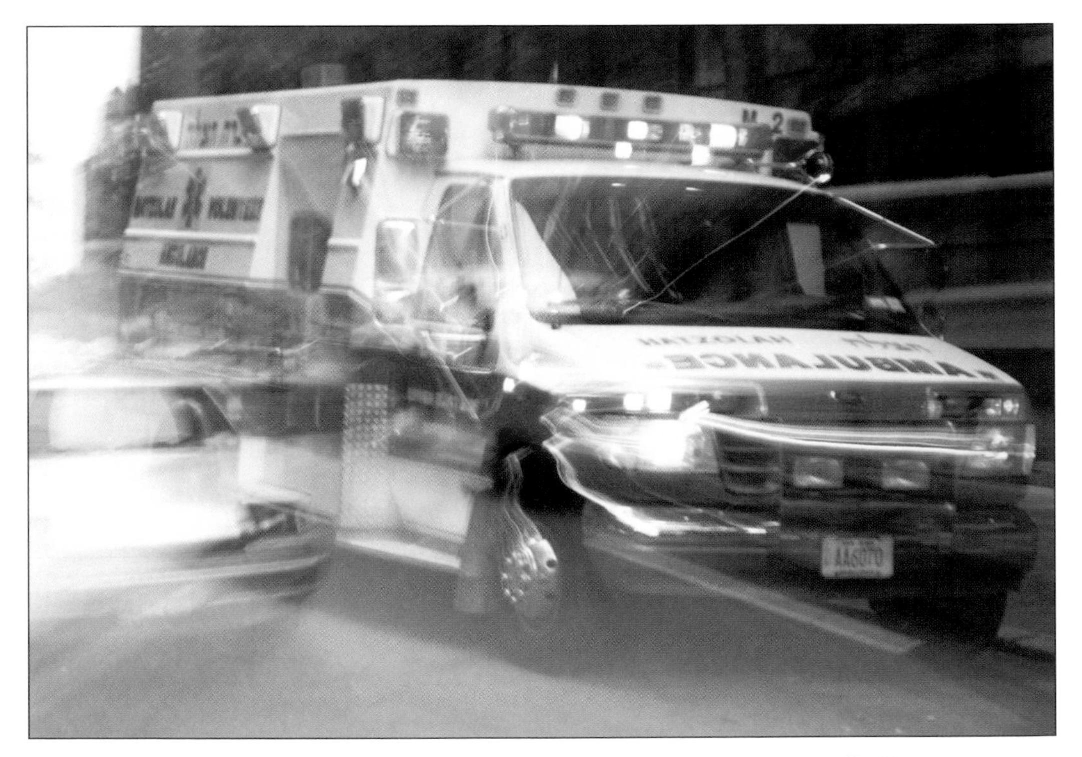

Figure 4 Ambulance response times are contingent upon anticipation of the geography of incidents.

- Single to multitiered demand-supply relationships
  - Single relationship: a system that solely considers locating ambulance bases according to demand distribution and accessibility
  - Multiple relationships: a system that simultaneously locates ambulance bases and hospitals
- Deterministic to adaptive or stochastic data input
  - Deterministic: optimising the layout of a cable TV network
  - Stochastic and adaptive: a dynamic system locating and re-locating ambulance bases considering a probabilistic distribution of demand and travel times and being able to accommodate changes in real-time

Another distinction derives from the underlying geometric concept of space, which can be continuous in one-, two- or three-dimensions, or, as for most practical applications, a network. These differences and their implications will be discussed later in this chapter.

3 The *p*-median problem

The *p*-median problem is a multifacility location-allocation problem that locates *p* facilities to minimise a global measure of cost of access. A solution to the *p*-median model has the properties that each facility is at the local median of its service area, that all demand is allocated to its closest facility, and that no re-location of facilities will decrease the global measure of cost of access. The basic model assumes proportionality between cost of access and distance. Hence, its objective is to minimise the system-wide sum of distances between demand and supply locations, weighted only by variations in the amount of demand at each demand location. The *p*-median model has central importance in location-allocation modelling. This is partly because the objective of increasing system-wide efficiency fulfils the needs of many practical applications, but also because the model is adaptable to representing a range of different and even multiple competing objectives. In fact, the covering models that were developed independently are very similar in structure (Hillsman 1984).

The focus on system-wide efficiency of the *p*-median model can, and frequently does, increase inequality of access throughout the study area because the global gain achieved through locating a facility near to or within a cluster of high demand often penalises those more isolated locations that constitute lower demand. Set-covering models aim to increase equity of access. Often, however, the choice between global performance and equity is not clear-cut but a matter of emphasis.

The weight or preference given to each goal may not even be clear to the decision maker in advance. The required, or an acceptable, balance between multiple objectives is often discovered only by exploring the solution space of the problem, something that SDSS are designed to facilitate. Planners of a retail chain, for example, may initially set out to cover the demand of an entire region by selecting appropriate branch locations. Whilst full regional coverage may be viewed as mandatory, the total number of shops is constrained by financial resources, and each branch requires a threshold number of potential customers. It may transpire that the initial model generates an optimal solution that violates at least one constraint. The decision-making process must then loop back to consider alternative retail strategies and policies, which may involve an entirely different department. The new options submitted to the spatial decision-making process may now include the establishment of a mail-order system. Complex considerations about consumer behaviour, such as whether or not the acceptance of a mail-order system by customers requires at least initial and occasional access to one of the branches, are beyond the scope of this chapter (but not beyond the scope of SDSS in general). The modified problem now requires a threshold of demand locations, uncovered by branches, which is larger than the unwanted residual of the first run. Thus, the objective shifts from coverage towards global efficiency, but as a subtle change of emphasis rather than a completely new approach. It is easy to imagine a decision-making process going through a number of such loops before settling on a solution.

Using a single, highly adaptable core model obviously has merits within the context of designing a SDSS. The rest of this section focuses on the p-median problem in network space and explores its adaptability. On a network, the occurrence of demand and supply are restricted to discrete locations—the nodes where links (edges) intersect. It can be shown that this restriction to nodes does not constitute a principal change to the model (Garey and Johnson 1979), although a subdivision of long links by artificial nodes may be necessary in certain practical cases. The following generic formulation is adapted from Densham and Rushton (1992).

Minimize $Z = \sum_{i=1}^{n} \sum_{j=1}^{k} w_i d_{ij} x_{ij} \qquad \forall i, j$ equation 1

Subject to $\sum_{j=1}^{k} x_{ij} = 1$ equation 2

$\sum_{j=1}^{k} x_{jj} = p$ equation 3

$K \subseteq N$ equation 4

$i, j, n, k, p \in IN; \qquad w, d \in IR^{+}$ equation 5

where

n is the number of demand locations, which is less than or equal to the number of nodes on the graph;

k is the number of candidate locations for facilities;

p is the number of facilities to be located;

w_i is the weight of demand at location i;

d_{ij} is the directed shortest path distance between demand location i and facility location j; and

x_{ij} is an allocation flag, 1 if demand location i is allocated to facility j, 0 otherwise.

In equation (1), Z is the sum of distances between all demand nodes and their allocated facilities (d_{ij}), weighted by the amount of demand at each node (w_i). The binary variable x_{ij} ensures that only distances between demand nodes and their allocated facilities are included in Z. Equation (2) forces each demand node to be allocated to only one facility. Equation (3) specifies the number of facilities to be located; it may take the form of an interval. Term (4) indicates that the set of candidate locations for facilities is either identical to or a subset of the set of demand nodes. The possibility of having no demand at a candidate node is covered by the option of setting its weight w_i equal to zero. Term (5) defines the variables and parameters as positive integer or real numbers respectively. It remains a computational implementation issue whether w_i and d_{ij} are forced to be integer as well. For the model itself this is irrelevant.

Distance itself is frequently too crude a measure of the cost of access. In certain cases, for example, problems in the IT and telecommunications domains, distance and accessibility may not be significantly correlated at all. Travel time on roads depends predominantly on congestion, road condition and speed limits, whilst the time spent turning at nodes depends on both from- and to-links, but not on their length. Therefore, the term d_{ij} must be understood as representing cost-distance, that is, any chosen appropriate measure such as travel time, price per instant of access, etc. It is normally calculated from properties of each constituent component of the internode connection. The calculation can also be carried out prior to running a solution algorithm. How the properties are obtained and stored remains a data issue that is external to the model. An alternative option is to obtain samples of the aggregate cost of access for each d_{ij}.

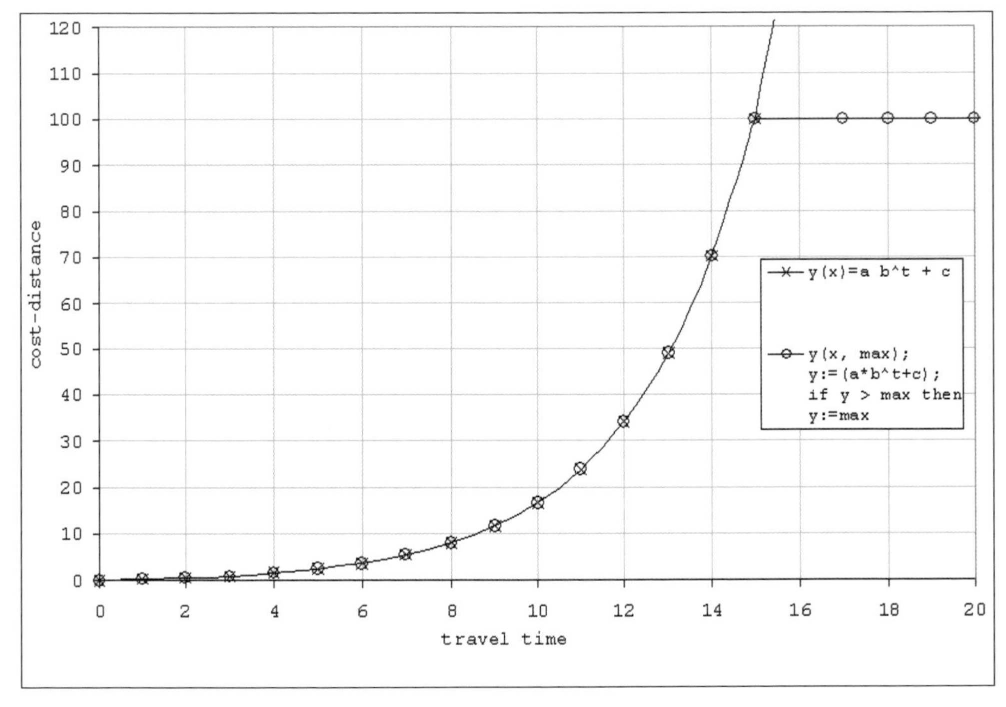

Figure 5 Exponential distance-cost functions suitable for ambulance response modelling

A more interesting consideration is that d does not need to be linear and can accommodate measures of probability. Figure 5 shows, as an example, two similar cost functions *d(t)* that would be suitable for a model of high-priority ambulance response. It reflects the fact that the risk of death or further health damage strongly increases with response times exceeding eight minutes, whereas the risk below this threshold is far less significant (Daskin 1987). The first function allows unconstrained exponential increase of cost that, under certain circumstances, can cause difficulties when running a solution algorithm. The second function therefore determines a maximum cost value beyond which the curve flattens, and assigns it to some maximum travel time, beyond which arrival is considered to be equally unacceptable. In this way, the original *p*-median model approaches the properties of a set-covering model.

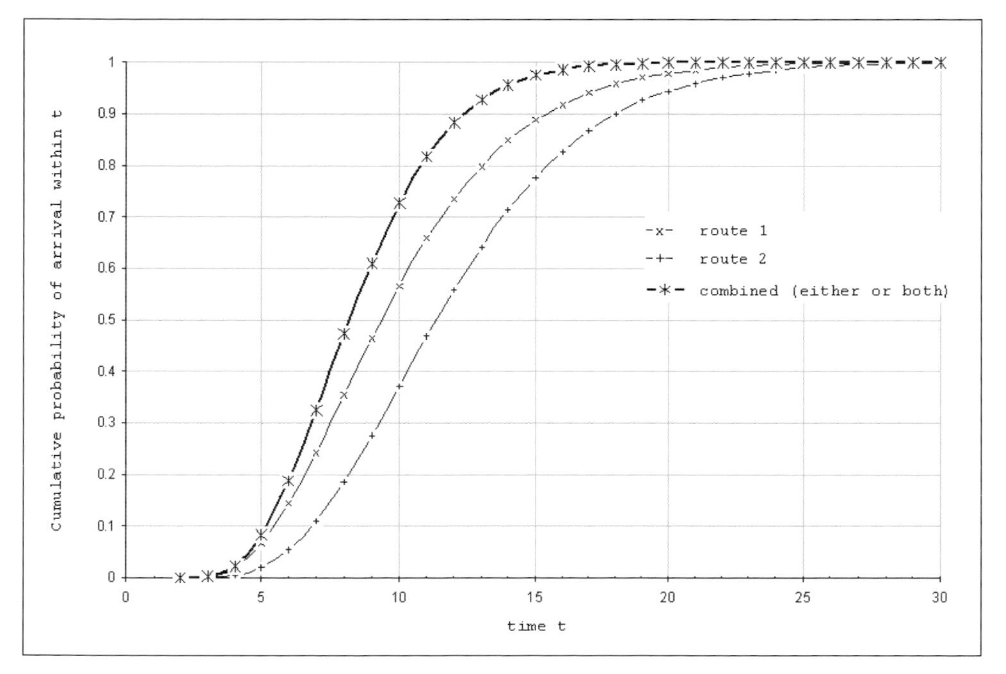

Figure 6 Cumulative Erlang distribution of travel time on two independent routes

Continuing with the example of emergency medical services (EMS), it can be shown that it is possible to account for uncertainty of travel time when specifying the value of d_{ij}. Stochastic internode travel time should be modelled by an Erlang distribution, which is also used for queuing models. So-called deterministic travel times are in fact mean travel times. Using the mean as the estimate for the expected value is an obvious, but not necessarily the most appropriate choice for modelling a critical emergency service such as ambulances. The decision maker may instead opt for a percentile as a measure to ensure the timely arrival of vehicles in most instances.

The implications of uncertainty of travel time are greater in a model that assigns one or more secondary ambulance locations to each demand node for backup, and models that account for the possibility of simultaneously dispatching multiple vehicles to an incident. Vehicles taking different routes are less likely to get stuck simultaneously in exceptional congestion than vehicles travelling the same route. This implies that multiple vehicles preferably should be dispatched from different bases (Daskin 1987). The combined probability that either the vehicle travelling route one or the vehicle taking route two (or both) will arrive within time t can be used as an additional objective component or constraint (figure 6).

The gain in the combined probability obviously depends also on any interdependence between delays occurring on the two routes. An estimate for this correlation can be obtained in a GIS based on the assumption of spatial autocorrelation. Figure 7 shows an example using the computationally efficient method of calculating the overlap ratio of smallest enclosing rectangles (SERs). Such an approach

Figure 7 Defining the SER-overlap ratio of two routes as a measure for congestion risk correlation

has the desirable side effect of favouring bases that are relatively distant to each other and, therefore, less likely to be simultaneously affected by overload.

Uncertainty also prevails with regard to the occurrence of discrete instances of demand. Only non-capacitated supply problems can be modelled with deterministic demand weights, w_i. Those may include genuinely non-capacitated problems such as, for example, electoral districting or problems where supply is sufficient to avoid queue build. EMS models often, and perhaps controversially, assume that the capacities of ambulance bases are sufficient and queues will not form. Capacitated models, on the other hand, can be represented through spatial queuing models. The occurrence of discrete instances of demand within some time interval normally can be assumed to be random, or Poisson distributed. In dynamic models, the Poisson-mean may change in time. The waiting time within the queue can then be modelled by an Erlang distribution, then used to include constraints on the maximum length of, or the maximum waiting time within, the queue. Non-spatial queuing models can be used initially to determine the required system-wide supply capacity, for example, the required total number of ambulance vehicles. Spatial queuing models, when allowed to form individual queues at each facility, find their application in simulation approaches to validate a solution. They are also used in dynamic models where the relocation of vehicles (facilities), either to increase the capacity of existing facilities or to increase the number of facilities, p, is triggered by reaching a threshold waiting time or queue length. Dynamic spatiotemporal variation of demand probability normally can be modelled by discrete stepwise adjustments of the (Poisson) mean. A sequenced succession of overlays representing the spatial distribution of demand can be used in a GIS to update the demand weights, w_i. The same method can also be applied to adjust link travel times to changing traffic situations.

4 Solution methods

Exact methods

Exact solution methods, such as integer linear programming (ILP), branch and bound techniques, and the Lagrange relaxation method have relatively restricted importance for applied location-allocation analysis. They have been successfully applied to reduced problem sizes within hybrid approaches and as validation methods for solutions previously found by heuristic algorithms. Generally, however, their scope is limited by inferior performance. Solution times tend to increase exponentially with problem size. Large configurations of several thousand demand and candidate nodes are common in real-world applications such as transport planning. Even with efficient heuristic methods, frequently hours are the adequate units to measure solution times (see, for example, Densham 1996); and there is no shortage of real problems that practically are not solvable with existing methods and computing resources. Therefore, developing more efficient heuristic solution methods remains a great challenge. The performance speed of this core step of locational analysis determines to what extent dynamic systems can be implemented, and to what extent complex and ill-structured problems can be addressed in a truly interactive and exploratory way of spatial decision making. In both cases numerous fast repetitions of the actual location-allocation algorithm is required.

Basic heuristics

Early heuristic solution methods include the established ADD (Kuehn and Hamburger 1963), Drop (Feldman, Lehrer and Ray 1966), Alternating (Maranzana 1964) and Interchange algorithms (Teitz and Bart 1968) and Goodchild and Noronha's (Goodchild and Noronha 1983) variant of Interchange. Basic heuristics form the backbone of later advanced methods, such as the global-regional interchange algorithm (GRIA: Densham and Rushton 1992), simulated annealing, tabu search, genetic algorithms and a number of hybrid approaches such as, for example, the method of heuristic concentration (Rosing and ReVelle 1997).

The ADD algorithm starts with an empty set of facilities and adds facilities to the solution, with an increment of one, until all p facilities are located. The Drop algorithm uses the opposite method: it starts with k facilities located and, one-by-one, drops the surplus of $(k - p)$. Maranzana's Alternating algorithm starts with some arbitrary configuration of p facilities, and then alternates between demand allocation to the nearest facility and facility relocation to the spatial median until a stable set of facility locations and service areas is found. These methods may terminate, however, with a solution that could be further improved by substituting a facility for a free candidate location.

The Teitz and Bart algorithm (TB), on the other hand, also starts with an arbitrary solution and then loops through iterations during which each free candidate is substituted for each facility, terminating when no improving substitution is found during an iteration. The original TB accepts the most beneficial out of p substitutions for each candidate, whereas Goodchild and Noronha's variant (GN) evaluates all $p * (k - p)$ options per iteration before triggering a swap, which reduces the amount of memory access.

A major drawback of both TB and GN is that they are naïve in the sense of considering any substitution, including those that, if human perception were fast enough to interfere with the process in real time, would be intuitively spotted as redundant. Another difficulty, common in principle to all heuristics, is the risk that the algorithm terminates at a local optimum that is inferior to the global. This tendency is reinforced by considering only pair-wise (1-opt) substitutions between facilities and free candidates. A 2- or n-opt exchange algorithm could possibly achieve further improvement beyond where a 1-opt algorithm terminates. Rosing and Hodgson (2002) offer some insight into the spatial patterns that cause multiple-node traps. The interested reader may also refer to (Schilling, Rosing and ReVelle 2000) for a detailed discussion of how network characteristics affect solution performance and quality. However, any general relaxation of the 1-opt constraint exponentially increases the number of combinations to be evaluated.

Advanced methods

While the 1-opt interchange method retains its central importance, research has concentrated on improving the efficiency and robustness of heuristics in five main ways:

1. Performing a more informed search of the solution space in order to reduce the number of transactions considered per iteration.
2. Introducing an element of controlled randomness to avoid or escape from entrapment in local optima.
3. Establishing rule sets for repeating a single or sequencing multiple algorithms.
4. Intelligent task division and subtask design to make use of the potential of parallel and heterogeneous processing and, in dynamic systems, system idle periods.
5. Exploiting the potential of visual-interactive modelling—focusing on the human-computer interaction aspects of locational analysis and decision making.

The GRIA algorithm effectively implements the first strategy. It achieves the reduction of possible exchanges to be evaluated by breaking the algorithm into a global drop-add and a local re-location component, using specially designed data structures that consist of demand strings, candidate strings and an allocation table. The relative savings in the computational load even tend to increase with the problem size. Church and Sorenson (1996) considered GRIA to be just as robust as the Teitz and Bart algorithm and argued further, that the two algorithms were at that time the only ones sufficiently well studied and established to be implemented in a commercial GIS. Both algorithms were incorporated in the spatial analysis module of ArcInfo™. Rosing, ReVelle and Schilling (1999), on the other hand, argue that GRIA does not necessarily terminate at a (local or global) 1-opt optimum.

None of the algorithms mentioned above addresses in itself the risk of entrapment in local optima. Unfortunately, this risk tends to increase with problem size. Tabu search is a meta-heuristic for guiding known heuristics to escape local optimality (see, for example, Glover 1986). The basic principle is to build a dynamic memory of those parts of the solution space that have already been investigated. Interchanges previously evaluated become marked as illegal and moderate increases in the objective function are accepted in order to avoid these illegal solutions.

Another way to eliminate locally optimal solutions is simply to repeat an interchange heuristic several times with different starting solutions in the hope that a sufficiently large solution set contains the global optimum. The key question is how large the set must be. Church and Sorensen (1996) propose a dynamic, probabilistic stopping rule, based on the observation that a global optimum tends to occur more frequently than any of the local optima. If a best solution occurs a given number of times, t (three, for example, as proposed in the work cited above), the decision maker can be reasonably confident that it is a global optimum. The actual threshold, t, to be adopted will obviously depend on how 'reasonable confidence' is defined and on how many runs can be afforded within a specific decision-making context.

Heuristic concentration (HC), a two-stage meta-heuristic proposed by Rosing and ReVelle (1997), takes a similar approach. However, rather than searching and counting occurrences of entire solutions, HC assumes that a sufficiently large set of 'good' solutions generated by some heuristic in the first stage contains all the facility locations of a global solution. However, it does not need to contain the optimal solution itself. According to the number of occurrences across all solutions, facilities are selected to form a concentration set (CS). On the same basis, the CS is further subdivided into one subset of now fixed facilities—those deemed beyond doubt to occur in the global solution—and a second subset consisting of the remaining locations to be evaluated at stage two of the process. Stage two can be an exact algorithm (Rosing and ReVelle 1997) or a sequence of more accurate heuristics than the one used at stage one (Rosing, Revelle and Schilling 1999). It is certainly worthwhile to explore alternative ways of building the concentration set, for example through using a generalized network (aggregation) at stage one. Another interesting question is whether 'quick-and-dirty' runs at stage one, not guaranteeing the occurrence of all globally optimal facility locations in CS, still tend to produce a good global solution. In many practical applications such as the dynamic location of emergency vehicles, time constraints are mandatory but 'good' suboptimal solutions can be acceptable.

Another approach implementing the concepts of controlled randomness and repetitive runs is GRASP (Klincewicz 1991). GRASP performs large numbers of runs of the fast but otherwise inferior Greedy algorithm to build a solution set. The process is randomised as it arbitrarily selects one out of a number of best candidates to be added to the solution at each iteration. (The original Greedy selects the best candidate and, hence, would normally only produce identical solutions.) The resulting solution set could also be used to extract a concentration set for HC.

Simulated annealing is a thermodynamic analogy. It maps the original problem into a system model of a material crystallising with decreasing temperature and entropy, tending to assume a lowest possible energy state. Part of the analogy is more or less intuitive: the energy equation of the thermodynamic system represents the objective function, the current state of the system is analogous to the current solution, and the lowest-energy state subsequently is the global optimum. The process starts at some high-energy state then temperature is lowered in small steps, and, at each stage, interchanges are evaluated. Solutions decreasing the objective function are accepted and those increasing it may be accepted with a certain probability that is the equivalent of the Boltzman factor in the original annealing problem. One effect is that the search begins as coarse global scanning of the solution space and then becomes more local, detailed and accurate as temperature decreases thus performing a systematic, informed search. Another is the controlled randomness introduced through the equivalent of the Boltzman factor. A similar search strategy can be also achieved through dynamic adjustments of the parameters of a tabu search, though somehow lacking the elegance of an analogy to a known natural process. The key questions of a simulated annealing approach are how to define the equivalent of temperature and how to ascertain the appropriate values for the decrement and probability factor.

Another analogy-based global optimisation method that has received much attention in location analysis is that of evolutionary algorithms (see, for example, Krzanowski and Raper 2001). This approach involves the competitive selection and development of a population of solutions under the influence of appropriate implementations of evolutionary and genetic concepts such as fitness, mating, crossover, mutation, chromosome and gene. If the search approach involves genetic analogies in the strong sense, namely the concepts of genes, chromosomes, crossover and mutation, it should be referred to as a genetic algorithm. An implementation of evolutionary principles in a more general and abstract sense is called an evolutionary algorithm or program (Krzanowski and Raper 2001). Implementations of those principles in location analysis can vary greatly, and it is beyond the scope of this chapter to give an in-depth discussion of those variants. The interested reader may, for example, refer to Delahaye (2001) who presents a case study of an evolutionary model of an airspace control sectoring problem.

Discussion of the concepts of task division and human-computer interactions has thus far been omitted in this chapter. Both actually represent another metalayer above the concepts that have been discussed until now. Subtask design has a vertical and a horizontal component. Vertical, or hierarchical task division aims at freeing the actual optimisation process from unnecessary burdens. For example, can and should route-finding and cost-distance calculation be carried out beforehand to provide an input for the search algorithm? In a dynamic system like ambulance relocation, these tasks can then run as a separate background process whenever a change to the input data occurs. Horizontal task division has as an objective to define independent components of the same task, which then can run on separate CPUs or nodes of a computer network. A discussion of implementation strategies can be found in Densham (1996). Human-computer interaction and visual-interactive location modelling follow a somewhat related principle (see Tobón and Haklay, this volume). Why should the computer resolve problem components that the user (with his or her superior cognitive power and the capability intuitively to comprehend complex problems) can resolve more efficiently and effectively? To illustrate this, one may conceptualise the problem of modelling the impact of a road accident and subsequent congestion at a major road crossing. One option is to employ sophisticated transport modelling techniques to calculate the impact on the network neighbourhood, likely to be based on a number of assumptions that are estimated user inputs anyway. The alternative would be that the user draws a freehand circle around the accident spot and applies some multiplier to the original link travel times. Arguably, the result can be just as good. A less tangible and more psychological effect is that decision makers tend to trust more in solutions found by a process they comprehend and in which they were actively involved.

5 Current research at CASA

A current research project at CASA investigates dynamic spatiotemporal location strategies for emergency vehicles, specifically ambulances. This research is building on the framework for location-allocation modelling set out in this chapter in the following important respects. A static distribution of emergency vehicles within a service area is in most cases inadequate, leading to suboptimal response times and inefficient use of resources. On the other hand, even marginally faster arrival at an incident can be vital for a patient's survival chances. A dynamic approach therefore aims to re-locate ambulances according to changes in the state of the system, which is given by the spatial distribution of demand probability, the road traffic situation and the current utilisation of the vehicle fleet. Ambulances in stand-by mode do not need to be located at a designated base station, but can be positioned anywhere within the service area. The only requirement is that each vehicle's position be known to the dispatch centre, and that the dispatcher has the technical means to rate its suitability for serving an emergency call to a given site. The permanent and partly unpredictable change of the state of the system in combination with a large and dense road network and the necessity of fast decisions constitute the actual challenge in providing spatial decision support.

One strategy investigated in this project is the implementation of the concept of model base management systems (MBMS). Rather than relying on a single static and monolithic location-allocation algorithm, this approach combines atomic elements and operations of known heuristics, stored and managed through the MBMS, to a bespoke meta-algorithm at run time. Such a bespoke algorithm can be expected to meet specific requirements of a given decision problem more accurately and, therefore, to perform faster. Often for example, the constraint on processing time is mandatory while, if necessary, a suboptimal solution may be acceptable (as long as it is a feasible and good—not markedly suboptimal—solution). Subsequently, the available time for the analysis is likely to determine the shape of the meta-algorithm. Part of the MBMS concept is that algorithms, once assembled, can be stored and efficiently retrieved when needed.

A second strategy, building specifically upon the advanced data storage and manipulation capabilities of GIS software, is an extended use of predefined and preprocessed scenarios. Most of the significant and system wide changes of the underlying data are, at least to some extent, predictable from empirical data and experience. For example, a rush hour situation that simultaneously affects incident probabilities, road traffic and fleet utilisation can be anticipated to recur in similar form on each working day at given times. The objective is to provide scenarios for any conceivable system state, thus reducing the need for real-time analysis as far as possible. This logically leads to the notion of a scenario base. The concept includes global as well as local scenarios. An example of a local scenario would be the impact of a major sports event. Individual scenarios can be combined into meta-scenarios, which again can be stored and retrieved on demand. The efficient management of the wide range of scenarios that can be created using GIS remains a challenge for human-computer interaction studies (see Tobón and Haklay, this volume).

A third strategy to be exploited within this project is intelligent task division and subtask design. One more or less obvious division is the separation of one process (and display), representing the current system state and supporting the actual dispatching and routing decisions, from another hosting the actual location-allocation optimisation. In practice, this most likely involves at least two separate workplaces and operators. The two main processes are interlinked in both directions. Dispatching and routing decisions are to be made in such a way that causing imbalances in the service provision is avoided (as long as alternatives exist), in awareness of not only the current but also the anticipated future state of the system. The closest vehicle is not necessarily always the best choice. On the other hand, the locational optimisation process requires constant knowledge of the current state. The optimisation process can be further subdivided into route re-calculation, scenario search and (possibly multiple parallel) runs of location-allocation algorithms. Route re-calculation is performed in order to supply the optimisation algorithm with precalculated internode travel times so that the latter can perform efficiently. Another aspect of route calculation, which however is beyond the current scope definition of the project, is that ad-hoc updates of the road traffic situation become available as samples rather than as comprehensive updates for the entire network. An approximation of the latter can only be deductively derived. Scenario search has as an objective to match the known current system state as closely as possible by a preprocessed scenario with a known location-allocation solution, assuming that this search is more efficient than seeking a solution from scratch.

A final strategy, pertinent to any decision support system and building upon a particular strength of GIS software as a core component of the SDSS, is to focus on efficient and effective human-computer interaction (Tobón and Haklay, this volume). Although user interface design is certainly not a key issue at a prototyping stage, emphasis on interactive decision making will remain a guiding principle.

The initial SDSS design approach is one of tightly coupled components using GisLink to add the required analysis functionality to the GIS. GisLink is a set of methods in Cadcorp SIS that make use of the Microsoft Windows® messaging system to allow Microsoft Visual Basic programs to communicate with SIS applications. GisLink is designed to facilitate a later transition to an embedded design using the Cadcorp SIS ActiveX control. At the time of writing, the work concentrates on the MBMS, which initially will comprise the Teitz and Bart algorithm, GRIA and an implementation of simulated annealing. As the project progresses, more information will become available online under casoco.casa.ucl.ac.uk/~torsten, where a detailed discussion of the EMS decision domain can also be found.

Acknowledgements

This research is sponsored by ESRC CASE Award S42200134072 (industrial sponsor: Cadcorp Ltd., Stevenage, United Kingdom).

Section V

GIS and community participation

A GIS is no better than the end use to which it is put, and users need to feel confident that GIS is efficient, effective and safe to use. Andrew Hudson-Smith, Steve Evans, Michael Batty and Susan Batty describe an innovative project in which residents of a large public-housing estate are able to use Internet-enabled GIS not only to visualise how their living environment is likely to change in the future, but also to question envisaged changes before they happen. In a wider sense, such use presumes that users are able effectively to interact with GIS, and in this context Carolina Tobón and Muki Haklay explore lay usage of GIS here. In both of these chapters, users are encouraged to experiment with data and to investigate scenarios. This theme is further developed by Theodore Zamenopoulos and Katerina Alexiou, who discuss our ability to marshal distributed knowledge sources in order to develop coordinated decisions through cooperative decision making.

Chapter 18

Experiments in Web-based PPGIS: multimedia in urban regeneration

Andrew Hudson-Smith, Stephen Evans, Michael Batty and Susan Batty

The Internet and World Wide Web are generating radical changes in the way we are able to communicate. Our ability to engage communities and individuals in designing their environment is also beginning to change as new digital media provide ways through which individuals and groups can interact with planners and politicians to explore their future. This chapter tells the story of how the residents of one of the most disadvantaged communities in Britain—the Woodberry Down Estate in the London borough of Hackney—have begun to use an online system that delivers everything from routine services about their housing to ideas about options for their future.

1 Introduction

Because GIS provide a visual medium in which two- and increasingly three-dimensional spatial information can be communicated on the desktop but also across the Web, there is a growing attraction to using GIS for public participation (see also Tobón and Haklay, this volume). PPGIS, as this movement has come to be called, is a natural consequence of the ubiquity of GIS and our ability to communicate maps and models in an intelligible visual form to those who have no expertise (or interest for that matter) in its technical basis (Kingston 2002). In this chapter, we illustrate how we have developed various aspects of this visualisation in an exciting and progressive example of public participation that is central to the process of urban regeneration in British inner cities (Rydin 1999). Our case study involves a 20-year programme of regeneration of housing in Woodberry Down, an area of deprived social housing in the London borough of Hackney, where public participation is regarded as a permanent part of the process of redevelopment. This is giving us a chance to develop some of the new technologies reported elsewhere in this book which will make PPGIS a reality (Evans and Hudson-Smith this volume; Tobón and Haklay this volume).

We begin with a brief review of new digital media and the way they are influencing different forms of communication. The Woodberry Down project represents one such form of communication, and we first sketch the background to the area, illustrating the critical problems of deprivation that dominate the regeneration that is taking place. We then provide a blow-by-blow account of how the online system we built has emerged. We detail the stop and go nature of the funding, the problems of ensuring that the residents and their representatives engage in its use, the way the community has been wired to ensure this, and the kinds of technical detail that are central to making sure the system delivers information in a robust and timely manner. One of the central issues that we bring out in our discussion is how such efforts help to push the frontier of what we are able to do technically, as well as identifying the problems of learning how best to exploit new technologies that are sustainable (Bullard 2000).

2 The new media for participation

Participation in planning requires information that is strongly dominated by visual media in the form of maps and pictures with text an important subset of such data. For online participation visual interfaces are essential, along with hardware in the form of computers and networks. Computers need to be powerful enough to process pictorial information, and networks need to have enough capacity to enable users to communicate quickly. This is the real bottleneck in using current systems. Most of those to whom online participation is geared do not have network access other than through traditional phone lines, and this limits the speed at which they can receive and transmit visual information. It is absolutely critical to develop systems with media that can be delivered quickly and successfully, and much of our own work in Woodberry Down is focused on developing systems that are workable and robust in these terms. Most systems developed to date are essentially passive in that information is delivered in one direction only—from the server to the client—with any interactivity on the client's part, the user, simply geared to choosing what information needs to be delivered. Acting on such information and offering feedback to informants is still quite rare, but, as we shall see it is essential if online participation is to move beyond a digital version of simply telling those affected what is planned (Laurini 2001).

Software, of course, is the key. Good software developed by those who serve information can turn that information into pictures and words that communicate the essence of an issue in the most effective way. Anything more than this depends on setting up the communication so that users can act on the information. Perhaps the easiest way is to provide users with data that they might manipulate themselves. For example, there are now many Web sites that deliver numerical data on planning issues that users can store locally and examine and manipulate offline at their leisure. The London borough of Wandsworth's planning applications site, where users can examine recent applications and decisions and map the data, is a relatively passive form of information delivery (www.wandsworth.gov.uk/gis/map/mapstart.aspx), although users do need to be able to interpret the meaning of such data (Tobón and Haklay, this volume).

The biggest problems occur when effective participation requires animated graphics, often more than maps, representing built environments through which users can move and fly. These kinds of virtual reality often require the user to decide what and where to navigate and in this sense are truly interactive. Web browsers have long been configured for such navigation, but there are still severe problems in delivering such information over standard phone lines. The dilemma, of course, is that the best information about a planning scheme usually requires such representations. Much of our technical work with Woodberry Down involves using and adapting software so that rapid fly-throughs and related manipulation of visual data are possible over the slowest and lowest capacity networks. An example of this kind of media is on CASA's Online Planning site at www.casa.ucl.ac.uk/online.htm. Recently there has been considerable interest in the use of GIS software to enable participation and in particular desktop GIS has been adapted to Web-based processing. Internet map servers deliver information processed on a central server to a client who activates the kinds of functions that require some knowledge of the problem. A good example is our Sustainable Town Centres site. This allows users to take layers of data that indicate various indicators of sustainability and to weight and combine these to produce an overall index used to rank-order different centres by the degree to which they are sustainable (see Thurstain-Goodwin this volume for an overview of the approach; and Evans and Steadman, this volume for a different project that is concerned with creating sustainability indicators). This online tool really moves beyond the realm of participation, for it is useful only to those who have an expert interest in town centres that requires, in turn, some expertise about the socio-economic functioning of land use and transport in cities (see www.casa.ucl.ac.uk/newtowns/index.html).

Truly interactive participation usually requires users at each end of the process acting in concert. Bulletin boards and their graphical equivalent in terms of white boards act in this way but require active responses and some commonality of interest to make the system function. As we will see in Woodberry Down, the bulletin board capability can be severely compromised if those who set it up are limited in responding to users by legal and other restrictions. Slightly shorter and sharper interactive responses between users at the server and client ends are contained in group decision making, in networked design studios, and in Internet systems that actively involve users in community design. For example, the Architecture Foundation www.architecturefoundation.org.uk) has developed a toolkit for engaging the community in urban design. The toolkit began as a passive Web resource where users follow the design process (see www.creativespaces.org.uk) and has developed into a much more interactive resource—The Glass House—which enables users to interact with various design options (see www.theglasshouse.org.uk). The Glass House uses state-of-the-art visualisation technologies that can be delivered quickly and at very low cost across the Web and has been developed by us in parallel with our Woodberry Down project.

There are many decision-making procedures usually fashioned for experts involving Internet-based communications. These have been developed on local area networks and are gradually being ported to the Internet. Invariably these involve some form of structured problem solving supported by various models and databases (Jankowski and Nyerges 2001), but, in general, such decision support systems are not suitable for the kinds of participation that we are involved in here. Of much more relevance are totally interactive systems in which there is no assumed hierarchy of users. Chat rooms and related forums are primitive examples of such communications, but where such interactive modes come into their own is through the idea of virtual worlds. Examples of these for design are rare, as the very notion of developing such visual representations where users can appear as avatars is highly exploratory, notwithstanding some notable examples which show considerable promise for enhanced methods of participation (Schroeder, Huxor and Smith 2001; Smith 2001).

3 Hackney and Woodberry Down: deprivation and regeneration

During the last decade, British local government has been dominated by problems of public (social) housing by and large created by those same governments two or more generations ago. The slum clearance programme and the re-housing of a very large proportion of the British population began in earnest in the 1950s, and many inner cities came to be dominated by high-rise dwellings under municipal control, built to rather poor standards and housing an increasingly deprived population. The run-down condition of this housing, primarily attributable to poor maintenance, has been exacerbated by the migration of the most active and able residents into owner-occupied homes elsewhere. Increased spatial polarisation of social groups has accompanied the erosion of the U.K. welfare state during the last 20 years. There has been a range of attempts to arrest the spiralling decline of many inner city areas, such as Woodberry Down, using a bewildering range of regeneration initiatives. Most of these initiatives involve frighteningly complicated sets of policies and instruments (Power 1998) and often require the financial underpinning of variants of the Private Finance Initiative, in which the private sector is encouraged to provide funds in return for long-term ownership of what has hitherto been public property.

There are 1370 housing estates in England that have been defined as 'deprived' and 112 of these or 8 per cent are located in Hackney, which is one of the poorest London boroughs. The best way of illustrating the environmental context is through the index of multiple deprivation (IMD), which is based on income, employment, health, education, housing and access, with child poverty identified as a critical subset of the income indicator. These six indicators are weighted as 25-25-15-15-10-10 and then aggregated to form the overall IMD. When mapped, they provide a picture of the relative geographical concentration of key problems

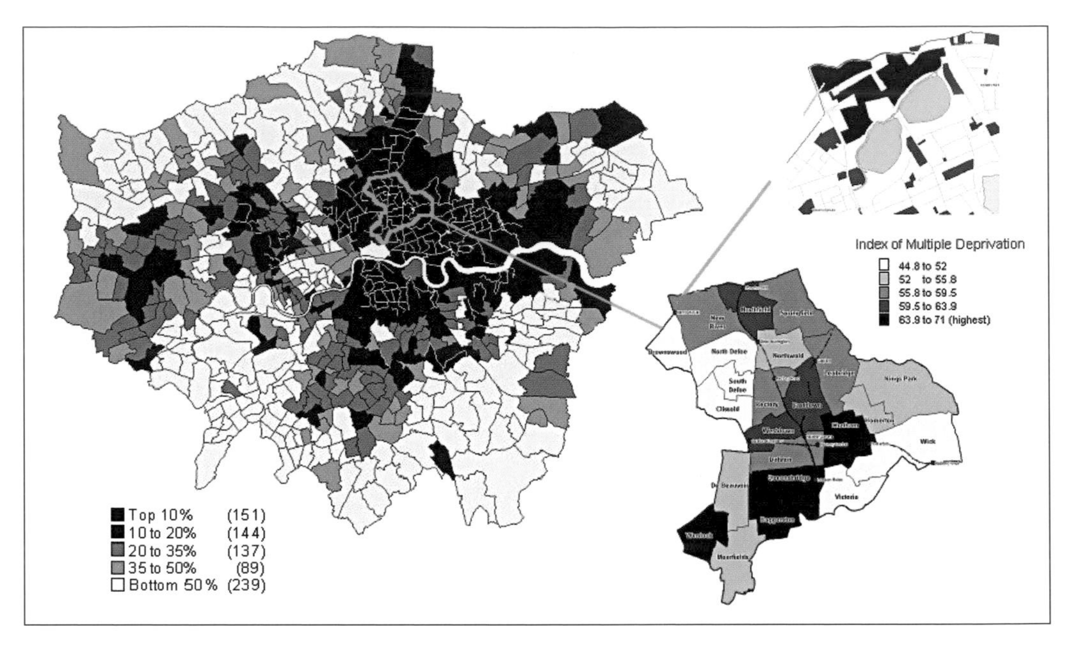

Figure 1 Deprivation in London, Hackney and the Woodberry Down Estates. The ward in which Woodberry Down is located is in the top 3 per cent of the most deprived areas in England and Wales, as measured by the 2000 Index of Multiple Deprivation

and problem estates in the country. Hackney is one of 33 boroughs in London, and in 2001 had a population of around 207 000. Forty per cent of its population belongs to ethnic minorities and 60 per cent of its housing is in the public or ex-public sector. As a municipality, Hackney is the second most deprived local government area in England, and it has the largest concentration of deprived estates in the land. All 23 of its wards are in the most deprived 10 per cent of all wards in England (where there are 8414 in total), 9 of these are in the top 3 per cent and the ward in which the Woodberry Down estates are located is one of these. The pattern of deprivation is shown for Greater London, for Hackney and for its estates in figure 1. In fact, the various housing blocks that make up Woodberry Down do not contain the most deprived households in the borough, but in terms of the housing indicator within the IMD, this is in the top half of 1 per cent of the worst housing conditions in England.

The Woodberry Down estates are in the Woodberry Down and Stamford Hill Single Regeneration Budget (SRB) area, and the renewal projects being financed from this source of funds are bid for competitively each year. In the wards that cover this area, more than 50 per cent of all households reside in public housing: if the stock that has been sold off is added to this, then it is clear that the area is

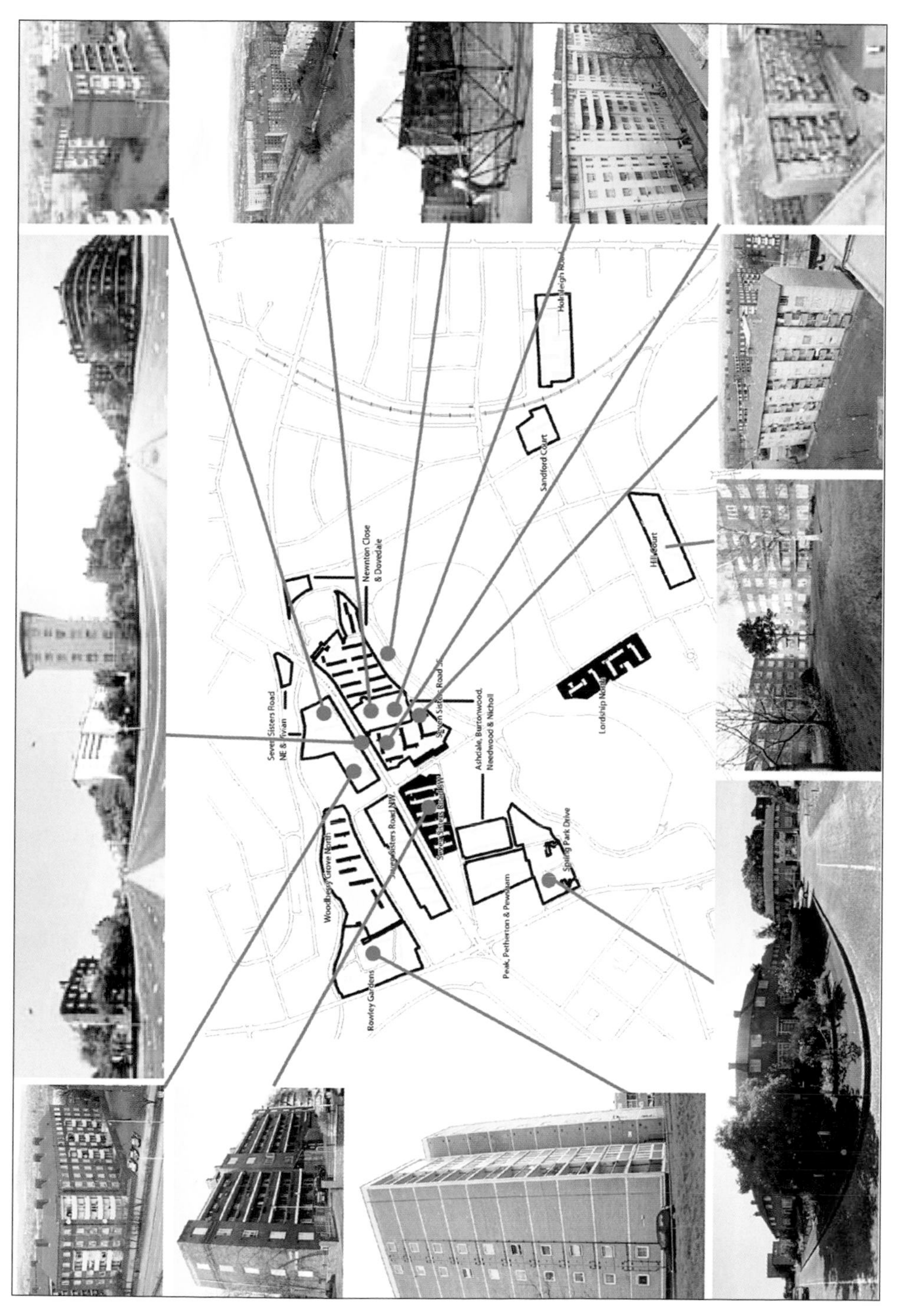

Figure 2 The estates that make up Woodberry Down

dominated by estates that are likely to be in need of considerable regeneration activity. We do not intend to develop an exhaustive analysis of the demographic profiles of the population, for it is clear enough that the populations housed in these areas lack basic amenities. The estates, in fact, tend to be residual sinks for the worst-off and for immigrants rather than being dominated by long-standing, ageing residents. There are problems associated with ageing, of course, but the principal problem centres upon poor housing conditions. To provide a quick visual impression of the kind of housing that we are dealing with, we show a collage of views around the 25 blocks that make up the estate in figure 2. Like so many illustrations, the real sense of how run down the area has become is hard to imagine from these photographs, although these pictures do capture a degree of the desolation of the environment.

The area to be regenerated is comprised of the estates shown in figure 2 and physically crosses various administrative and historically integrated, ethnic neighbourhoods. The first housing blocks were developed in the late 1940s by Forshaw as part of his and Abercrombie's vision for London. The form of the blocks is inspired by the Bauhaus, even appearing a couple of years ago in the film *Schindler's List*. The oldest blocks are listed. There are around 6000 residents in 2500 housing units of which some 29 per cent are owner occupied. The Woodberry Down Regeneration Team (hereafter called WDRT) has divided the locality into 18 distinct geographical areas, although for purposes of resident consultation, these are currently aggregated into 14. There is considerable confusion with respect to tying the official statistics, noted above, to what actually happens on the ground, and local surveys reveal that in these estates, the white population comprises just less than 40 per cent of the population, and that there are predominant black and Turkish populations. These estates permeate the area of Stamford Hill, the largest concentration of orthodox Jewish population in the United Kingdom.

The Woodberry Down project began in 2000 with the establishment of an on-site team and the beginning of negotiations for a Single Regeneration Budget proposal for some £25 million, which was subsequently successful. Currently much of the project is dominated by the negotiation of a Private Finance Initiative to fund the lion's share of the cost, which is estimated at some £160 million over 10 years. However, the project did not get off to a good start. The WDRT was located on-site in public offices that had hitherto been a local library, and the change of use to the team's HQ/centre generated considerable hostility amongst the local population. The team (WDRT 2001a) reported: 'Local residents are still angry that not only was their library taken away but also that the centre is, to many of them, not providing any tangible benefit or service to the estate. The WDRT believes that this is not because of the fault of the resident managers but due to the conception and delivery of this project' (page 11). In fact, what this issue reveals is that there is already substantial community participation and representation in the area that the entire project is attempting to draw on in managing the regeneration.

In the area, there are nine tenants' and residents' associations with another two in the process of forming. There are six estates' committees serviced by Hackney Council and these meet quarterly. The Stamford Hill Neighbourhood Committee meets nine times a year and is attended by council officers and local councillors. The council's housing stock in Stamford Hill is managed by the Paddington Churches Housing Association, and there is a monthly tenants' panel that discusses management issues. The Estates Development Committee (EDC), set up to represent the regeneration of the estate, cuts across these. It currently has 27 members whose role is to liaise with the WDRT and to represent the views of those affected by the considerable disruption that will occur as the regeneration gets under way. The process of online participation has been both motivated and endorsed by the EDC and the WDRT, and the Web site reflects the close involvement of this committee.

The WDRT have spelt out very laudable and ambitious aims for the project on behalf of the council and the community (WDRT 2001b). These involve conscious bottom-up consultation and involvement at all levels. In the first year, 12 community meetings were held, with all households being contacted and the various meetings formally involving over 20 per cent of the residents in the area. WDRT's vision includes a strong wish to generate a sustainable estate development in which social and ecological threats are minimised. There is also the explicit goal of taking account of new technologies to deliver services and involve the population. These goals are consistent with the national government's 'modernising Britain' campaign. The worrisome aspect of the project, like most such initiatives in Britain at present, is that it is beset by different kinds of financial bargaining. These continually threaten the scheme by throwing it off course in terms of timing and diverting valuable resources to open-ended and inconclusive debates about showing 'best value for money'. We are currently three years in, £25 million has been committed, £135 million has still to be negotiated and signed off, designs have still to be prepared, and there is nothing to show for any of this on the ground where it counts. We believe that the Web resources we have developed go at least a little way to pushing what is clearly a tortuous process forward, and to these we now turn.

4 Development of the Web resources

The decision to develop an online method for participation emerged in early 2000 from a series of related projects that involved projects in Hackney. The catalyst in many ways was the Hackney Building Exploratory, a community-based initiative that enables local communities to learn about their local environment and to participate in ideas about making it more liveable. The Exploratory is located in an old school within the borough and is full of fascinating models and maps of the community, built professionally from standard materials, as well as informally by children and adults as part of their educational visits. In 1999, we began to develop a series of digital exhibits to complement the material exhibits. These exhibits have enabled local residents to examine planning information using the latest ways of visualising development by gaining access to this media across the Internet. This led to the direct development of educational software that let visitors to the Exploratory find out about the local community using GIS, digital panoramas of street scenes within Hackney, different types of housing within the borough, and patterns of deprivation and disadvantage within the East End of London (Batty and Smith 2001). Computers were located in the Exploratory and were an instant hit with children who form a very large proportion of visitors. A Web-based version is available at www.casa.ucl.ac.uk/hackney.

The Exploratory was also involved with the Architecture Foundation, a charitable trust devoted to promulgating good architectural design which has a strong community influence supported by the leading architects and planners in the United Kingdom. It also had good contacts with Hackney borough whose GIS team was actively seeking ways of extending the relevance of their work through other digital media such as 3-D visualisations. Moreover, the Architecture Foundation was organising the British entries to Europan 6, a competition for young architects of which an entry based on one of the sites in Woodberry Down was chosen. In mid-2000, the Architecture Foundation and the Exploratory also began to explore funding for a wider London-based project involving online community design, and we were involved in proposing various extensions to projects that we had already developed as a basis for this. As part of these proposals, the development of online Web-based resources for public participation in Woodberry Down emerged, where the crucial issue was the development of multimedia content in sufficiently intelligible form for residents to make use of the resources in thinking about future design options for the community.

The WDRT decided to fund the project in late 2000 after they became convinced that we had fast enough multimedia methods to deliver visual content to the site. We had been perfecting these techniques using an area of central London around the BT Tower, adjacent to our offices, and in an application for British Nuclear Fuels at Dounreay where they were mothballing part of their reactor. The Architecture Foundation act as brokers for the project and, in early 2001, the WDRT laid out preliminary ideas for the structure of the Web site. A rough draft of the site was made, and we then began to meet with residents to illustrate what might be done and to test the extent to which the medium that we were proposing was acceptable to individuals who had only the most basic IT skills.

The structure of the site is divided into four different areas. First, there is information, mainly in the form of text, about the process of regeneration and this occupies at least half the site. Data in the form of reports can be downloaded from this area, but the main focus is on explaining what is happening with the regeneration process and informing the residents about how the developing situation will affect their own homes. The second area is mappable information supported by panoramas; this is currently exploratory in intent, but eventually will be used for residents to get some feel about what the future of the area might hold. This material makes use of fast multimedia and currently portrays the area as it stands. Residents may not get much from this as they already know their area, although readers of this chapter will find it very useful; it illustrates what is possible with respect to seeing the area physically as communicated over low-capacity bandwidths. The other two parts of the site are much more interactive. The third part is a bulletin board of fairly standard form that enables anyone who is registered to post comments. The fourth is the most experimental and currently shows how different physical options for the future can be viewed in 3-D. The four options currently available can be manipulated to show how the existing form of the site can be changed. The manipulation involves only zoom, pan and move capabilities so far, but once developed, will allow residents to engage in their own designs and post their schemes to the WDRT and other groups who have an interest in the future of the estate.

Initially our idea was to use an Internet map server to deliver maps online that residents could query. However, now and then, it has not really been possible to use typical map servers for the kind of purpose needed here. Residents do not want to query a map but they do want to see visual information in 2-D and 3-D very quickly. They need to be able to do this over standard telephone lines. Thus although ESRI (United Kingdom) donated a copy of ArcIMS map server, we quickly moved to much faster and simpler media, developing and using freeware/shareware based on various software products developed by Viewpoint (www.viewpoint.com). In fact, the development of the Web site and the testing of different media in hands-on form with the EDC represented one of the high points of a series of resident meetings which, for reasons not directly related to our project, were rather acrimonious. The principal motivation for the residents' meetings was to deal with more pragmatic issues involving the process of regeneration but they were also used to demonstrate our virtual site.

A particularly innovative feature of this project was the decision to directly engage the resident representatives in the EDC in the design of the Web site. As part of the overall funding, monies were set aside to purchase enough computers and Internet access to put all representatives online with access located in their own dwellings. The representatives agreed to use their access to engage their wider community in the participation process, yet this decision was rooted in problems. The notion of a public authority providing residents with free computers, the fact that their usage could not be controlled, the requirement that representatives would engage those who they represented in their own homes—all these were highly controversial and debatable issues. The notion that if representatives did not use their computer, they would be taken from them also presented difficult issues. As a result, the computers, once purchased, remained in a warehouse for six months before the council agreed to their release. To an extent, the idea that homes would be wired when those very homes would then be demolished or refurbished went against the grain. Yet it raises a far-reaching issue—that to replace physical infrastructure one may need to add to that infrastructure before the replacement takes place.

The database was constructed over a four-week period in the early spring of 2001. A massive number of panoramas were photographed at roof and ground level, and these form the various visual sequences that are embedded in the Web site. Zoomview and related products from Viewpoint were used for fast animation, zooming and panning of the aerial photographic coverage of the estate, which is used as the basic locational referent. Essentially these products generate views using a data-streaming technique called 'pixels on demand' in which a scene is divided into a large number of small pieces, with each piece delivered being dependent on the pan and zoom within the given window that is selected. The scene is quickly refreshed to produce the greatest detail but the user has a clear idea of what the overall scene looks like while this process is going on. The Viewpoint Media Player (VMP) is required, but this is now common on many machines and comes bundled, for example, with America Online. It can be downloaded over a standard phone line in a couple of minutes, and the request to do so is always activated when a Viewpoint scene is generated. The software also enables IT designers to layer information and to link the scene to other Web-based software such as Flash. We decided very early on that VRML would generate 3-D file sizes far in excess of what might be handled by a basic user, and thus the focus of software is no more elaborate than the fast graphics that can be read by VMP. In fact, to develop the site we had to collaborate with Viewpoint who were quite literally writing elements of their software while we were using it (see Evans and Smith, this volume).

The initial site design was meant to run until April 2001 but because the graphics design team that produced a first draft of the site and the Woodberry Down logo were slow to start, a working prototype was not available until the summer. Four versions were developed during these months and this entailed detailed collaboration between WDRT and the Architecture Foundation over many issues. Moreover, involving the residents was painful at times. For example, although the EDC are central to the design of the Web site, representatives wanted their addresses to remain anonymous in case they were identified as having free computers, and their homes were then burgled. This fear might seem fanciful to readers in other countries but in this part of East London, it is well-grounded. The site was finally launched in November 2001 along with the exhibition of designs submitted as part of the Europan 6 competition.

5 The structure of online participation at Woodberry Down

As we implied earlier, many online resources for participation are one way; that is, user interaction is passive, based rarely on anything more than e-mail questionnaire and comment forms. However, in Woodberry Down, interactivity—two-way communication between providers and users as well as between users themselves—is central to participation, and the Web site is thus configured to contain various comment forms, bulletin boards, animations, fly-throughs and pictorial manipulations. As the Web site is continually under development and will evolve with the process of participation and the schedule of regeneration, we will soon add sketching facilities, as well as policy forums for online debate. The structure we have designed is strongly orientated towards low-level but

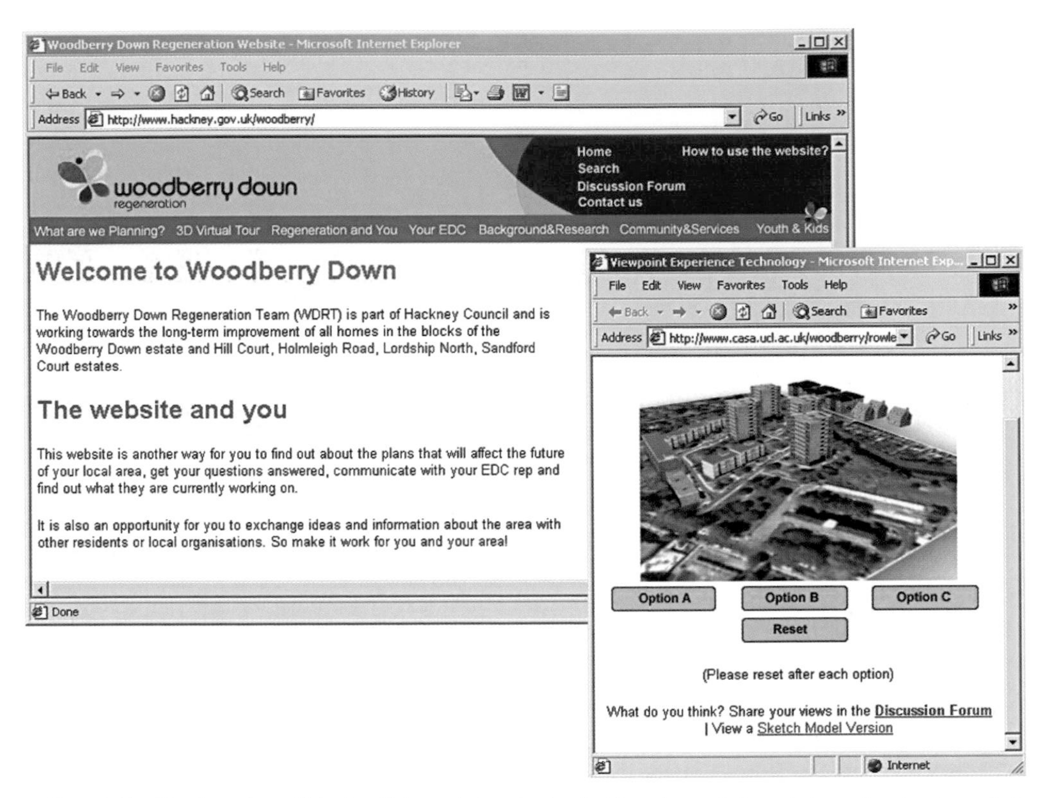

Figure 3 The Woodberry Down Web site with inset window showing the Viewpoint media for exploring various housing options

comprehensive interaction, is geared to online discussion and has a clear focus on community design. Professional experts and the community are the target users and providers, although political representatives are also likely to feature in its use.

The Web site has a particularly simple organisation. Essentially there are four main types of information: textual information about the entire process of regeneration and the site itself, services, and related facilities; multimedia as maps and panoramas about the various component housing blocks which make up the estates; design options reflecting the kinds of designs that might be developed for the site; and a discussion forum which enables users to interact with the WDRT concerning any aspect of the regeneration process. Textual data forms the vast majority of information that the site is able to deliver and this is accessed as pages through various drop-down menus accessible from the home page. These menus cover seven topics: What We Are Planning, 3-D Virtual Tour, Regeneration and You, Your EDC, Background and Research, Community and Services, and Youth and Kids. We show a version of the home page in figure 3 containing the design option that we will now examine, as an inset, and which enables any user to access the home page of the site while also opening up other windows from the site itself.

What We Are Planning gives access to four pages: the vision for the future; the partnership that will enable the site to be developed through various private finance initiatives yet to be chosen; the first stage of the works with access to the decant status of the various housing blocks; and the planning brief. The process of regeneration is plagued by esoteric terminology and acronyms and under the menu associated with **Regeneration and You**, there is a section of frequently asked questions (with answers) and a jargon buster that defines the various terms used by officials, such as 'basic credit approval'. There are links to the decant status page and to housing advice via links to other housing agencies from associated pages. Under **Community and Services** there are links to housing management advice and local services, all leading to their own pages. There is also a section here that lets users provide the WDRT with information about local events. **Background and Research** provides a brief history of the area as well as key documents, referred to as Yellow Books, about the regeneration that can be downloaded as Acrobat® PDF files. The use of PDF files illustrates the sorts of problems that we have had to grapple with: PDF readers are free, whereas documents set up in Microsoft Word require the appropriate software, which is not free. Yet PDF is a much less-intelligible format for the average user.

Extensive information about the EDC is accessible from the **Your EDC** menu, which gives information about the constitution of the committee, how often it meets, what it does and its local representation. Pages dealing with **Youth and Kids** are under construction and currently simply display graffiti and such like in the environment. As the site is under active development, visual information about the existing site and future plans are contained under the **3-D Virtual Tour** menu which lets users select from 104 blocks, load pannable and zoomable aerial photographic maps, and thence select digital panoramas of different parts of the site thereby giving a feel for what the place is like now. The 3-D Virtual Tour uses the Viewpoint media introduced in the previous section. When the user zooms in on an area of the map, a panorama is loaded and, using a sequence of point and click, this panorama can be opened from a spherical window, and the user can get some sense of the physical conditions and space of the housing in that area. Currently this facility is, as implied, a tour in that it simply illustrates what is possible, but in time we intend to integrate this into the sketch-planning capability that we

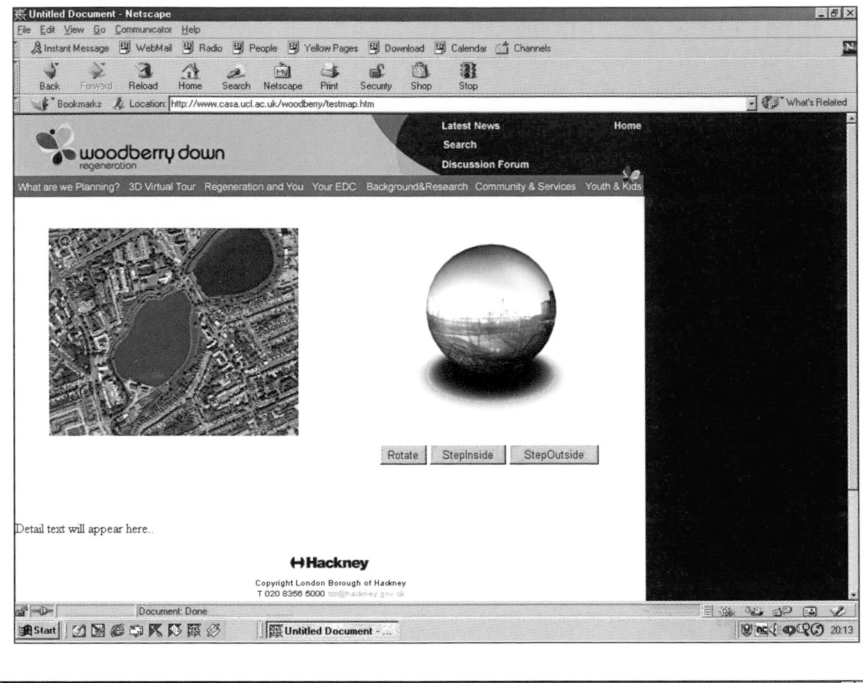

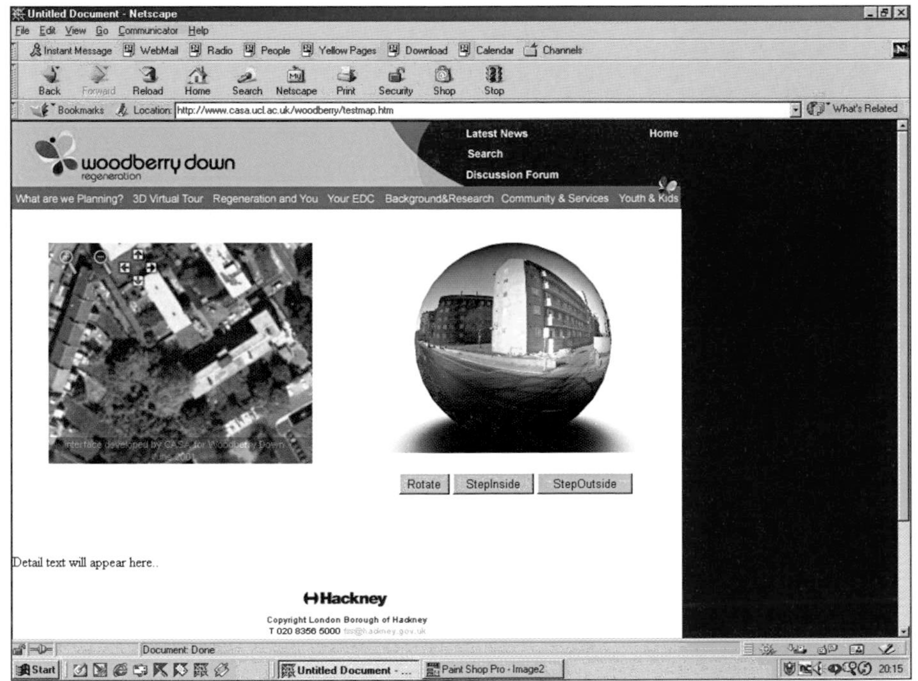

Figure 4 Accessing different housing areas, loading panoramas and moving about

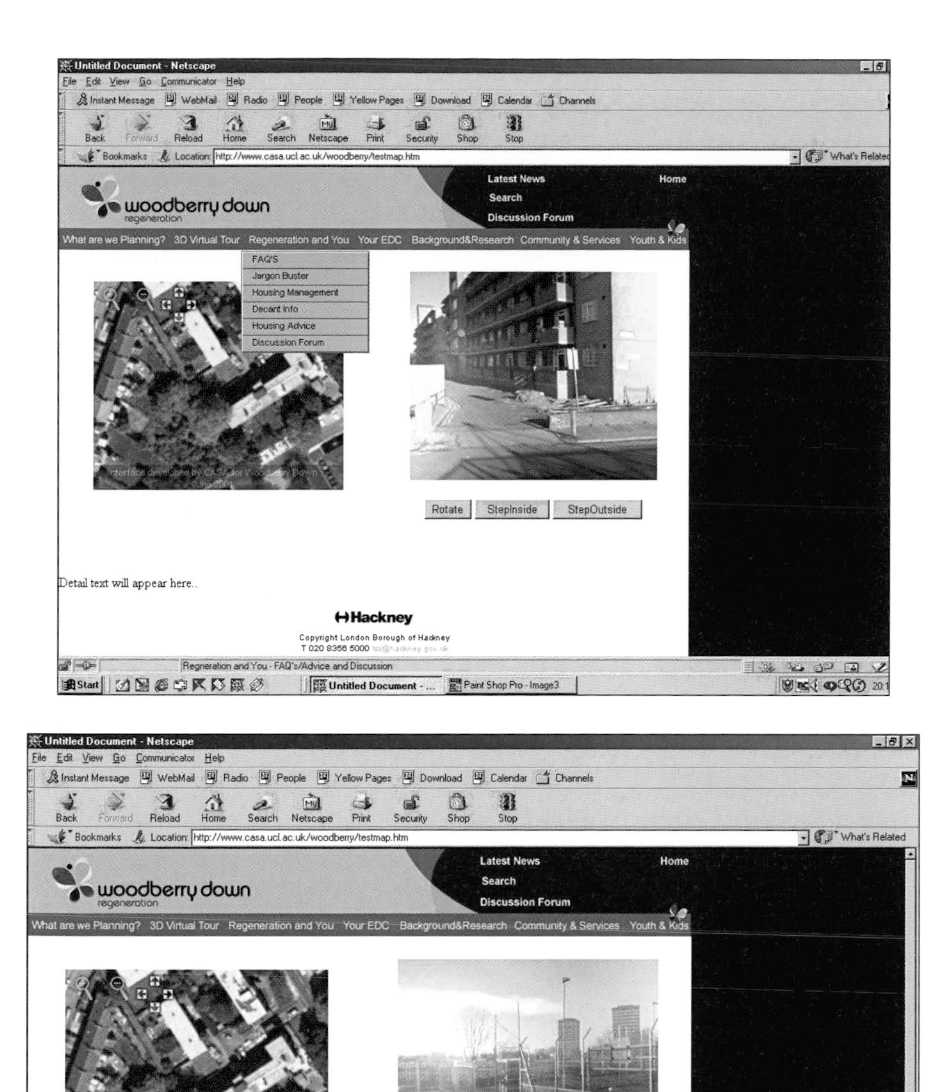

Figure 4 (cont.) Accessing different housing areas, loading panoramas and moving about

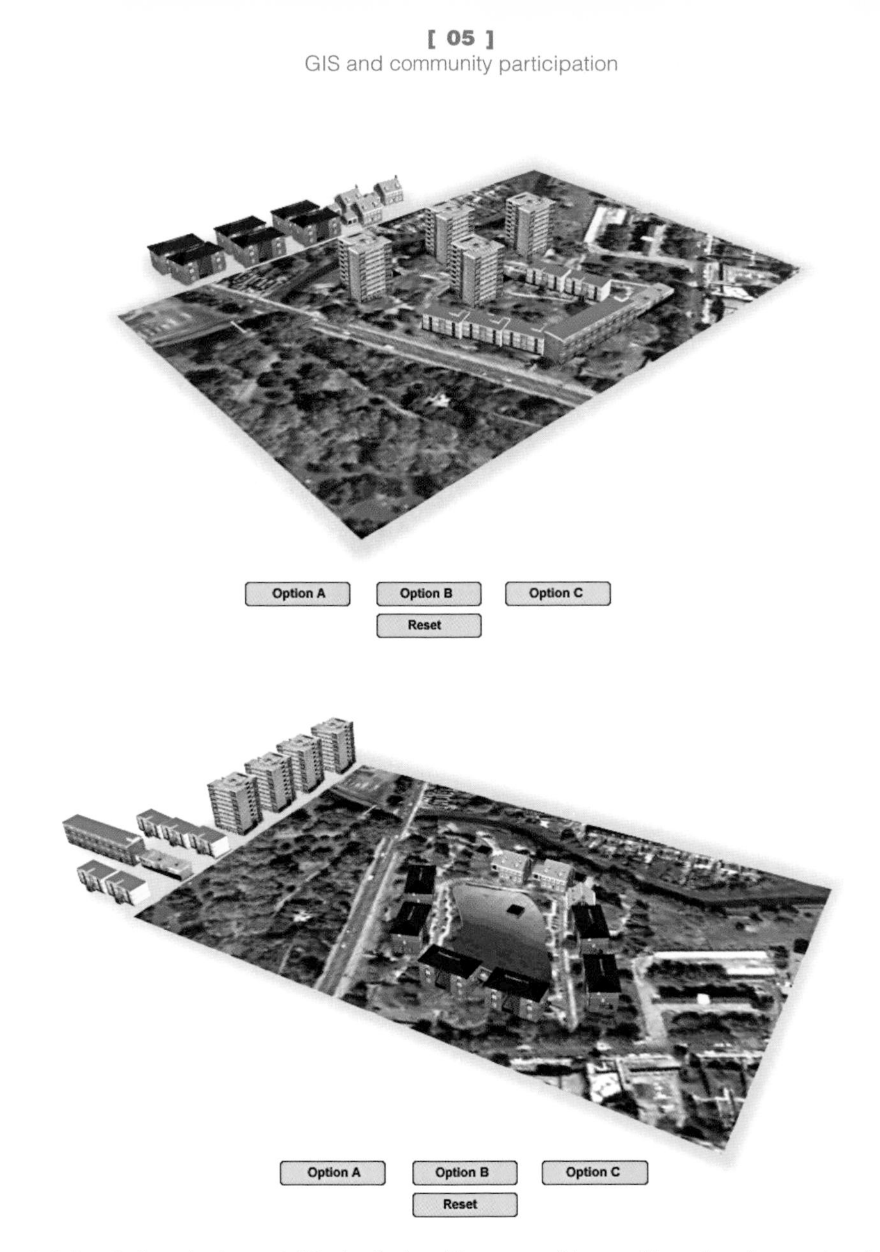

Figure 5 Options for the redevelopment of Rowley Gardens. When a user clicks one of the options, the current configuration of housing at A above moves to the side of the map and new housing options automatically assemble themselves in B, C and D

are developing in another area of the site. In figure 4, we show typical examples of the visual panoramas and zoomable map layers that can be accessed for all housing blocks on the site.

The sketch-planning capability, which will go online once the design options stage is underway, is currently accessible under the Wired Communities menu

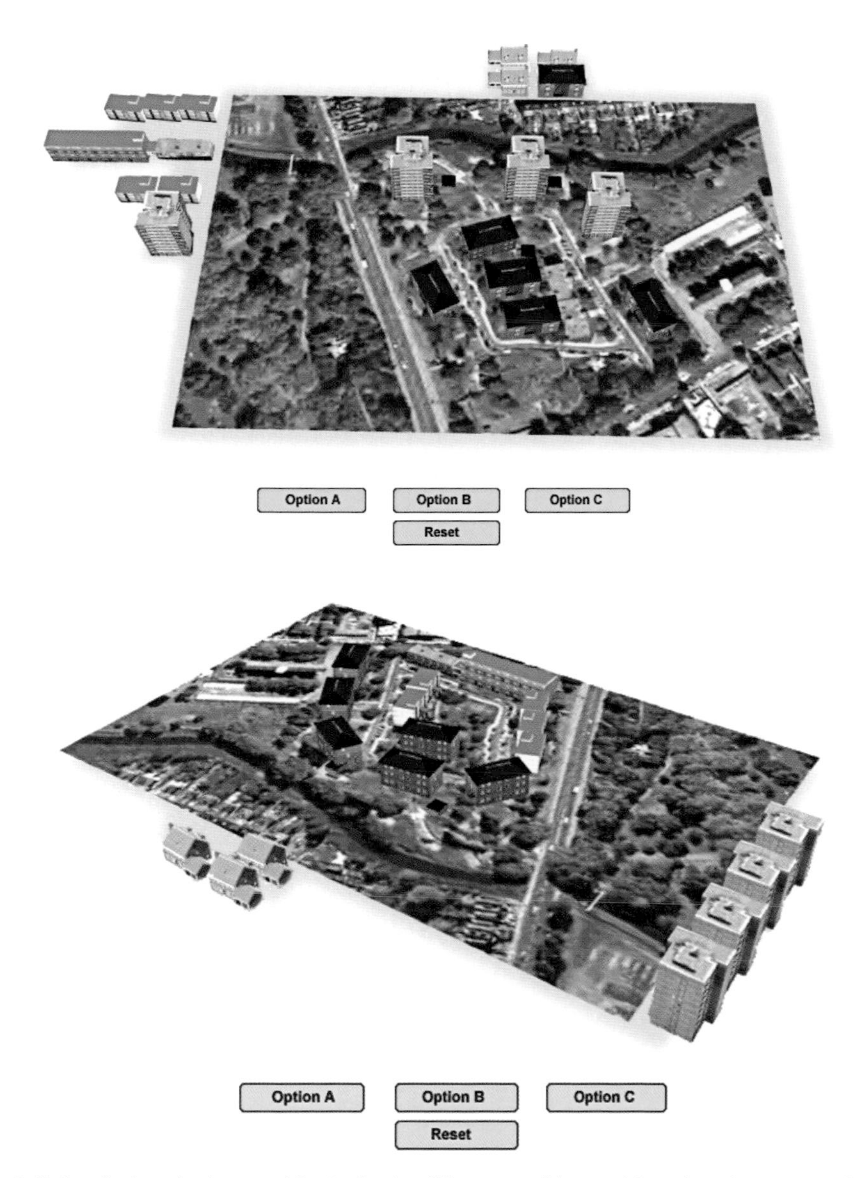

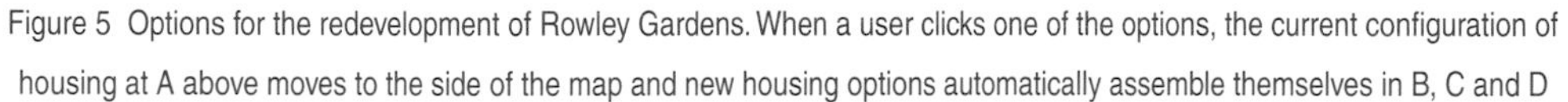

Figure 5 Options for the redevelopment of Rowley Gardens. When a user clicks one of the options, the current configuration of housing at A above moves to the side of the map and new housing options automatically assemble themselves in B, C and D

item that appears in the drop-down menu **Background and Research**. So far, we have only developed typical options for Rowley Gardens; there we present three options to enable the user to see the present configuration of housing blocks and to test three alternative designs that can be explored in 2-D and 3-D. The initial screen shows the existing housing which is composed of a mix of high-rise and low-rise blocks. When activated the three options replace the existing buildings, thus giving the user a sense of how the estate might look. We need to do much more to make this effective, but the tools are being developed and we are encouraged by the fact that residents are excited by these possibilities. We show the existing housing and three options in figure 5.

The last feature we note involves the bulletin board, or discussion forum. This feature went online in March 2002 and immediately residents began to post notes to the board. One of us acts as moderator and manages the site but once material is posted, then the WDRT needs to be involved immediately. As many of the postings relate to services to be provided by the council as part of their role as landlords of public housing and as many of the messages are critical, it was ultimately resolved that the WDRT were barred from responding for fear of litigation. This is a major obstacle to the very notion of participation, and it shows no sign of being resolved. It further reinforces the general feeling amongst many residents that local government is hostile, remote, uncaring and even malicious, and this

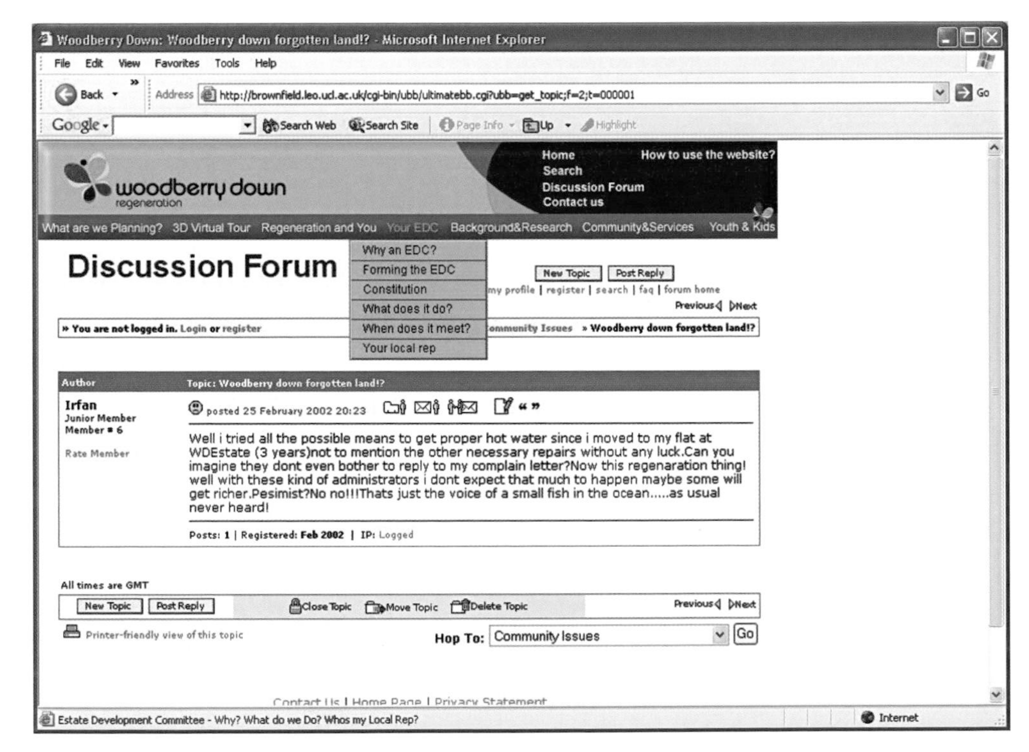

Figure 6 The discussion forum with a 'typical' message

does not bode well for the process. In figure 6 we reproduce a typical message from a user to give an idea of the power of discussion as well as the nature of the argument.

6 Conclusions: what next?

One of the key issues emerging from this chapter is the need to develop participation that is truly interactive. This is no more or less than two-way dialogue where providers of information respond to users and where users respond to providers in an ongoing collaborative process. This is easier said than done. Most participation schemes tend to be short-lived and somewhat passive: information is provided and sometimes meetings held to engage those affected, but any dialogue often takes place too late in the process for any effective action on the part of the community. In online participation, and in the smaller but distinct subset of activity in providing mapping to the community (in PPGIS for example) the challenge is even greater as the medium is unfamiliar. The nature of this medium is such that there needs to be active use and provision. Often participation schemes are financed as one-off ventures, but in the case of Web-based dissemination, it is essential to provide the greatest funding for ongoing development and maintenance of the media.

The development of any Web site needs to be a continuing affair, as it is a portal for information that is continually changing. It might even be argued that if the context is to provide those affected with information where none has been provided before, then there are better ways than the Web. Delivering leaflets door-to-door is the classic example of public participation. But where there is the need to involve the community, then Web-based methods are attractive in that they can incorporate responses, and they can chart and communicate changing circumstances. In the Woodberry Down project, the single biggest problem is not convincing the WDRT or the community of the need for such continuing involvement. It is finding continuing sources of funding for this participation, given that this is but one strand in a much wider portfolio of participatory activities. The very existence of the regeneration team in the local community is a kind of participation and this online venture has to compete with all the alternatives and complements. Currently the project is awaiting further funding with the Web site maintained and updated by ourselves as a labour of love.

Yet there is the real prospect that the work will continue as more and more residents acquire computers and as much of the rest of the world gets wired. It is important to note the mission of the WDRT when they say: ' . . . the WDRT believes that it needs to build a trusting working relationship—a real partnership with residents. It is certain that when it comes to involving the local community in the regeneration, the quality of their involvement in the process may well be more important than the final outcome of many key decisions' (WDRT 2001a: 18). Online participation is a central construct in achieving these goals.

Human-computer interaction and usability evaluation in GIScience

Carolina Tobón and Mordechai (Muki) Haklay

The primary concerns of human-computer interaction (HCI) and usability evaluation are supporting the goals and tasks that users wish to achieve with a computer system, thus ensuring that the system and its functionality are fit for purpose. These considerations are centrally relevant to GIS applications. Moreover, there are useful synergies to be realised from integration of HCI and GISc research. This is illustrated in this chapter through two distinct applications in which usability techniques are adapted and applied to learn about user requirements and tasks.

1 Human-computer interaction issues in GIS applications

The first commercially available computers that appeared in the 1950s were very large and expensive machines that could only be operated by specialists 'familiar with the intricacies of offline programming using punch cards' (Preece et al 1994: 4). Technological advances, however, have dramatically changed this situation by decreasing both the cost and size of computers. The advent of the silicon chip, for instance, allowed the miniaturisation of circuits and the development of powerful computers with large storage capacities. This also facilitated the innovation of the personal computer (PC) in the 1970s, which after the 1980s came to be used by a wide variety of people who are not computer experts or programmers but who utilise computers for a vast range of applications. The success of the PC, however, has also been made possible by improved understanding of the ways in which humans interact with computers, since this has enabled the design of systems that support a larger user population with the broadest range of requirements. Human-computer interaction (HCI) was the term adopted in the 1980s to describe the field of study concerned with these issues.

The aims of HCI are to create systems that provide functionality appropriate to their intended use and which are 'good enough to satisfy all the needs and requirements of the users and other potential stakeholders' (Nielsen 1993: 24). These people, however, may vary in their computer literacy skills, world views, cultural backgrounds or domain knowledge. Thus, it is important to understand the ways in which people use computer systems in particular settings if system design is to support users in an effective and efficient manner. Furthermore, users expect computer systems to be useful in achieving their goals not only in terms of the appropriateness of the functionality they may provide, but also in terms of how well and easily such functionality can be operated (Nielsen 1993; Preece et al 1994). Usability is thus a key concept in HCI and broader notions of system acceptability.

The concern with usability issues within geographic information science (GISc) has paralleled developments in HCI, but not without a time lag. The initial concern with GIS usage mostly dealt with data management—handling large files, formats and interoperability—and getting the system to perform the manipulations of information from overlay analysis to network tracing. GIS usage was initially confined to large organisations and was performed by professional users such as engineers and drafting technicians. However, by the mid-1990s the technology had diffused much more widely to the desktop and a new generation of non-GIS expert users began to use them in their daily routines. This raised concerns about how GIS were used and how they could accommodate different user needs. Interest in GIS metaphors and interface design began to emerge from a need to support the developing range of user requirements and tasks, some time after HCI became formalised as a research field and usability emerged as a major research area.

Research on HCI and usability issues in GIS has improved our understanding of user behaviour in a range of user settings (Medyckyj-Scott and Hearnshaw 1993; Nyerges et al 1995). Nevertheless, GIS still require users to have or acquire considerable technical knowledge in order to operate the computer system (Traynor and Williams 1995). This presents major obstacles to non-expert usage since the interface encapsulates a language, world-view and concepts that support the system's architecture rather than the user's world-view. These issues have led to interest in the cognitive aspects and psychological dimensions of user interaction with GIS. Research themes include the ways in which human cognition influences GIS use (Nyerges et al 1995), how people think about geographic space and time (Egenhofer and Mark 1995), and how spatial environments might be better represented by computers and digital data.

This chapter focuses on the application of HCI and usability evaluation techniques within GISc research. We argue for research into the usability of GI applications to underpin GISc, and we see this as providing important insights into the design of improved GIS applications. Two case studies are discussed in order to illustrate our arguments. Section 2 describes a study designed to understand how GIS might be used to encourage greater public participation in planning. Four usability evaluation methods were combined and adapted to appraise user tasks in this context. A similar objective lies behind the evaluations described in section 3. A number of usability evaluation techniques are considered in order to learn about user tasks of a more exploratory nature. Results from the evaluations are particularly relevant to the design of systems that support very broad user goals and semi-structured or ill-defined tasks, such as in exploratory data analysis (EDA) and geographic visualisation. The final section concludes with some thoughts about further integration and the developing synergy between HCI and GISc research.

2 The relevance of usability evaluation for Public Participation GIS (PPGIS): case studies in Wandsworth, London

The proliferation of information and communication technologies (ICTs) during the 1990s opened up avenues for public access to, and participation in, planning processes. As Hudson-Smith et al (this volume) argue, the use of the Internet and the multimedia capabilities of the World Wide Web (WWW) can increase public participation in planning and by so doing, make such processes more democratic and relevant to local residents who are directly affected by planning decisions in their area. This topic has received much attention within the GISc research community during the mid- to late-1990s. This interest is illustrated in the National Centre for Geographic Information and Analysis (NCGIA) Initiatives 17 and 19—Collaborative Spatial Decision Making (Densham, Armstrong and Kemp 1995) and GIS and Society (Harris and Weiner 1996), respectively. The latter stemmed from growing concerns about the social implications of GIS and the use of GIS to empower disadvantaged and marginalised groups. These provide some of the main research issues in Public Participation GIS (see Craig, Harris and Weiner 2002; Smith and Evans, this volume). The diverse applications of GIS in public participation settings entail its use by non-experts and occasional users, and thus successful PPGIS requires that complex computer technology be made more accessible and easy to use by such users. PPGIS is also concerned with increasing the access to and use of GIS by people who bring a diversity of knowledge, technical capabilities and cultural perspectives to decision-making processes. To enable this participation, GIS must provide users with a positive experience of the technology. In this context, the next section discusses a case study aimed at understanding the tasks that users may attempt in public participation settings.

Applying usability evaluation methods within PPGIS

The planning department of the London borough of Wandsworth collaborated with CASA to explore using GIS to foster public participation in planning. As the focus for discussions on PPGIS we initially explored the issues of redevelopment of previously used (brownfield) sites and the actions of local amenity groups and individual residents. The case studies involved discussing a proposal for high-density developments of luxury homes on the banks of the Thames that were of concern to local residents (figure 1). Residents were concerned about the lack of provision for affordable homes and the environmental impacts of the

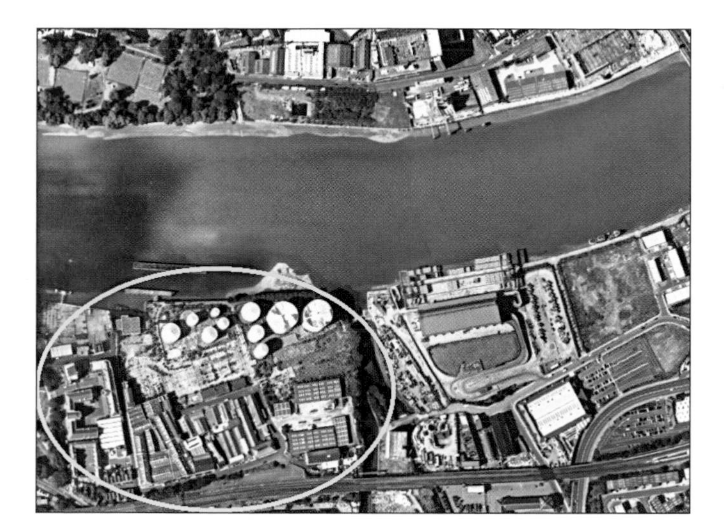

Figure 1 Former Shell repository—a site for development of luxury homes on the banks of the Thames

development, for instance, traffic congestion and pressure on local services such as schools, recreational facilities and libraries.

Two workshops were organised to explore these issues. For the first, we recruited 14 people all of whom were active members of the Wandsworth community. For the second, we recruited nine participants who had objected to a planning application in Wandsworth borough during the previous 12 months. Both workshops were structured in three parts: an introductory plenary session, a practical hands-on session, and a focus group discussion. The introductory session outlined the basic features of the GIS and the database that was compiled for the workshop[1]. In the second session, participants worked from a free-standing PC in groups of two or three with a GIS 'chauffeur', a person familiar with the GIS and the data content of the system. The chauffeurs demonstrated some of the basic tasks and then encouraged participants to take the mouse and keyboard to navigate their own way through basic operations of the system (figure 2). The hands-on session continued for 90 minutes, followed by a break and an hour-long focus group discussion to elicit the experiences and opinions of the participants. All the discussions during the workshop were recorded and transcripts prepared[2].

While the substantive element of the workshop was to explore the use of GIS within public participation settings, HCI techniques played an important role and were integrated into the research methodology. Four HCI and usability evaluation techniques were used in these studies. The first was the use of chauffeurs, which has long been an established practice in computer-supported collaborative work (Nunamaker et al 1991). In essence, the chauffeur acts as a

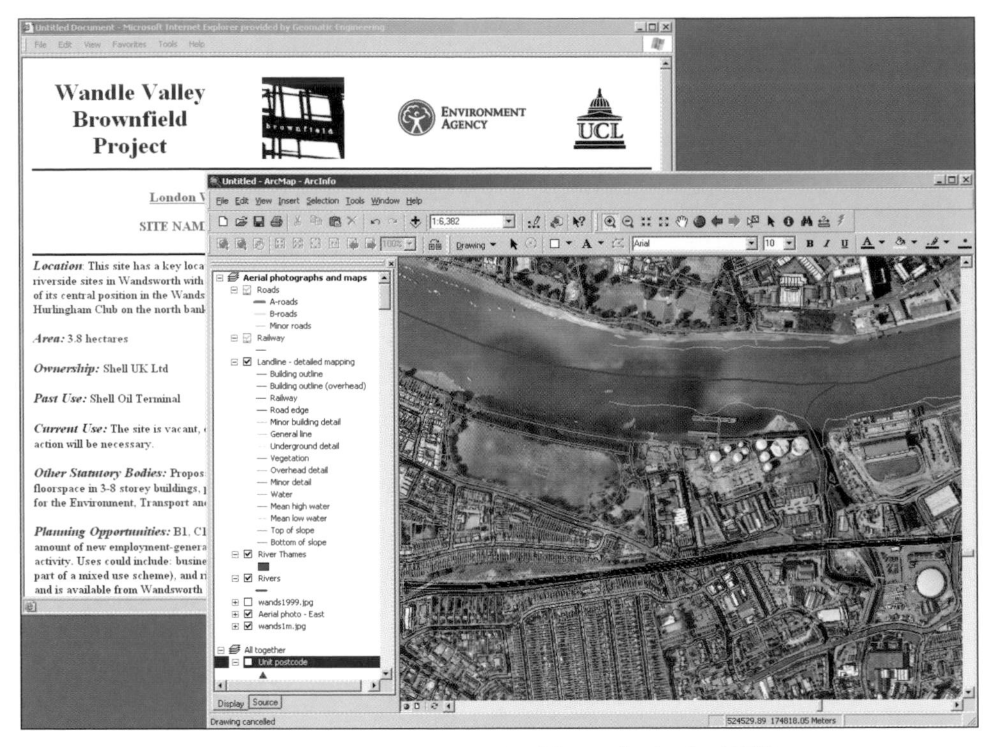

Figure 2 Participants explored datasets in the GIS, as well as in related HTML pages.

mediator between those who need to use the GIS but lack the technical know-how and the system. Hence, the chauffeur 'drives' the system on behalf of the participants who in this case varied in computer literacy from the novice to the experienced. In terms of PPGIS research, the use of chauffeurs reduces the technical complexities that the participants experience when working with a GIS, as a professional assistant is always present.

For the purpose of analysing and understanding how users performed tasks with computers, sessions were recorded, including the audio and the computer screen. This provided contextual information about why users performed as they did. In other words, it allowed us to identify what participants viewed on the screen during the sessions and to relate these images to the topics that were discussed. Videotaped sessions can also be used to time users on different tasks, to evaluate how participants accomplish them, or simply to allow multiple views of the users' experience with the system. In the context of PPGIS studies, the session recording enabled the researcher to analyse the sequence of events that led users to make a specific comment about the system or to obtain a certain result.

Figure 3 Participants worked in groups aided by a GIS chauffeur who drove the software

Within usability evaluations, the practice of asking participants to verbalise their experience with the software is intended to identify various cognitive activities or user problems with the system. For instance, this method elicits the understanding of users' expectations of system behaviour when using specific functionality. By arranging the participants into groups, we also encourage them to discuss their thoughts and capture their attitudes while using the system (see figure 3). This third technique was useful within our PPGIS studies as participants provided clear examples of their difficulties with respect to the concepts embedded in the GIS.

In usability evaluations, participants are expected to complete a set of tasks that are representative of the actual work potential users would attempt with the system. Information such as quantitative measures of user performance—the time it takes to accomplish a task or participant success and failure rates—is recorded from each task to aid in the evaluation of a system's usability. Within these case studies, tasks were used to guide the participants through a set of activities designed to provide them with an experience with the software and explore PPGIS research issues. For example, towards the end of the hands-on session we asked participants to add new information into the GIS which was not part of the database provided. One group of participants added information about local playground facilities, commenting on the need to include such facilities in the local authority planning agenda. The PPGIS literature stresses the importance of capturing local knowledge and the ability to integrate it into a GIS. Thus, by asking participants to accomplish these tasks without a clear definition of what type of information they were expected to fill in, we encouraged them to discuss the type of information that they would like to capture, as well as the value of that information.

Within the PPGIS context, GIS has a pivotal role as a tool for supporting and enhancing public participation. The integration of usability-evaluation techniques with traditional qualitative research methods, such as transcript analysis and the use of focus groups, was successful. It also provided a means to investigate non-expert GIS user interaction with GIS software and spatial data and to understand some of the types of tasks that users wished to accomplish. Many geographic applications, however, require greater flexibility than is provided by GIS, especially in terms of supporting dynamic and interactive data investigation. Such is the case in EDA (exploratory data analysis) of spatial data and geographic visualisation, where users aim to explore a dataset in order to facilitate the process of hypothesis formulation and knowledge construction concerning the information the data may contain. The following section discusses the usability evaluation techniques that enable us to learn about user tasks in this context.

3 Usability evaluation for defining geovisualisation tasks

Geovisualisation is in important respects a cognitive activity, the purpose of which is to gain understanding of a geospatial dataset by representing or encoding data in some graphical form that can reveal otherwise hidden information (MacEachren 1995). Geovisualisation tools or systems are frequently used to explore and sift information from large or complex geographical datasets. Geovisualisations have at least two components: dynamic and interactive computer environments; and users whose domain knowledge is key to the data exploration and hypothesis formulation process. The computer system must, therefore, allow users to manipulate and represent data in multiple ways so that they can investigate 'what if' scenarios or formulate questions that prompt the discovery of useful relations or patterns. In this manner, the geovisualisation environment is intended to support a process of knowledge construction which is 'guided by [the user's] knowledge of the subject under study' (Cleveland 1993: 12).

Nevertheless, the abductive nature of data exploration and knowledge discovery in geovisualisation that aims to 'discover patterns within the data while simultaneously proposing a hypothesis by which the patterns might have come to be' (Gahegan 2001: 275) implies that tasks and goals are often ill-defined, making them difficult to formulate and even measure. Formal usability evaluation techniques, however, require structured tasks or scenarios to assess user performance or success in accomplishing them. Two evaluations were conducted with the purpose of gathering information about the support requirements of geovisualisation tasks. The first was carried out using a commercially available system. Findings from this study guided the formulation of user tasks for evaluating a second geovisualisation environment. These two studies are discussed next, along with the combination of analysis used as a means of understanding the extent to which potential users could utilise computer-based (geovisualisation) tools to accomplish their work effectively and with ease.

Gathering information from users: study 1

The first usability evaluation was carried out using a commercially available tool comprised of a visualisation system (DecisionSite Map Interaction Services by Spotfire: Boston, Massachusetts) and a lightweight GIS data viewer (ArcExplorer™ from ESRI: Redlands, California) coupled as a plug-in to DecisionSite. Figure 4 illustrates the environment. To the left-hand side of the image is the GIS data viewer with the map at the centre, a context window and layer control at the bottom, and a toolbar at the top to link the two systems. The main limitation of this environment is the lack of a dynamic link between the two components. Hence, when data are selected on the map in ArcExplorer, the operation is not reflected in the other visualisations in DecisionSite (right-hand side) or vice versa. Instead, once a visual operation such as a selection is made, for instance in DecisionSite, a button in one of the tool bars has to be pressed for the selection to be reflected in the map.

It was anticipated that users would have difficulties in exploring spatial data using this environment. Thus, a first usability evaluation was designed to obtain as much information as possible about the ways in which the limitations of the environment would restrict user understanding and exploration of the data. For this purpose, nine participants were asked to perform four tasks. The first two tasks were open-ended questions designed to allow the users to become familiar with the environment while revealing how they attempted to solve the task in hand. Users were allowed to ask questions and were encouraged to explain problems encountered and to engage in discussions about the information they were obtaining. In the last two tasks, users were not allowed to engage in discussions and were required to obtain more precise answers about the dataset being explored. Despite the aforementioned limitation of the environment, the

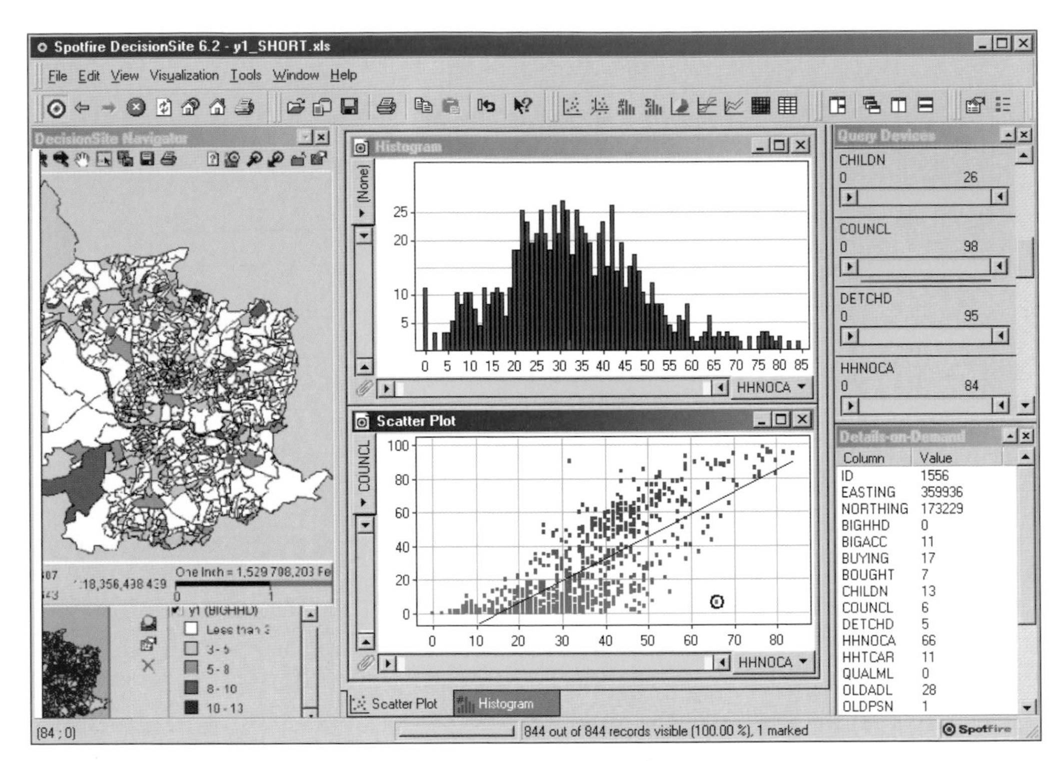

Figure 4 DecisionSite Map IS

evaluation was nevertheless successful in gathering information about how users attempted these tasks and their difficulties or successes in doing so and also in discovering other problems that required system support.

The participants who were invited to take part in this study were expected to have some experience with handling data in a GIS or in other systems where graphical displays are used for spatial-data manipulation. Participants included GIS professionals and students of various ages and nationalities, as well as planning students with different levels of GIS experience. The lack of a dynamic link in the software, however, made it difficult for all users to understand the full functionality of the coupled system. At least two participants did not understand that the data represented in all views were from the same dataset. It took two other users the whole session to grasp this. Two more failed to comprehend the information that the environment was providing and, therefore, did not see the utility of the linked software. Nevertheless, seven out of the nine users commented positively about the advantages of combining flexible and interactive tools for attribute value exploration with the mapping capabilities of a (lightweight) GIS and were excited about the possibilities for hypothesis formulation that tools of this nature might support.

Developing an environment to appraise user tasks: study 2

A second environment, in which DecisionSite was coupled to ArcMap in ArcGIS, was therefore created in order to solve the problem of dynamic linkage described above. Figure 5 shows the environment developed for the second evaluation where the selection or highlighting of features in any view is reflected in all others. It was an objective of this study to obtain further information from the users about the functionality that environments used to visually explore geographic vector data should include, rather than defining a priori what such functionality should be. For this reason, DecisionSite was coupled to a fully fledged GIS, and the evaluation was designed to give participants as much flexibility as possible in tackling tasks and in using the functionality they deemed necessary to solve these tasks.

Tasks for this evaluation were defined according to three factors: the extent of the geographical area that needed to be investigated, the number of attributes or variables to explore, and the type of visual operation users had to perform. The first two factors had two levels: either one or many areas or attributes to be explored. The visual operations required users to identify geographical area units with particular characteristics, locate where a particular phenomenon occurred, uncover associations between attribute values, and compare geographical areas where some relation between attribute values occurred (Knapp 1995). An experiment was designed so that combinations of the three factors (spatial extent, number of attributes investigated and visual operation) at their various levels (as defined previously) defined 16 tasks of differing complexity (see Tobón forthcoming for a detailed discussion). Twenty participants with varying degrees of GIS expertise completed two sessions on two different days when they had to solve all 16 tasks. Although the tasks they were presented with during each session were different in terms of the particular data attributes that they explored, the types of tasks were the same. Therefore, this experimental design allowed two observations to be recorded for every user and for every task type. This made possible the investigation of learning effects of the software environment, as well as comparison of user performance before and after the users gained some confidence in using the software or acquired knowledge about the data.

Performance measures, such as the time it took respondents to complete a task, were recorded. Analysis of variance (ANOVA) was then used to study the changes

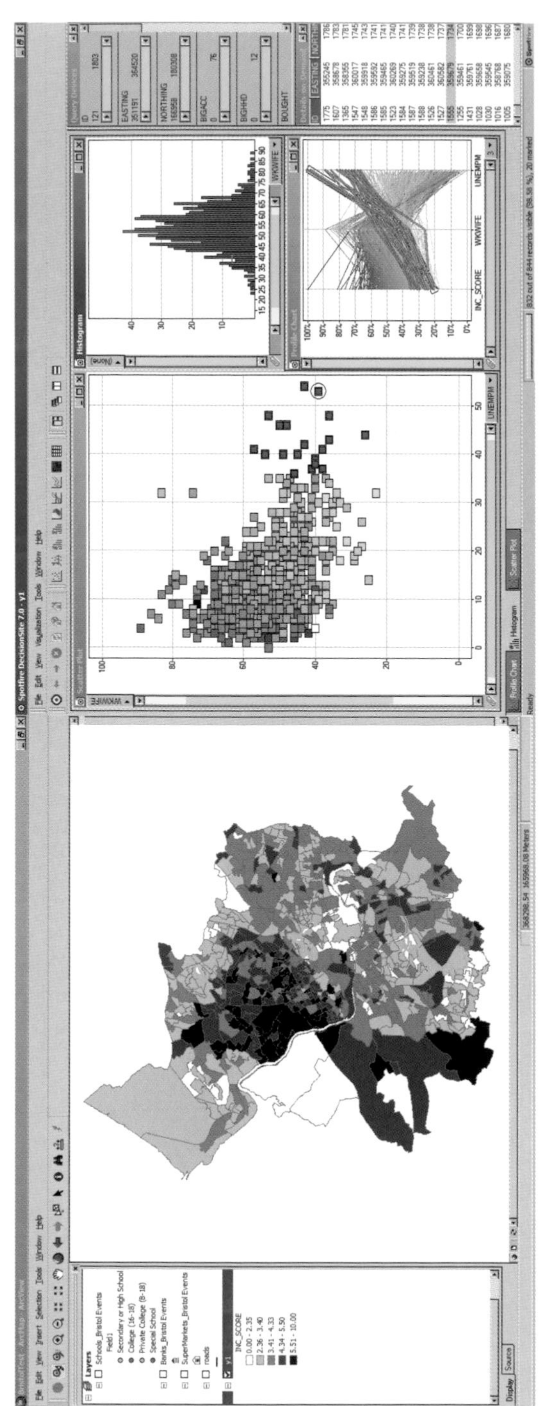

Figure 5 ArcMap environment on the left and DecisionSite environment on the right

in performance from users' experiences with the tasks. The technique makes it possible to detect the effect of varying the level of each factor on the performance measure. It also makes it possible to detect interactions between the factors, or in other words, to assess the incremental effect of each factor at varying levels of each of the other factors in order to account for overall user performance (see Cox 1958 for a detailed discussion).

Both the main and the interaction effects were likely to be important. This is because the time required to complete a task was expected to depend on the types of visual operation attempted, the spatial extent of the problem area and the number of variables involved in solving the task, as well as on the combinations between these factors. The results of the statistical analysis confirmed this to be the case, and further qualitative information gathered during the experiments helped explain the particular needs of users for supporting each type of task. Thus, the statistical evidence was complemented with interviews and questionnaires that provided information about user perceptions and experience of the environment. Interviews also made it possible to obtain detailed explanations about how users went about solving tasks or about the problems that they encountered. This adds weight to and enriches the quantitative evidence. Users also answered a questionnaire designed to obtain information about the perceived usefulness and ease of use of the environment (Davis 1989).

Questionnaires provide a quick and inexpensive measurement instrument in this kind of study. They also generate useful quantitative data, particularly if measurements are recorded on a Likert scale[3], which can be used as an appropriate yardstick with which to measure each user's perceptions. Questionnaires administered in this second study made it possible, for example, to establish that although the environment was perceived as being very useful to obtain relevant information and speed up user response once it was mastered, the software was not seen as easy to use, at least initially.

To summarise, a number of usability techniques were tailored and combined with appropriate experimental designs to learn about some of the types of tasks that geovisualisation applications need to support. These methods can be used to evaluate not only a system or prototype in terms of some aspect of its usability, but also to test research concepts, such as how well a particular visualisation works for certain data types.

4 The synergy of HCI and GISc research

The functionality provided by off-the-shelf GIS is becoming more powerful and sophisticated, and their use by people who are probably not GIS experts is increasing continuously (Longley et al 2001: 360). Furthermore, as Schietzelt and Densham (this volume) discuss, GIS are frequently part of larger systems for supporting decision making and analysis when using georeferenced data. These circumstances create a tension between the rich yet complex environment that GIS can offer and the proliferation of applications that an ever-wider range of users can develop. Hence, an ongoing engagement with HCI and usability aspects of GISc is necessary for the success of GIS to be made relevant to the growing user community.

The two examples discussed in this chapter illustrate how usability evaluation techniques can be adapted to a variety of settings and applications. These methods can be tailored to evaluate GIS applications in order to detect problematic, missing or incomplete functionality, as well as to measure user perceptions about specific aspects of a system or tool. The first case study discussed the investigation of user tasks in a PPGIS context by applying usability evaluation methods to elicit some of the results. The second study also investigated user task definition in a geovisualisation context where user goals were rather broad and tasks ill defined. The latter study in particular is a good example of evaluations where adopting or adapting usability techniques may not necessarily be an easy task in itself because of the complexity of the environment being evaluated or because of the difficulty in defining user tasks. Nevertheless, there is a myriad of available techniques for every stage in the design, development and deployment of an application and a large body of HCI literature with examples of when and where these techniques are appropriate. This makes it possible to identify the type and quality of information that it is realistic to expect users to obtain. Incorporating these methods into GIS applications may improve the likelihood of them being fit for purpose and hence adopted by its potential users.

The integration of HCI and GISc research can provide important insights for both communities, as there appears to be a synergy between the two. In this chapter we have only identified the tip of the iceberg and there are many other issues which present major challenges in this area. We have discussed mainly desktop-based GIS applications and were more concerned with user tasks rather than with usability problems of the GIS interface. The application environments in which GIS are used have changed dramatically since the GIS community first started thinking about usability. In particular, the proliferation of GIS in ubiquitous mobile devices—from handheld computers to in-car navigation system and mobile telephones (see Li and Maguire, this volume)—raises a wide spectrum of new challenges. This provides exciting and fertile ground for a renewed interest in a combined HCI and GISc research agenda.

Endnotes

1 In both workshops we used an off-the-shelf GIS package (ArcView from ESRI) which also provided multimedia access to specially designed Web pages or existing Web sites.

2 For a full description of these workshops and the substantive outcome see Boott et al (2001).

3 See, for instance, Trochim (2002) and trochim.human.cornell.edu/kb scalgen.htm for a simple description of how they are designed, administered and analysed.

Chapter

Structuring the plan-design process as a coordination problem: the paradigm of distributed learning control coordination

Theodore Zamenopoulos and Katerina Alexiou

Structuring the plan-design process entails investigating design decision making in distributed human-computer systems. The motivation for the research reported here is to develop tools that can improve our ability to foresee and generate coordinated decisions in distributed and cooperative decision-making environments. In this chapter, we set out our detailed research hypotheses and we present a model and tool that learns by interacting with distributed knowledge sources (human or artificial) and uses this knowledge to generate spatial plans. In this context, the generation of spatial plans becomes a process of continuous learning and adaptation based on interaction. Finally, we illustrate some simulations built to test the model in a virtual environment prior to its application within a human-machine decision-making network. We conclude with a discussion of preliminary results and directions for future development.

1 Introduction

Plans are fundamental instruments of change within society in general and in architectural design and urban development in particular. Because of this, the development of methods and tools to study and support the designing of plans is of paramount importance. For that reason, computational models have been used for some 50 years now and in multifarious ways (Batty 2001b). Some models address plan designing by following the **analytical tradition**, which implies that if we get to know the world around us better, then we will know how to change it. In this sense, computer models are used as methodological tools to support and reflect theory building and hypothesis testing. Such models are developed to computationally reconstruct phenomena or realities and, thereafter, support explanations, predictions or evaluations. On the other hand, computational models have also been developed following the **design tradition**. In this framework, models are employed as methods or tools that directly attempt to resolve design or planning problems and support the execution of design and planning tasks. The main concern is the computational construction of useful 'inputs' to design and planning activity that work as problem solvers, collaborators, intelligent assistants, exploratory devices and so on. Three interrelated directions of research drive the development of computational models: the first relates to the study of problem-solving strategies, methods and techniques (see for example Schietzelt and Densham, this volume; Gero 1985); the second relates to the study of the design and planning object or artefact and its representation (see for example Steadman 1983; Batty and Longley 1994); and the third relates to the study of design and planning knowledge and its acquisition and representation (see, for example, Candy 1998; Rubenstein-Montano 2000).

Each of these approaches typically concentrates on the inner coherence and effectiveness of the computational models or the effectiveness of the humans using these models and, thus, underestimates the design potential that may be created by humans and machines working together in networks. This implies the need to look at design neither as a human or artificial process, nor as a social process alone, but as a process governed by the interaction of humans and artificial constructs within complex socio-technical systems. This also implies a need to develop and evaluate models as part of this larger complex, able to work together with humans and their environments, and adapt their behaviour according to this interaction.

In this chapter, we present a study that aims to discover some efficient conditions to enable the emergence of plan-designing abilities in human-model networks, and in distributed and cooperative design and planning systems in particular. We study these questions and conditions by developing a distributed and adaptive design model and introducing it within a human-computer cooperative network.

More specifically, we present a model/tool that learns through interaction with distributed knowledge sources (human or artificial) and then uses this knowledge to generate spatial plans. In this context, the generation of spatial plans is a process of continuous learning and adaptation based on interaction. We take a hypothetical urban development assignment that concerns housing, retail, and open-space development within a virtual city. In urban development the simultaneous and continuous generation of both architectural and urban plans is paramount. Additionally, requirements and targets are typically distributed among different teams and vary in time according to the emergence of new situations. The proposition of this research is that design and planning processes are by their nature distributed, and, therefore, coordination is of paramount importance for the 'constructive' ability of the design/planning system. Coordination refers to the ability of the design system to address current problem formulations and at the same time guide the overall network to anticipate and exploit new opportunities (new problem formulations and new solutions). The plan-design problem is formalised as a coordination problem and is addressed by distributed learning control methodologies.

2 What is a plan and what is plan designing?

In order to discuss the plan-design process, we need first to have a clear and unambiguous view of what a plan is, what kind of knowledge/information it represents, and what kind of processes typically characterise the formation of plans. In one sense, and if we restrict ourselves to typical physical and spatial planning problems, plan designing involves specifying a spatial configuration (in terms of land uses, distribution of volumes, and rooms) that reduces or resolves conflict between different goals (see Batty 1984 for a general discussion). However, plans aim not only to address current needs, but also to create future opportunities (Hopkins 2001). Hence, plan designing involves specifying those configurations that in parallel can anticipate and formulate future conflicts (opportunities) and eventually suggest new configurations.

This broad definition might imply different interpretations regarding the function and scope of plans. An interesting classification is found in Hopkins (2001: 34-42), which summarizes how a plan works in five ways: as a design (a fully worked out outcome); a vision (an image of possible outcomes); a strategy (a set of decisions that set a contingent path); a policy (if-then rules for actions); and an agenda (a list of things to do). In each instance, plans contain and represent information about decisions to be implemented (in a solution space), performance attributes (in a performance space), requirements and expectations (in a problem space), as well as interdependencies among these spaces, usually in a time-critical manner. The relations between designing processes and plans, and the system to be designed, vary between disciplines and according to the nature of the system to be designed (for example, if it is a building or a city). This variation reveals different interpretations of the plan-design process.

Nonetheless, plan designing tends to be perceived as a purposeful, constrained, decision-making and knowledge-constructive process, where search within a given set of decision variables, exploration of alternative problem spaces, co-evolution of the problem and solution space, and learning are significant driving forces (Maher 2000). We argue that learning control methodologies are well suited to this definition of plan designing. The aim of learning control is to explore and adapt problem and decision formulations so that they follow time-variant expectations and performance constraints. Learning is a function that captures, represents and restructures interdependencies among design problems, decisions and performance attributes. In parallel, learning is a function that represents and improves the ability of the system to control the design process and, therefore, generate and make purposeful and feasible decisions that can lead to better problem formulation and decisions.

Plan designing often also involves a need to cope with—and why not exploit?—the fact that design knowledge and processes are distributed. This assumption is applicable not only because plans are usually formed collectively by communities, interdisciplinary groups, or in general by distributed decision makers, but also because even expert reasoning might be considered to be fragmented between diverse purposes, criteria (constraints), scales and methods. The distribution of decision making also implies a conflict resolution or coordination problem. Coordination has been discussed in the context of organisational theory and multi-agents systems (MAS) and relates to the question of how a distributed system can synchronise its activities. In the following, we will formalise the plan-designing problem as a problem of coordination between distributed agents (human or artificial) that convey individual knowledge, individual targets and methods. We term these self-interested agents. The control abilities discussed above represent the coordination abilities of the individual agents.

Three hypotheses (conditions) underpin our plan-designing framework. The first is that the ability to anticipate and generate spatial plans for ill-defined problems will be improved by distributing the control process to an open structure of individual agents. The second is that the abilities of distributed agents to synchronise their activities and the consequential emergence of coordination can represent the generative or creative features of the plan-design process. Finally, the third hypothesis suggests that domain knowledge cannot be defined for ill-defined problems and distributed processes a priori, so some learning mechanism needs to be devised to capture this knowledge and effectively use it to generate plans.

The main hypotheses: distribution, coordination and learning

The first hypothesis considers plan designing as a complex process that involves multiple decision-making agents (human or artificial) that are sources of diverse and often conflicting knowledge and that express individual views and goals. Hence, design decisions are seen as a result of the constructive abilities found in distributed systems and organisations, and cannot be attributed solely to any individual mind or action. This alludes to the ability of distributed systems to generate alternative plans out of partial and incomplete knowledge, based on the interaction/communication between multiple agents. Plan designing as distributed decision making implies that the normative activity of change is set under the weight of a collective dynamic, which also underlines the fact that plan formation processes must be studied both in positive and normative terms (see Batty 1984).

In this sense, it is relevant to note that we have moved from using computational models and machines as automatic design devices to using computational models that support the generation of designs through user interaction. In the context of multi-agent design this interaction is distributed over networks as can be documented by current interest in collaborative design and planning and Computer Supported Cooperative Work (CSCW). The hypothesis of the distribution of decision making suggests that knowledge is also distributed, not only because plans are collectively formed by communities (or multidisciplinary groups), but also because even expert reasoning is fragmented according to diverse goals, criteria and evaluations.

Naturally, in the context of distributed decision making, plan design involves searching for configurations that reduce or resolve conflict between distributed goals. Broadly speaking, we can distinguish three typical structures in distributed systems. The first appoints a collective function that needs to be optimised for the sake of 'social welfare'; the second leaves the dynamic between the involved parts to determine the distribution of welfare; and the third distributes welfare equally between the involved parts. In the decision sciences formal definitions include concepts of bargaining, negotiation, conflict resolution, social choice, consensus or cooperation. Similar approaches have been developed in the context of artificial intelligence (Ossowski 1999), and some relevant examples in operational research can be found in Batty (1984).

In the research reported here, plan designing is viewed as a coordination problem, in the light of distributed decision making and conflict resolution. Coordination is extensively discussed in the context of organisational decision making (Malone and Crowston 1990) and is a recurring issue in the literature on distributed artificial intelligence and multi-agent systems (for example, Jennings 1996; Ossowski 1999). Whether talking about actors or agents, human or artificial, coordination is what makes them act as a distributed system and reach solutions on the basis of managing interdependencies between individual requirements. In the following we will introduce the idea of coordination as a learning control problem. Learning corresponds to a process of capturing interdependencies among decision variables, while control corresponds to a process of using this knowledge to generate control actions (plans) that meet time-variant individual targets, despite endogenous uncertainties or exogenous disturbances introduced by distributed agents. In this context, creativity and innovation lies in the possibility of unforeseen solutions emerging through agent interaction and learning.

Finally, the third hypothesis relates to the question of how to incorporate domain or descriptive knowledge about the system that is to be designed within the model. Very often domain knowledge is seamless with the proposed model. For instance, facility planning has been extensively addressed with respect to studies on user behaviour, resulting in building models (for example, gravity models) that represent this behaviour. Therefore, the design of optimum location-allocation plans strictly depends on this predefined formulation of user behaviour. On the other hand, in MAS this knowledge is distributed to local agents, and global patterns of behaviour emerge by local interaction (see Torrens, this volume). In this way, knowledge about system behaviour is not defined a priori but rather emerges as the collective design process progresses. In case-based design (CBD) systems, domain knowledge is represented in the form of design cases that are dynamically adapted and updated (see Maher and Pu 1997). In both MAS and CBD systems, learning is an implicit function of the system that supports the maintenance, re-use and adaptation of knowledge. Starting from the recognition that learning plays an important role to improve the quality of design knowledge through time, several theories and models have been developed that suggest different ways to interpret the function and scope of learning within a design system (for example, Grecu and Brown 1998).

In our research, learning has a dual function and meaning: first, to capture and restructure interdependencies among the problem, solution and performance spaces so as to improve understanding about the domain problem and reduce uncertainty; and second, to improve the generative ability of the design system towards solutions that fulfil time-variant expectations and performance constraints and reduce conflict. Here we will use distributed neurocontrol methodologies (see Hrycej 1997; Šiljak 1991) as a paradigm for plan generation based on learning.

Control

The discipline of control emerged from cybernetics (Wiener 1948), and since then the ideas of control have been applied in various scientific areas: in engineering, economics and the physical and social sciences. Control systems are systems in which a controller interacts, by way of one or more controlling variables, in order to influence the state of a controlled object (also called 'plant'). Some of the typical ingredients of control problems are: the controlled objects have their own dynamics so their outputs cannot be changed instantaneously; the range of available adjustments is usually constrained; characteristics of the controlled object are not known with certainty; the state of the controlled object may be affected by uncontrolled, external and unpredictable inputs (known as exogenous variables); desired values for the state of the controlled object are often exogenous and unpredictable; current values of the state of the controlled object are uncertain; and measurements carrying information about current states are noisy (Jacobs 1993: 1). The main idea behind adaptive control systems is feedback—the controller attempts to steer the controlled object by adjusting a controlling variable (control signal) on the basis of measuring the difference between a desired value and the actual value of a controlled variable. Exogenous sources may disturb the feedback signal, but the controller needs to achieve its objective despite uncertainties about the characteristics of the control object and about the exogenous variables.

We believe that control seems particularly suited to the constrained, goal-oriented activity of design, specifically where design problems are characterised by complexity, uncertainty and non-linearity. We are interested here in adaptive learning control. Neural networks (NN) display many desirable characteristics required to address such problems: they are capable of learning by experience (from the data); they have the ability to map non-linear functions; they do not require deep understanding of the process or the problem being studied; they have the ability to generalise; and they are robust in the presence of noise (Kecman 2001). Neural networks have been extensively used in adaptive control, and there are many different NN-based control methods. Next we will elaborate a formal model for plan designing using control-based methodologies, and, in particular, we will formalise the problem of coordination among self-interested agents.

3 Structuring the plan-designing process

In the research reported here, structuring the plan-designing process entails investigating design decision making in distributed human-computer systems by focusing on the three main hypotheses already discussed concerning distribution, coordination and learning. The motivation is to develop tools that can improve our ability to foresee and generate coordinated decisions in distributed and cooperative decision-making environments. The following section explains the structure of the proposed design and planning system and clarifies how plans are built, how they work, and what kind of information they represent.

Human agents, artificial agents and plans

Plans are built on distributed domain problems or partial proposals developed and controlled by agents (human or artificial). For instance, a trivial location and space layout problem may involve various groups of agents: a group that defines an appropriate location, another that designs a suitable distribution of volumes, a third that configures a potential spatial distribution of rooms, and a last one that is involved in the structural engineering of the building. Agents' proposals are considered to be partial not only because they convey domain-specific knowledge about the design problem, but also because these proposals are incomplete and change over time according to changing situations and new knowledge gained in the process.

Plans are devised by human and artificial agents in a 'collective space'. In the simulations presented here, artificial design agents are introduced by users on the basis of a purpose or domain problem, but, in principle, artificial agents should be able to control their own definition and population. That is to say, agents should be able to adapt and change themselves but also generate new agents in order to resolve particular problems. The interface between human and artificial agents is configured by objects (which are collections of variables) that are embedded in a Virtual Reality (VR) world. These objects are dynamically identified and modified by agents. We should note that human actors, their computational constructs, and the way they manipulate objects form the 'knowledge' or 'reasoning sources' for the artificial agents. Thus, artificial agents learn through the modification of objects in the VR world (figure 1). The relationships between plans, objects and agents are shown in figure 2. Here, plan descriptions are built around aggregated objects introduced by users. In the centre, the class **Object** is an interface among users (or their digital counterparts) and the agent society. Objects are built on three classes of information: **Function**, **Behaviour** and **Structure**. Each user can set and edit a number of objects that are organised in different groups according to their purpose. Each group that is an instance of the class **Object** corresponds to an **Agent**. Agents are artificial entities that act as controllers.

Figure 1 illustrates that plans are devised in a collective space by agents (human and artificial) that each control parts of the overall description. The artificial agents (on the right of this figure) are introduced by users for a particular purpose or to resolve a particular domain problem. The interface between human and artificial agents is configured by objects (here, cuboids) that are embedded in a VR world. The ways in which human agents (or their computational models)

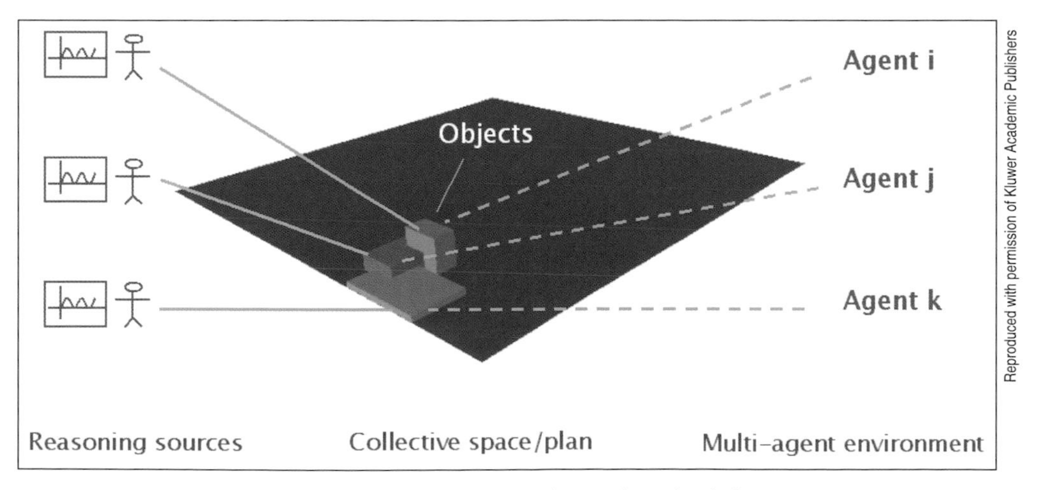

Figure 1 How plans are built (see text for explanation)

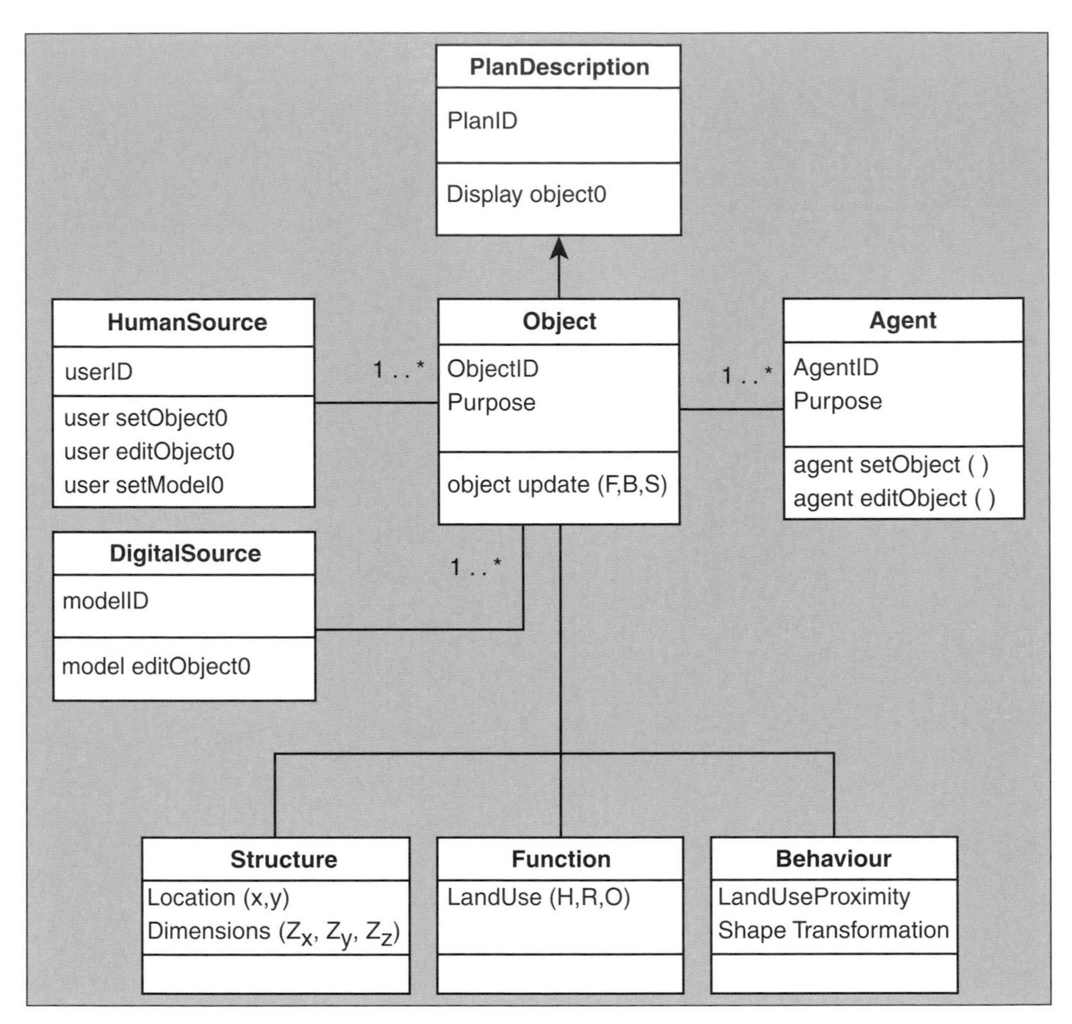

Figure 2 A Universal Markup Language (UML)-like diagram outlining the basic structure of the model

modify the objects in the VR world provide the reasoning sources for the artificial agents.

For the simulation described in this chapter, we use three objects (initially in the form of three cuboids) located in a hypothetical virtual city, which represent the preliminary development goals for a housing unit, a retail facility, and an open space. The extent to which the overall model for plan generation is working autonomously from human operators depends mainly on the degree to which formal models are incorporated as domain knowledge sources. Also, the granularity of the objects may determine the scale to which we study the design artefacts and the depth to which we manipulate their characteristics through human-model interaction.

The function-behaviour-structure framework

The objects within the VR environment are built on three classes of information: functional, behavioural and structural (FBS). The implications of the FBS framework for plan designing have been extensively discussed in the literature in a variety of different contexts (for more information on the FBS framework see, for example, Gero 2000). Here we will restrict our attention to the ways in which this framework is adapted in the context of our model.

Structural information specifies the components of the proposed plan, their attributes and their relations. For the simulations presented in this chapter, structural information depicts the physical components of the objects and their topological relations. So, for instance, for an object A_H (housing), structural information includes location $[x, y]$, volume dimensions $[z_X, z_Y, z_Z]$ and relations with other objects such as distance to other facilities—like retail and open space *[dr, do]*—and adjacency with other buildings to north, south, east and west. Behavioural information specifies how each object reacts to changes in its state and to its environment. Behaviour is a description of structural change of the design objects in order to reach their intended functions. For instance, behaviours describe land-use attractiveness (the tendency of land uses to be attracted to—or repelled by—other facilities) or cost (according to land value and floor-area ratio). We should distinguish between structural behaviour, which is directly derivable from structure (for example, the cost of a building is related to its total floor area) and expected behaviour, which is derived from the function (for example, there is a minimum floor area threshold for a housing unit).

Finally, we consider that functional information represents the teleology and purpose of the proposed objects expressed as land use—in our case housing, retail and open space. Those three classes of variables are linked together by processes that transform one another; namely the processes of synthesis, analysis, evaluation, formulation and reformulation (Gero 2000). In the next section, we will discuss these processes by adopting a control-based approach.

The important point in this framework is that it gives us the possibility to represent, within plans, information about the problem or requirement space (functional information), the solution space (structural information), and the performance space (behavioural information). This last class of information is

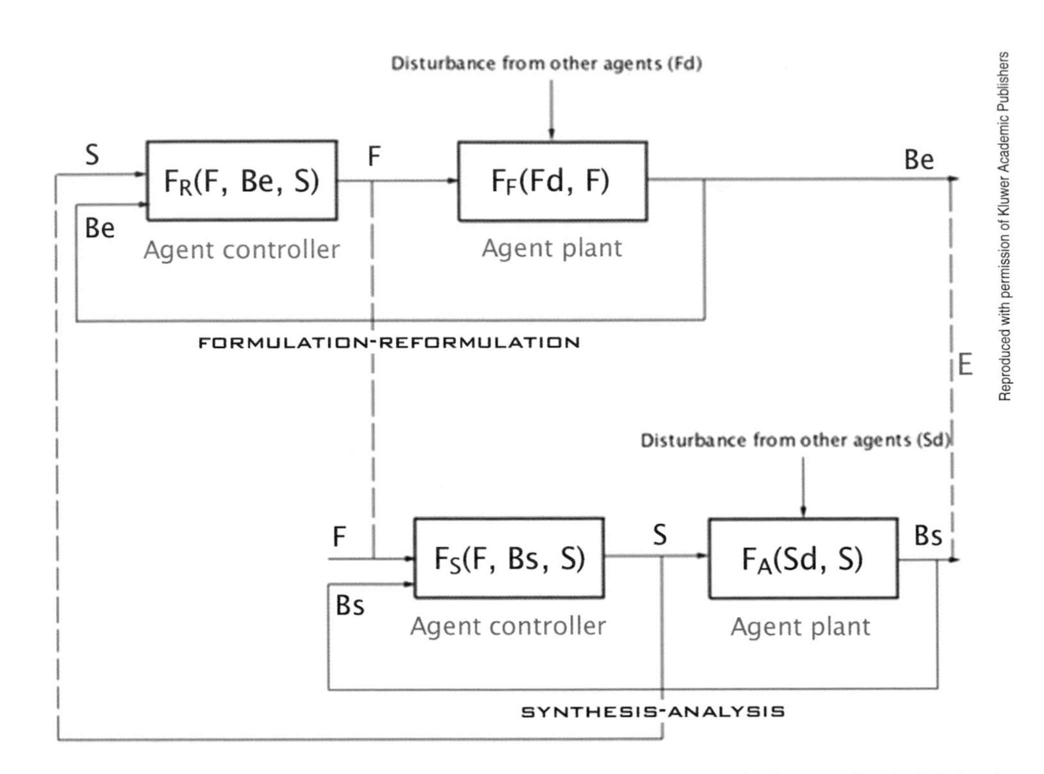

Figure 3 The structure of an artificial design agent. (Notation: *B* = Behaviour, *F* = Function, *S* = Structure, F_A = Analysis function, F_S = Synthesis function, F_F = Formulation function, F_R = Reformulation function and *E* = Evaluation)

critical, as it represents high-level knowledge and plays an important role in linking function and structure and providing an evaluation mechanism.

The structure of an artificial design agent

Artificial design agents are autonomous software entities that generate domain-specific (and partial) proposals that constitute inputs to the human-model designing network. The artificial design agents learn through the interactions that take place in the collective (VR) space and adapt their behaviour according to this knowledge. Their function is to steer the plan-designing process towards (scalable) decision and problem formulations that may lead to the coordination of their distributed requirements. Artificial agents do not have global knowledge about the whole system and there is no central coordination mechanism. In effect, coordination is an emergent property of the system, triggered by the process of learning and self-adaptation of agents at the local level.

Even though there are several different control-based formulations that might be reasonable for coordination problems (for another example see Alexiou and Zamenopoulos 2001), here we formalise coordination as a self-regulating process that aims to satisfy temporal targets, despite external disturbances that might cause conflict. Each agent carries out two combined control-based activities: the first alludes to a synthesis-analysis-evaluation route and the second alludes to an evaluation-formulation-reformulation route as we show in figure 3.

The objective of each agent is to find a suitable path of structures *S* that leads the behaviours *Bs* to follow a reference (expected) behaviour *Be* despite uncertainties and despite exogenous disturbances *Sd* produced by other agents' decisions. The expected behaviour Be is defined by a reference model, which is developed following a similar control process. The objective in that case is to find the appropriate functions *F* that lead the expected behaviour *Be,* to follow a reference structural behaviour *Bs* despite uncertainties and despite exogenous disturbances *Bd.* Hence, the desired performance of the synthesis-analysis system is evaluated (denoted by *E* in figure 3) through the reference model (formulation-reformulation) which is defined by its input-output pair *{F, Be}.* The control system attempts to make the plant model follow the reference output *Be* asymptotically as seen in the limit equation (1), where ε is a positive integer:

$$\lim_{t \to inf} | Be\text{-}Bs| < \varepsilon \qquad \text{equation 1}$$

To sum up, what we call synthesis is a control process that generates solutions (structures) so that time-variant expectations can be satisfied. Reformulation is the control process that aims to redefine the problem formulation (function) so that the expectations developed for a design object respect limitations posed by the domain knowledge. Evaluation is the process of measuring the degree of matching between the two control systems. The control signal $S_t,\ldots,S_{t+n}$ produced by this combined control process might be intepreted in three ways: as a set of partial actions (solutions) that build a global solution in time; as a course of complete actions (solutions) that the agent has to follow in order to reach its targets; or as a set of actions that regulate a given set of variables in order to satisfy time-variant targets. The latter option was implemented for the model presented here. The different intepretations of the control signal essentially correspond to different kinds of plans. For instance, the first paradigm is an approriate intepretation of plans as designs or visions. The second is appropriate for plans as strategies,

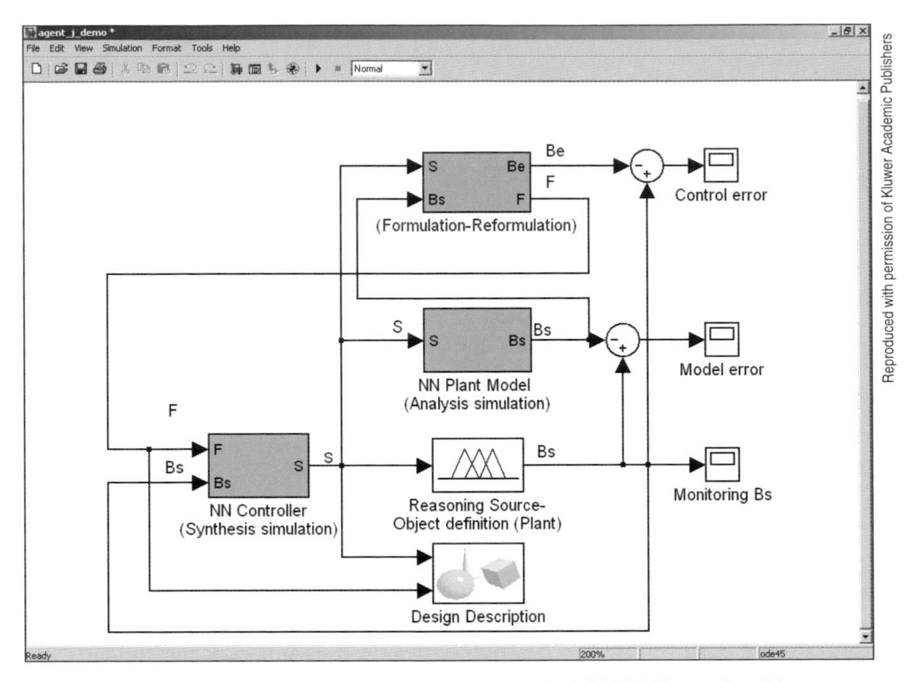

Figure 4A The control model of a generic agent in MATLAB-SIMULINK, overall model

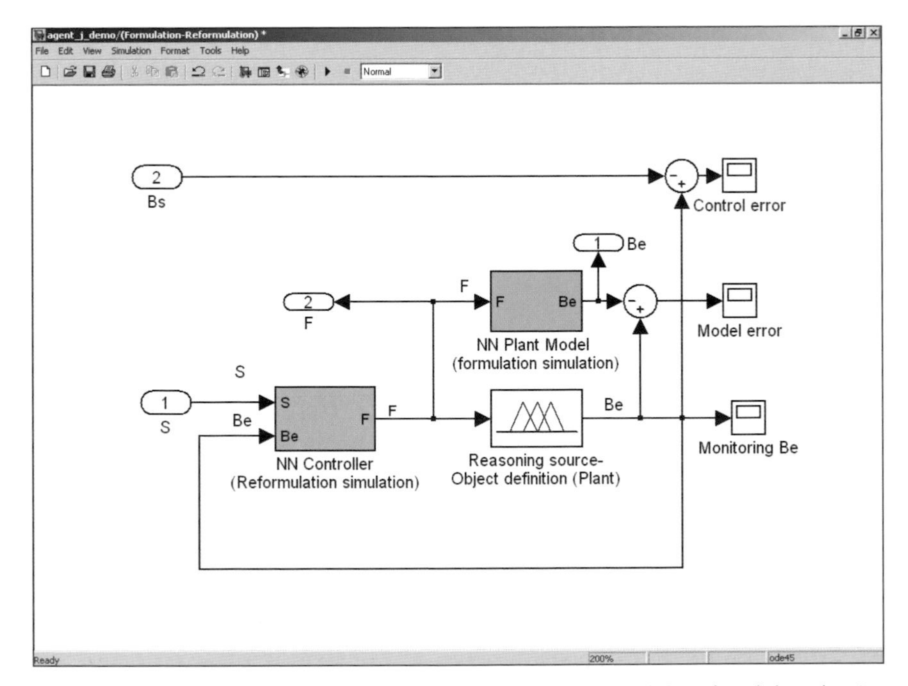

Figure 4B The control model of a generic agent in MATLAB-SIMULINK, formulation-reformulation subsystem

policies or agendas, and the third is a good paradigm for the interpretation of plans as designs.

4 Simulation and experimentation

The proposed plan design system is a prototype developed in a MATLAB-SIMULINK (MathWorks, Inc: Natick, Massachusetts) environment. This is a high-level programming environment that enables us to develop and experiment with the model by simulating its function and behaviour in a distributed plan-designing situation before introducing it in a real application.

How the proposed model works

The typical architecture of an artificial agent is shown in figure 4. It is developed in a SIMULINK (MathWorks, Inc: Natick, Massachusetts) environment where each box in the figure represents a subsystem of the overall system. We are experimenting with adaptive back-through control architectures. These structures typically use two neural networks: the controller (the system that controls) and the plant model (a model of the system to be controlled). First, the plant model is trained to approximate the reasoning sources (human operators or their computational models), by learning, either online or offline, the input-output patterns of FBS attributes. Then, these patterns are used 'backwards' as a guideline for the controller (Kecman 2001). Each box in figure 4 represents a subsystem of the entire model. The controller and the plant model are two subsystems that contain feed-forward neural networks used to represent the control abilities. The reference model box is another subsystem built on a similar control architecture, which represents the formulation-reformulation process (figure 4B). The reasoning source (or plant) box represents the system to be controlled and can be a human operator or a formal model. The system is connected with a VR sink that is used to visualise the evolution of the design description.

To model the reasoning sources, we have experimented with mathematical formulations that model the behaviour of objects (like motion, shape transformation and costs) based on state-space methodology, as well as with fuzzy systems. As an example, fuzzy systems are built on the basis of the fuzzy IF-THEN rules shown in figure 5A, which for example may represent expected proximity among land uses according to their volume dimensions shown in figure 5B. This is a simplified example that demonstrates behavioural expectations regarding the proximity of land uses expressed by a virtual participant in the design process.

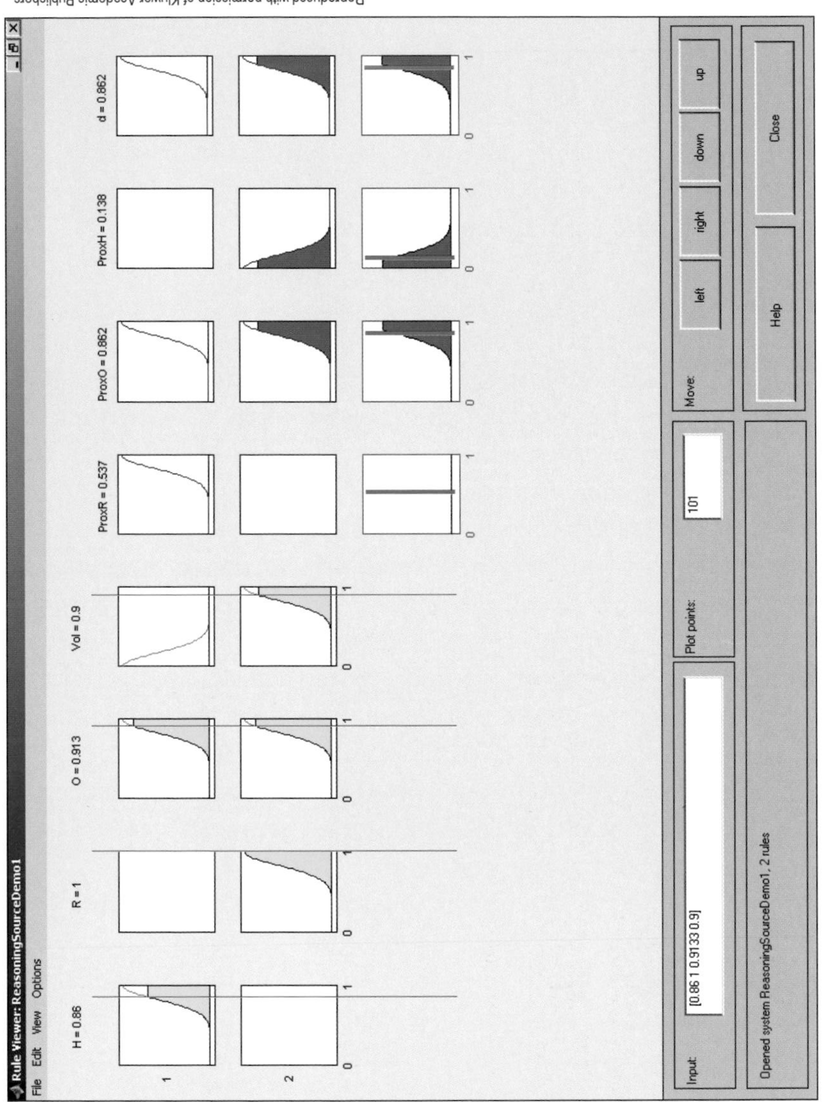
Rule Viewer: ReasoningSourceDemo1
File Edit View Options
H = 0.86
R = 1
O = 0.913
Vol = 0.9
ProxR = 0.537
ProxO = 0.862
ProxH = 0.138
d = 0.862
Input:
[0.86 1 0.9133 0.9]
Plot points:
101
Move:
left
right
down
up
Opened system ReasoningSourceDemo1, 2 rules
Help
Close

Figure 5

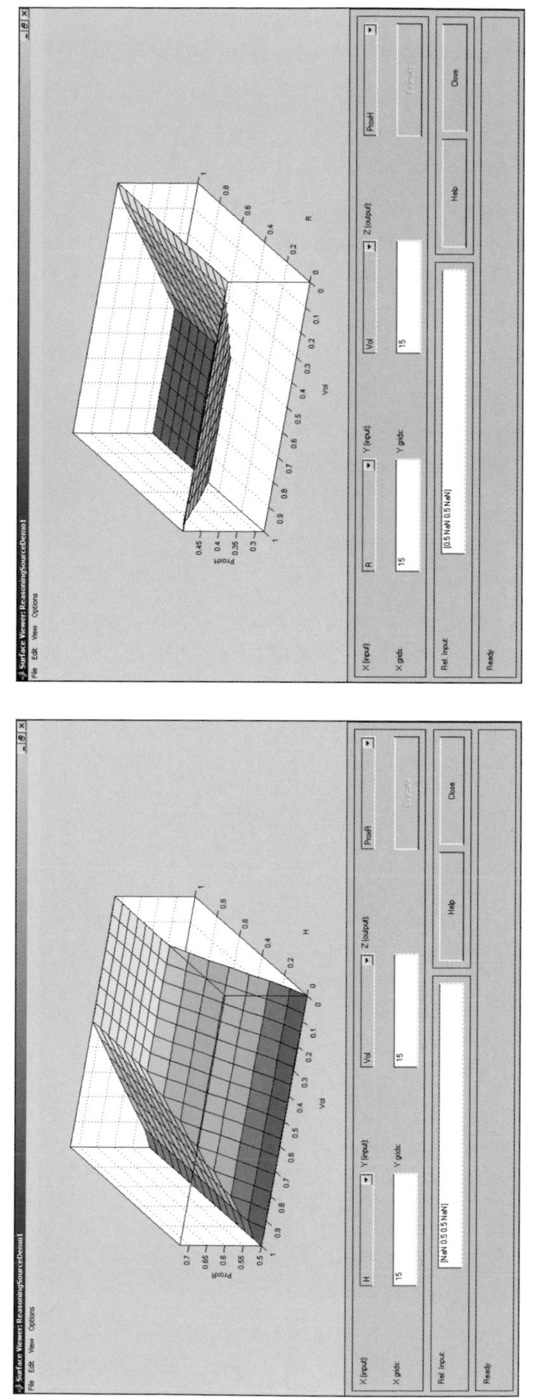

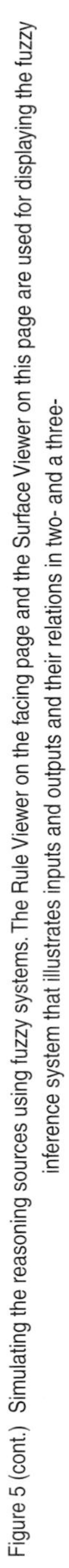

Figure 5 (cont.) Simulating the reasoning sources using fuzzy systems. The Rule Viewer on the facing page and the Surface Viewer on this page are used for displaying the fuzzy inference system that illustrates inputs and outputs and their relations in two- and a three-dimensional space

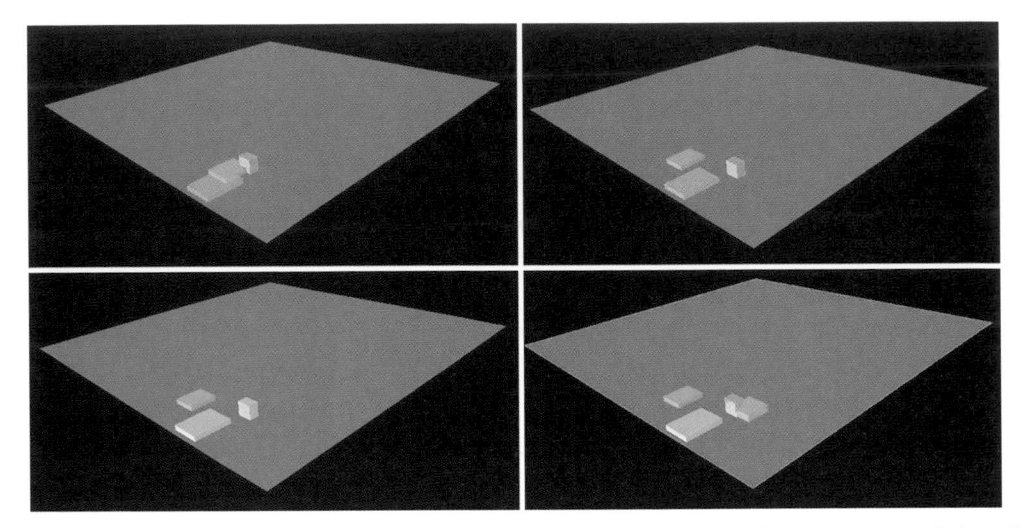

Figure 6 A conflict situation: A. In the series of pictures at the top, the housing facility (blue object) tends to move towards the retail unit (yellow object). The retail agent has a different expectation and moves away from the housing object. B. The blue agent may change its structure and add 'volume' (bottom left) so that it satisfies its expected behaviour, or it may change its functional description (bottom right) in order to adapt to the actual situation

The virtual reality toolbox offers the possibility to visualise the evolution of the design-decision space. We can directly retrieve and manipulate the location and shape variables of the three objects and view the conflict as it evolves in the three-dimensional space (see figure 6). For more information about the simulation see Alexiou and Zamenopoulos 2002.

Reflections on some preliminary simulations

Conflict in the current version of the model is introduced as disturbance in the control terminology. Conflict may arise because agents control parts of the overall description, and there might be variables affecting the performance of the agents that cannot be anticipated. Agents need to find appropriate/alternative solutions (functions or structures) that cope with uncertainties and conflict. To illustrate this in more detail, we present a conflict situation. In figure 6A, the agent that controls the blue housing unit, has developed an expectation to move close to the retail facility. In contrast, the agent that controls the yellow retail facility has developed an expectation to move far from the housing unit. This conflict situation can be resolved in two ways. Given that the retail cuboid stays unchanged, the agent that controls the housing unit might:

a. change function and hence expectations for the plan-design process (through the reformulation process) and adopt a mixed land use, in order to follow its structural behaviour and keep far from the existing retail unit in figure 6B (bottom left), or
b. radically change its structure and generate another cuboid that could facilitate retailing in close proximity to the housing, and therefore satisfy its original expectation in figure 6B (bottom right).

The process of artificial generation of plans based on learning control is a process of self-adaptation of agents that leads to coordination of their distributed descriptions. The simulations presented are still under development, so little can yet be claimed regarding the overall performance of the system and, hence, the validity of our hypotheses. Validating such systems which are generative in nature is, nevertheless, an important objective. One approach is to subject the resulting plan descriptions to evaluation by domain experts. The other approach, followed so far, is to stage different conflict scenarios, like the one described in figure 6, and to review the rationality of the results in each specific case.

In general, we suggest that plan designing based on human-model interaction (or, in the example shown in figure 6, based on interaction among human simulators and artificial agents) seems to cope well with the closed-loop iterative phenomena that may appear in human systems. This means that such models can be effective in avoiding entrapment in loops produced by the re-occurance of two or more conflicting views. This capability may be attributed to the problem reformulation process: it prompts humans to explore a wider range of alternative decisions and formulate conflict resolution scenarios in a timely manner. On the other hand, the ability of the system to reformulate problem definitions, and relax certain constraints might be problematic if the agents do not realistically reflect real human preferences. Weak learning performance on the part of the artificial agents could lead to a ceaseless problem reformulation loop. Similarly, generalisations produced in the learning process may be innefficient in coping with conflicting or alternative preferences expressed by individual humans. In each of these instances, the system's efficiency is primarily related to its learning ability, and much attention should be paid in devising and using the appropriate learning algorithms.

The architecture of the model is modular and we anticipate that it will be possible in the future to apply it to different problems and in different settings. Thus, we also envisage the possibility of incorporating this model into design and planning support systems. Harris and Batty (1993: 193-4) set out some requirements for planning support that correspond to two principal needs in planning:

first, the search for good alternative plans by way of an informed trial-and-error process; and second, the evaluation of the consequences of the generated alternatives. We suggest a third requirement: that of co-exploration and co-evolution of the way problem and alternatives are formulated and scaled into more complex decisions. This may also suggest an alternative approach to the way computers are used as support tools. The principle of complementary action of computers and humans is the predominant approach in both the design and planning decision support systems literature. Harris and Batty (1993: 195), for example, stress the importance of an appropriate clear division of labour between planners (and designers) and computers. In the system presented here, this division is not always clear. On the one hand, artificial agents play a clearly complementary role in enchancing the human ability to make decisions in complex environments, and to produce coordinated plans. On the other hand, artificial agents are trained to directly reflect human reasoning and knowledge, so the generation of coordinated plans is in fact an emergent characteristic of the human-model network. The benefits and drawbacks of this approach remain to be seen in future research in CASA, when we hope to test the above simulations using real human-computer networks.

Acknowledgements

Figures 1, 3, 4 and 5 are reproduced with kind permission of Kluwer Academic Publishers. Original publication: Zamenopoulos T, Alexiou K 2003 computer-aided creativity and learning in distributed cooperative human-machine networks. In *Digital design: research and practice, Proceedings of the 10th International Conference on Computer Aided Design Futures (CAAD Futures 2003)* :191-201. Chiu M-L, Jin-yeu T, Kvan T, Morozumi M, Jeng T (eds). The research reported here was partly supported through awards by the U.K. Engineering and Physical Sciences Research Council (EPSRC).

Zamenopoulos, T. and K. Alexiou. 2003. Computer-aided creativity and learning in distributed cooperative human-machine networks. In *Digital design: research and practice,* edited by M.L. Chiu, T. Jin-Yeu, T. Kvan, M. Morozumi and T. Jeng, 191-201. Proceedings of the 10th International Conference on Computer-Aided Design Futures (CAAD Futures 2003).

Chapter 21

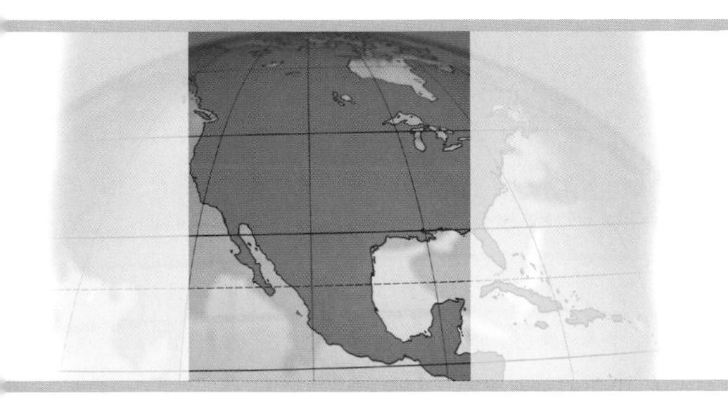

Researching the future of GIScience

Michael Batty and Paul A. Longley

To conclude this set of research snapshots of state-of-the-art GIScience, we reflect on the massive decentralisation that has taken place in world society in the last 30 years in terms of policies, opportunities and our ability to communicate with one another. Good GIScience is about using our understanding to adapt the ways in which we carry out our scientific investigations and improve the kind of findings that we are able to generate. We identify a series of theoretical imperatives from the themes we have developed here: disaggregation and the treatment of the individual in space, temporal dynamics, the simulation of spatial behaviour, networking, visualisation, communication, and public participation using GIS. We then sketch the practical problems of extending theory in applied contexts and appraise some of the ethical concerns that are coming to dominate this field: problems of ownership of data, copyright and the limits posed by changing conceptions of ownership which are changing the way we are able to share and communicate spatial information.

1 Where do we go from here?

It is no coincidence that as the miniaturisation of computer hardware has continued apace with the resultant invasion of digital media into every aspect of social life, the world has moved significantly from a paradigm in which centralised control from the top down has been replaced by a much more decentralised, bottom-up approach to organisation. The impact of this is evident throughout all the contributions we have gathered together here, from ways in which we now communicate information and expertise that was once garnered behind closed doors to ways in which we now deploy that information in real time. This decentralisation has been accompanied by the quest for much richer, more realistic and more applicable forms of knowledge that manifest themselves here in new approaches to representation and simulation at the finest geographical scales and temporal levels. This wave of change towards more decentralised thinking is as much a response to the limitations of past scientific practices as it is of new technology (see Goodchild and Longley 1999). Top-down strategies are now widely regarded as being rather insensitive ways of understanding and changing our systems of interest and as being democratically suspect in local contexts. However, representations of change that are intrinsically important at the scale of the individual must be cognisant that the rights of the same individuals are sacrosanct.

What is clear, however, is that this sea change in the way we think about the world has only just begun. Its course will run much longer, and we should expect that by the end of the next decade, by 2010 and certainly by 2020, the world will be a much more decentralised place than it is today. For one thing, there will be countless more millions of individuals, an increasing proportion of which will be educated to a standard where they will able to grasp the opportunities for social and economic interaction that a decentralised world offers. The impact that this will have on GIScience is, of course, difficult to predict, and, although a consequence could be more of the same—more of what has been reported in this book—it is more likely that there will be even more radical changes to the way we deal with geographic theory and its application than we can ever envisage through incremental thinking. Nevertheless, in trying to chart our path to the future and, more specifically, to define the research frontiers where many of the contributors to this volume will be working, we can at least draw out some critical issues that will become increasingly important in the next decade. We will organise these around three themes: theoretical imperatives which will change the very science that we are working with; practical problems that will dictate the way we build applications and the end users for whom we will build them; and ethical concerns which will determine the way we respond to problems and influence the GIScience we wish to further.

2 Theoretical imperatives

The drive from 2-D to 3-D GIS, which is illustrated in several chapters here, poses particular theoretical questions involving extension of the standard functionality of GIS into the third dimension. This raises rather specific problems. Much of the theory of 2-D space, for example in the urban and environmental realms, has been developed with only two dimensions in mind. Although there are standard functions of accessibility, buffering, spatial overlay using algebras, and related techniques that have been well developed for 2-D digital maps, their extension to 3-D is in its infancy and the problems posed differ. We suggest that this particular challenge will be hard to meet and slow to resolve despite the fact that the extension of GIS into the third dimension is likely to become routine within the next decade. For example, the remote-sensing techniques described by Smith (this volume) will be routinely used to ground and render the kinds of models illustrated by Hudson-Smith and Evans (this volume), while the CAD techniques described by Grajetzki and Shiode (this volume) will also be integrated fully with the multimedia and 3-D GIS used in the construction of virtual cities. In short, the visualisation capabilities of these developments are likely to be entirely realised as we indicated in the application of multimedia to public participation in Woodberry Down (Hudson-Smith et al, this volume), whereas the theoretical extension of spatial analysis into 3-D is likely to remain problematic. It is almost as if new theories of the 3-D realm in cities and related environments are required before the analytic capabilities of GIS can be extended from analyses such as those presented by Batty and Shiode (this volume). The theoretical challenges are great, although the existence of new data sources and routine visualisation will provide the springboard from which these can take place.

Perhaps more fundamental are the forces driving spatial theory towards local events, more detailed geographic scales, and temporal dynamics. Proprietary GIS software is often ill equipped to change the representation of space from bounded regions existing at a cross section in time. Events, the form and behaviour of which is dictated by dynamics, are hard, if not impossible, to represent within current standard software (Torrens, this volume; Barros and Alves-Junior, this volume). GIScience remains equally deficient, although these limitations have been noted for a long time. The problem is as much one of finding appropriate conceptualisation as it is of finding the requisite data to validate models of systems whose dominant features are revealed in their dynamics. The idea of the individual or the agent (Batty, this volume) is hard to represent in GIS especially when such a focus requires scientific understanding of the processes through which agents use, occupy and mould space. Although GIScience, as we have implied here, encompasses a broad church, there is little in contemporary spatial theory and analysis which augurs well for the development of consistent theory in this direction. Although several chapters have dealt with individuals and their behaviour, particularly the chapters by Torrens (this volume) on the land development process using cellular automata and agent-based systems and by Batty (this volume) on local scale movement in the context of pedestrians, the ways in which these models have been operationalised owes little to GIS software and is, at most, complementary to such efforts. Moreover, the chapter by Batty and Shiode (this volume) on more macro dynamics that lack an explicitly spatial component are more part of the development of regional science than GIScience. Whether or not such efforts are part of this science is, however, a fruitless debate, for what constitutes the science is what we define it to be and the research questions that drive it. Nevertheless, there are severe problems in developing the dynamics at fine levels of granularity where the agent rather than the aggregate is the prime concern.

Finer scales based on individual events, rather than aggregate patterns and the dynamic processes which define associated behaviour, are hard to incorporate in GIS, but so too is interaction. This is quite problematic in that interaction and connectivity are the defining issues of the emerging bottom-up network paradigm which is widely regarded as the most promising way of looking at complex systems of all kinds (Watts 2003). Despite promising advances in spatial interaction theory a generation or more ago, a vacuum in this research was created because the theory was cast in a static equilibrium mould. This is largely inconsistent with ideas about dynamics. Spatial dynamics, networks, and individual agents and events—all key ways of articulating our current concerns for urban and environmental systems—are largely absent from current theories which comprise the portfolio of GIScience. Our prediction is that although these will become the cutting edge, for the reasons already elaborated, progress will be slower than we would like it to be.

The advent of better data, of course, can both provide us with better support for our existing theories and force us to develop better ones. Better theory in turn usually forces us to collect better data. In questions where dynamics and individuals are paramount, then what we are able to do depends on the context that drives us to improve our theories and the technologies that enable us to collect better data. In agent-based dynamic modelling, for example, both of these forces will reshape parts of GIScience in the next decade. New local remote sensing (Smith, this volume), new kinds of geodemographics (Webber and Longley, this volume), and new methods of synthesising data from diverse sources (Longley et al, this volume) will force the pace, and we should see more appropriate theory emerging in the next decade. But the relativism of the context is important to note. In one sense, the questions that we are now posing are very different in kind from those that drove theory development two or more decades ago and that were reported in the first reader on spatial analysis edited by Berry and Marble (1968) at the heyday of the Quantitative Revolution in geography. Our current concern for the individual and the way we communicate our science to those likely to be affected by it is quite consistent with the highly decentralised societies that have emerged in the last 50 years where the individual reigns supreme. Our concern for participation and visualisation is reflected in the work of Hudson-Smith et al (this volume), Tobón and Haklay (this volume), and Lloyd et al (this volume).

Technology is hardly theory per se, although the process of miniaturisation will continue apace for the foreseeable future and will be reflected by much more routine uses of GIScience and GIS in mobile situations—on handheld devices rather than on the desktop or across fixed networks. Wireless will surely become the order of the day, and this scaling down to the routine and to the individual will again change the theoretical questions that will be addressed. For example, we already see a much greater concern for the relationships between space and behaviour, and it is easy to predict that this concern for human and individual behaviour is likely to change GIScience quite radically in the next decade. It has already begun, and it forges the link between new models of the individual, the agent and human behaviour (Li and Maguire, this volume).

3 Practical issues

In this book we have not said very much about the policy process per se, largely because our principal concern is with the science that supports such processes rather then the precise manner in which we articulate such behaviours. We have, however, given many examples of how GIScience might be used in planning and design, and we have assumed a complex, bottom-up context in which a wide range of connected interests are the recipients—the actors who manipulate and use the information that our models and techniques generate. Nevertheless, many of the techniques that we are using to advance and extend GIS are relevant to the ways in which we theorise the planning process. For example, Zamenopoulos and Alexiou (this volume) use neural network techniques from AI (artificial intelligence) to structure the coordination of actions which define the design process, concentrating on the way models can be used to generate and, thence, evaluate alternative scenarios. Not only does this approach build on the network paradigm that pervades this epilogue and will dominate the next decade, but these models are consistent with the ways in which our science is used in participation. In parallel fashion, the various techniques for visualisation discussed by Thurstain-Goodwin (this volume) in the context of spatial data surfaces and by Lloyd et al (this volume), in terms of the structure of spatial data, also support contemporary planning process as key examples illustrating how we adapt science to spatial context.

The practical problems of making such science operational, hence useful and usable do not only concern the way it is implemented. It is important to translate such theories and methods into techniques and algorithms that bring the science to fruition. For example, optimisation techniques based in location-allocation models that distribute facilities to optimal locations, thereby minimising some cost function or maximising some benefit-cost ratio, have been widely developed over the last 30 years. However, these models still remain difficult to use, largely because the solution spaces that characterise their structure do not easily yield to techniques for finding optimality. Accordingly, the focus in this area is on finding new techniques to speed to the optimisation process, indeed, even to find ingenious formulations of such problems in which solutions become identifiable. Schietzelt and Densham (this volume) explore these issues and show how intractable problems can be made tractable and, thence, can be embedded in wider solutions that reflect the decision support needed to enable successful conclusion to the problem-solving process. In a seemingly more abstract mode that is nevertheless essential to practical implementation, de Smith (this volume) shows how we need to transform space into a form which is meaningful, not only for understanding, but also for better implementing many of the more routine GIS techniques that are currently embodied in software.

Strategic planning of urban growth, and the accommodation of sprawl and peri-urbanisation, requires a science that complements contemporary GIS and urban dynamics models. Sprawl in Europe is underpinned by the search for indicators that are useful to planners and decision makers, and many of the visualisation techniques adapted to GIS in this book are relevant to the planning of urban development. Besussi and Chin (this volume) explain how this is being fashioned through supra-national policy research associated with the European Union. Barros and Alves-Junior (this volume) discuss similar issues in rapid-growth situations in Latin American cities and develop techniques for simulating urban change based on the agent-based and CA models discussed early in this book. More specific adaptations of standard GIS technology are dealt with by Evans and Steadman (this volume) who illustrate how well-developed cross-sectional static models of land use and transportation can be communicated using GIS in desktop and Web-based settings, where the object of translation from models to spatial structure is based on indicators derived from such models. These relate to the techniques developed by Lloyd et al (this volume) and Thurstain-Goodwin (this volume) which enable specific data outputs from interpolated and aggregated surface indicators to be communicated effectively to decision makers.

4 Ethical concerns

The massive drift towards decentralisation seen in the last 20 years in western societies provides at least the potential for much wider participation than hitherto. There are greater opportunities for involvement in a whole range of social activities, particularly the planning of local environments. This sets the context for widening the process of enfranchising not only the public at its grass roots but also politicians, decision makers, various special-interest groups, and, indeed, professional experts who themselves cannot hope to cover all aspects of their profession. Communicating information through new visualisation technologies is central to this process. Our example in Woodberry Down where the organisational infrastructure for participation is being complemented by our information technologies is a typical example of the substantial progress that is being made in this direction there and in many places (Hudson-Smith et al, this volume).

Issues about the ownership of information have taken on new meaning since the information revolution began two or more generations ago, and these issues have become increasingly significant. The information industries have seen great strides in the last decade in sharing information, in the development of new standards, and inter-operability between different hardware, software and even organisational environments. But the key issue remains ownership. This is compounded by the kinds of decentralised technology through networks that have appeared, the classic example being Napster, the system used to swap music freely on the

Internet. Information can be costly to produce but is usually very cheap to reproduce, and a central question is whether, if you have purchased information, you can share it with whomever you want to. The detail of this issue in particular instances is far from being resolved, as ever-newer and still less-centralised new technologies further assert the right of the individual. This kind of debate has enormous implications for using networks for the dissemination of information, and it is central to the participation debate. Much information used in GIS is proprietary in that costs are levied when it is purchased and sometimes when it is used. When this usage broadens to the wider community, copyright issues come into play, and this is dramatically limiting the diffusion of expert information and advice through GIS in situations where the basic data are proprietary. For example, again in our Woodberry Down example (Hudson-Smith et al, this volume), the restrictions on the use of Internet GIS software that arose out of copyright issues was an important problem when determining what software we could use to implement the system. The use of census data was also subject to similar limitations, despite the fact that the project was very definitely a public domain 'labour of love' with no profit being incurred at any stage.

There are of course major ethical issues concerning the collection of social as well as environmental data. Again, a key issue is whether or not others have the right to collect data about us without our permission. Clearly we have laws which assign government this right, but is a retailing chain or a banking institution ethically correct in compiling and then selling data about its customers if those customers have not given approval in the first place? Is the government in breach of its duty to its electors if it sells information about them, even back to those very electors themselves? These are some of the issues that are becoming critical as the information age gathers pace. An excellent example of this is the kind of data that are collected by remote devices such as laser scanning and closed circuit TV. Many British city centres, for example, are now completely covered by such TV, and police use it routinely to monitor crime. However, there are many private agencies and firms using footage without any mandate by the public at large. Our rights are confused in this area so, for example, although we may call for new methods of monitoring movements at the local scale in order that we can build the sorts of pedestrian models that Batty (this volume) presents, it is not clear if this is legally acceptable and, of course, the debate about the ethics of such collection has only recently emerged.

Lessig (2001) argues that the 'information commons' that we have come to accept as being the public domain, are gradually and surreptitiously being eroded as we march headlong into the information age. There are, of course, signs of opening up everywhere, but there are also signs of clamping down, not least in the wake of September 11. Copyright law is being extended into areas that traditionally have been the preserve of the individual, and much of the technology that we are involved in is being fashioned to aid and abet this process, however unwittingly. We consider that these issues will be central to GIScience in the next decade. What we can do in theory will as much be dictated by what we can do in terms of getting access to data that we know are available but not necessarily to us, as by what we would like to do and what we can do. The network revolution has brought this issue into focus, but the wireless revolution, which is only just beginning, will throw it into stark relief. We anticipate that data and information will become increasingly important issues in GIScience, and we are likely to see much greater activity in these areas than in new developments in theory. Our mid-term predictions for this science is that if another book like this is to be written in a few years, then it will likely be more concerned with data and how we can access them, albeit with those data being associated with new and different theoretical directions. Wired and wireless networks will be central to the way we develop and use our technologies, while the quest for participation, for communication, for the visualisation of our science will proceed apace.

Bibliography

3D Net Productions. 2002. Web city office towers 3D world. Accessed 2/9/2002. www.officetowers.com/florida/index.htm

Adamic, L. A. 1999. Zipf, power-laws, and Pareto—a ranking tutorial. Accessed 2/9/2002. ginger.hpl.hp.com/shl/papers/ranking/ranking.html

Ainley, P., J. Jameson, P. Jones, D. Hall and M. Farr. 2002. Redefining higher education: a case study in widening participation. In *Access and participation in higher education,* edited by A. Pajuska and A. Hayton, 89-105. London: Stylus Publishing.

Aitken, S. C. and R. Prosser. 1990. Residents' spatial knowledge of neighborhood continuity and form. *Geographical Analysis* 22:301-25.

Alexiou, K. and T. Zamenopoulos. 2001. A connectionist paradigm in the coordination and control of multiple self-interested agents. In *Proceedings of the 2nd National Conference Input 2001: Information Technology and Spatial Planning, Democracy and Technologies,* edited by G. Concilio and V. Monno. Politecnico di Bari, Dipartimento di Architettura e Urbanistica: 4. www.casa.ucl.ac.uk/NeWGis/A_Connectionist_Paradigm.pdf

____. 2002. Designing plans: a control based coordination model. CASA working paper 48. Centre for Advanced Spatial Analysis, University College London. www.casa.ucl.ac.uk/working_papers/paper48.pdf

Allen, P. M. 1997. *Cities and regions as self-organising systems: models of complexity.* Reading, United Kingdom: Gordon and Breach.

____. 1998. Evolving complexity in social science. In *Systems—new paradigms for the human sciences,* edited by R. Altman and W. Koch. Berlin: Walter de Gruyter.

Almeida, C. M. D., M. Batty, A. M. V. Monteiro, G. Câmara, B. S. Soares-Filho, G. S. Cerqueira and C. L. Pennachin. In press. Stochastic cellular automata modeling of urban land use dynamics: empirical development and estimation. Computer, environment and urban systems. Available online January 2003.

Amato, P. 1970. A comparison: population densities, land values and socioeconomic class in four Latin American cities. *Land Economics* 46:447-55.

Anas, A., R. Arnott and K. A. Small. 1998. Urban spatial structure. *Journal of Economic Literature* 36:1426-64.

Arnold, D. 1974. *Der Tempel des Königs Mentuhotep von Deir el-Bahari I. Architektur und Deutung.* Mainz: Verlag Philipp von Zabern.

Aurigi, A. and S. Graham. 1998. Cyberspace and the city: the virtual city in Europe. In *A companion to the city,* edited by G. Bridge and S. Watson, 134-47. Cambridge, Mass.: Blackwell.

Badawy, A. 1954, 1966, 1968. *A history of Egyptian architecture.* Vol. 1, Giza, 2-3. Berkeley: University of California Press.

Bailey, T. C. and A. C. Gatrell. 1995. *Interactive spatial data analysis.* Harlow, United Kingdom: Longman.

Barkan, J. D., P. J. Densham and G. Rushton. 2001. Designing better electoral systems for emerging democracies. Department of Geography, University of Iowa. Accessed 9/9/2002. www.uiowa.edu/%7Eelectdis

Barr, S. L. and M. J. Barnsley. 1999. Improving the quality of very high-resolution remotely sensed land-cover maps for the inference of urban land-use information recognition approaches. In *RSS 99 earth observation: from data to information.* ISPRS, Cardiff, United Kingdom: University of Wales.

Barrett, C. L., R. J. Beckman, K. P. Berkbigler, K. R. Bisset, B. W. Bush, K. Campbell, S. Eubank, K. M. Henson, J. M. Hurford, D. A. Kubicek, M. V. Marathe, P. R. Romero, J. P. Smith, L. L. Smith, P. E. Stretz, G. L. Thayer, E. Van Eeckhout and M. D. Williams. 2001. TRansportation ANalysis SIMulation System (TRANSIMS). Portland study reports. Los Alamos National Laboratory Reports LA-UR-01-5711, 5712, 5713, 5714, 5715. Los Alamos, Calif.: Los Alamos National Laboratory.

Barros, J. and F. Sobreira. 2002. City of slums: self-organisation across scales. CASA working paper 55. Centre for Advanced Spatial Analysis, University College London. www.casa.ucl.ac.uk/working_papers/Paper55.pdf

Batty, M. 1984. Plan design as committee decision making. *Environment and Planning B: Planning and Design* 11:279-95.

____. 1998. Urban evolution on the desktop: simulation using extended cellular automata. *Environment and Planning A* 30:1943-67.

____. 2000. Geocomputation using cellular automation. In *Geocomputation,* edited by S. Openshaw and R.J. Abrahart, 95-126. London: Taylor and Francis.

____. 2001a. Exploring isovist fields: space and shape in architectural and urban morphology. *Environment and Planning B* 28:123-50.

____. 2001b. Models in planning: technological imperatives and changing roles. *International Journal of Applied Earth Observation and Geoinformation* 3:252-66.

____. 2001c. Polynucleated urban landscapes. *Urban Studies* 38:635-55.

____. 2002. A decade of GIS: what next? *Environment and Planning B: Planning and Design* 29:157-8.

Batty, M., C. Chapman, S. Evans, M. Haklay, S. Kueppers, N. Shiode and P. M. Torrens., eds. 2000. *Computer visualisation for the Corporation of London.* 3 vols. and executive summary. London: Centre for Advanced Spatial Analysis.

Batty, M., C. Chapman, S. Evans, M. Haklay, S.Kueppers, N. Shiode, A. Smith and P. M. Torrens. 2001. Visualizing the city: communicating urban design to planners and decision makers. In *Planning support systems,* edited by R. Brail and R. Klosterman, 405-43. Redlands, Calif.: ESRI Press.

Batty, M., H. Couclelis and M. Eichen. 1997. Special issue: urban systems as cellular automata. *Environment and Planning B* 24 (2).

Batty, M., J. Desyllas and E. Duxbury. 2002. The discrete dynamics of small-scale spatial events: agent-based models of mobility in carnivals and street parades. CASA working paper 56. Centre for Advanced Spatial Analysis, University College London. www.casa.ucl.ac.uk/working_papers/Paper56.pdf

____. Forthcoming. Safety in numbers?: modelling crowds and designing control for the Notting Hill Carnival. *Urban Studies.*

Batty, M., B. Jiang and M. Thurstain-Goodwin. 1998. Local movement: agent-based models of pedestrian flows. CASA working paper 4. Centre for Advanced Spatial Analysis, University College London. www.casa.ucl.ac.uk/local_movement.doc

Batty, M. and S. K. Kwang. 1992. Form follows function: reformulating urban population density functions. *Urban Studies* 29:1043-70.

Batty, M. and P. A. Longley. 1994. Fractal cities: a geometry of form and function. San Diego, Calif.: Academic Press.

Batty, M. and P. Longley. 1997. The fractal city. *Architectural Design* 67 (9-10: Profile 129):74-83.

Batty, M. and A. Smith. 2001. Virtuality and cities: definitions, geographies, designs. In *Virtual reality in geography,* edited by P. Fisher and D. Unwin, 270-91. London: Taylor and Francis.

Batty, M. and P. M. Torrens. 2001. Modeling complexity: the limits to prediction. *CyberGeo* 201. www.cybergeo.presse.fr/ectqg12/batty/articlemb.htm

Bellman, C. and M. Shortis. 2000. Early stage object recognition using neural networks. *International archives of photogrammetry and remote sensing* XXXIII:Supplement B3.

Benenson, I. 1998. Multi-agent simulations of residential dynamics in the city. *Computers, Environment and Urban Systems* 22 (1):25-42.

Benenson, I., I. Omer and E. Hatna. 2002. Entity-based modeling of urban residential dynamics: the case of Yaffo, Tel Aviv. *Environment and Planning B: Planning and Design* 29:491-512.

Benenson, I. and P. M. Torrens. Forthcoming. Geosimulation: Object-based modeling of urban phenomena. *Computers, Environment and Urban Systems.*

Berger, M. 1999. Mobilité résidentielle et navettes domicile-trevail en Ile-de-France. *Espace, populations, sociétés* 2:207-17.

Berry, B. J. L. and D. F. Marble, eds. 1968. *Spatial analysis: a reader in statistical geography.* Englewood Cliffs, N.J.: Prentice-Hall.

Birkin, M. 1995. Customer targeting, geodemographics and lifestyle approaches. In GIS *for business and service planning,* edited by P. A. Longley and G. P. Clarke, 104-49. Cambridge, United Kingdom: GeoInformation International.

Birkin, M., G. P. Clarke and M. Clarke. 2002. *Retail intelligence and network planning.*

Chichester, United Kingdom: John Wiley & Sons, Inc.

Bizarre Creations. 2001. Metropolis Street Racer and Project Gotham. www.bizarrecreations.com

Blank, A. and S. Solomon. 2001. Power laws and cities population. Accessed 2/9/2002. xxx.tau.ac.il/html/cond-mat/0003240

Bonabeau, E., M. Dorigo and G. Theraulaz. 1999. *Swarm intelligence: from natural to artificial systems.* New York: Oxford University Press.

Boott, R., M. Haklay, K. Heppell and J. Morley. 2001. The use of GIS in Brownfield redevelopment. In *Innovations in GIS 8: Spatial Information and the Environment,* edited by P. Halls, 241-58. London: Taylor and Francis.

Bracken, I. 1989. The generation of socioeconomic surfaces for public policy making. *Environment and Planning B* 16:307-25.

Bracken, I. and D. Martin. 1989. The generation of spatial population distributions from census centroid data. *Environment and Planning A* 21:537-43.

Brail, R. and R. Klosterman, eds. 2001. *Planning support systems.* Redlands, Calif.: ESRI Press.

Brakman, S., H. Garretsen, C. Van Marrewijk and M. van den Berg. 1999. The return of Zipf: towards a further understanding of the rank-size distribution. *Journal of Regional Science* 39:183-213.

Breheny, M. and R. Rookwood. 1993. Planning the sustainable city region. In *Planning for a sustainable environment,* edited by A. Blowers. London: Earthscan.

Brimicombe, A. J. 2002. GIS: where are the frontiers now? Proceedings GIS 2002. Bahrain. 33-45.

Brody, H., M. R. Rip, P. Vinten-Johansen, N. Paneth and S. Rachman. 2000. Map-making and myth-making in Broad Street: the London cholera epidemic, 1854. *The Lancet* 356 (9223):64-68.

Brunsdon, C. 1995. Estimating probability surfaces for geographical point data—an adaptive kernel algorithm. *Computers and Geosciences* 21:877-94.

Brunton, G. and G. Caton-Thompson. 1928. *The Badarian Civilisation and predynastic remains near Badari.* London: Publications of the British School of Archaeology in Egypt and Egyptian Research Account.

Bullard, J. 2000. Sustaining technologies? Agenda 21 and U.K. local authorities' use of the World Wide Web. *Local Environment* 5:329-41.

Burrough, P. and R. McDonnell. 1998. *Principles of geographical information systems.* Oxford: Oxford University Press.

Burstedde, C., K. Klauck, A. Schadschneider and J. Zittarz. 2001. Simulation of pedestrian dynamics using a two-dimensional cellular automaton. *Physica A* 295:507-25.

Burton, E. 2000. The compact city: just or just compact? *Urban Studies* 37:1969-2001.

____. 2002. Measuring urban compactness in U.K. towns and cities. *Environment and Planning B* 29:219-50.

Camagni R, M. C. Gibelli and P. Rigamonti. 2002. Urban mobility and urban form: the social and environmental costs of different patterns of urban expansion. *Ecological Economics* 40:199-216.

Camazine, S., J. L. Deneubourg, N. R. Franks, J. Sneyd, G. Theraulaz and E. Bonabeau. 2001. *Self-organization in biological systems.* Princeton, N.J.: Princeton University Press.

Candy, L. 1998. Representations of strategic knowledge in design. *Knowledge-Based Systems* 11:379-90.

Carroll, G. R. 1982. National city-size distributions: what do we know after 67 years of research? *Progress in Human Geography* 6:1-43.

Center for Spatially Integrated Social Science (CSISS). 2001. Location-based services specialist meeting. Workshop Web site. Accessed November 2002 www.csiss.org/events/meetings/location-based/index.htm

Center on Social and Economic Dynamics. 2001. Ascape 1.9 (software). Washington, D.C. www.brook.edu/es/dynamics/models/ascape

Cervero, R. 2001. Efficient urbanisation: economic performance and the shape of the metropolis. *Urban Studies* 38:1651-71.

Champion, A. G. 1995. Analysis of change through time. In *Census users' handbook,* edited by S. Openshaw, 307-35. Cambridge, United Kingdom: GeoInformation International.

Chorley, R. and P. Haggett, eds. 1967. *Models in geography.* London: Methuen.

Chrisman, N. 1997. *Exploring geographic information systems.* New York: John Wiley & Sons, Inc.

Church, R. L. and P. Sorensen. 1996. Integrating normative location models into GIS: problems and prospects with the p-median model. In *Spatial analysis: modelling in a GIS environment,* edited by P. A. Longley and M. Batty, 167-83. Cambridge, United Kingdom: GeoInformation International.

Clark, C. 1951. Urban population densities. *Journal of the Royal Statistical Society. Series A.* 114:490-6.

Clarke, D., ed. 1977. *Spatial archaeology.* London: Academic Press.

Clarke, G. P. and M. Clarke. 2001. Applied spatial interaction modelling. In *Regional science in business,* edited by G. P. Clarke and M. Madden. Berlin: Springer.

Clarke, K. C. 1998. Visualising different geofutures. In *Geocomputation: a primer,* edited by P. A. Longley, S. M. Brooks, R. McDonnell and W. MacMillan, 119-137. Chichester, United Kingdom: John Wiley & Sons, Inc.

Cleveland, W. S. 1993. *Visualizing data.* New Jersey: Hobart Press.

Coombes, M. and S. Raybold. 2000. Policy-relevant surface data on population distribution and characteristics. *Transactions in GIS* 4:319-342.

Cooper, L. 1963. Location-allocation problems. *Operations Research* 11:331-43.

Cox, D. R. 1958. *Planning of experiments.* Chichester, United Kingdom: John Wiley & Sons, Inc.

Cox, J., D. Fell, M. Thurstain-Goodwin. 2002. *Red man green man: performance indicator for urban sustainability.* London: RICS Foundation.

Craig, W. J., T. M. Harris and D. Weiner. 2002. *Community participation and geographic information systems.* In Community participation and geographic information systems, edited by W. J. Craig, T. M. Harris and D. Weiner, 3-16. London: Taylor and Francis.

Dal Pozzolo, L., ed. 2002. *Fuori cittî, senza campagna.* Milan: Franco Angeli.

Darby, H. C., ed. 1963. *An historical geography of England before A.D. 1800.* Cambridge, United Kingdom: Cambridge University Press.

Daskin, M. S. 1987. Location, dispatching, and routing models for emergency services with stochastic travel times. In *Spatial analysis and location-allocation models,* edited by A. Ghosh and G. Rushton, 224-68. New York: Van Nostrand Reinhold Co.

Davis, F. D. 1989. Perceived usefulness, perceived ease of use, and user acceptance of information technology. *MIS Quarterly* 13:319-40.

Day, T. and J. P. Muller. 1989. Digital elevation model production by stereo-matching spot image pairs: a comparison of algorithms. *Image Vision Computing* 7 (2):95-101.

de la Barra, T. 1989. *Integrated land-use and transport modelling: decision chains and hierarchies.* Cambridge, United Kingdom: Cambridge University Press.

____. 2001. Integrated land-use and transport modeling: the TRANUS experience. In *Planning support systems: integrated geographic information systems, models, and visualization tools,* edited by R. K. Brail and R. E. Klosterman, 129-56. Redlands, Calif.: ESRI Press.

de Smith, M. J. 1981. Optimal location theory-generalisations of some network problems and some heuristic solutions. *Journal of Regional Science* 21:491-505.

Delahaye, D. 2001. Airspace sectoring by evolutionary computation. In *Spatial evolutionary modeling,* edited by R. Krzanowski and J. Raper, 180-202. New York: Oxford University Press.

Densham, P. J. 1991. Spatial decision support systems. In *Geographical information systems: principals and applications,* edited by D. J. Maguire, M. F. Goodchild and D. W. Rhind, 403-12. New York: John Wiley & Sons, Inc.

____. 1994. Integrating GIS and spatial modelling: visual interactive modelling and location selection. *Geographical Systems* 1:203-19.

____. 1996. Visual interactive locational analysis. In *Spatial analysis: modelling in a GIS environment,* edited by P. A. Longley and M. Batty, 185-205. Cambridge, United Kingdom: GeoInformation International.

Densham, P. J., M. P. Armstrong and K. K. Kemp. 1995. *Collaborative spatial decision making: scientific report for the specialist meeting.* Santa Barbara, Calif.: National Center for Geographic Information and Analysis.

Densham, P. J. and G. Rushton. 1992. A more efficient heuristic for solving large p-median problems. *Papers in Regional Science* 71:307-29.

Department of the Environment. 1996. *Revised PPG6: town centres and retail developments.* London: HMSO.

DiBiase, D. 1990. Visualization in the Earth Sciences. *Earth and Mineral Sciences Bulletin* 59:13-8.

Diggle, P. J. 1985. A kernel method for smoothing point process data. *Applied Statistics* 34:138-47.

Dijkstra, J., J. Jessurun and H. J. P. Timmermans. 2002. A multi-agent cellular automata model of pedestrian movement. In *Pedestrian and evacuation dynamics,* edited by M. Schreckenberg and S. D. Sharma, 173-80. Berlin: Springer-Verlag.

Dodge, M. 1999. Finding the source of the amazon.com: examining the hype of the 'earth's biggest bookstore'. CASA working paper 12. Centre for Advanced Spatial Analysis, University College London. www.casa.ucl.ac.uk/amazon.pdf

Doll, C. N. H. and J. P. Muller. 1999. The use of radiance calibrated nighttime imagery to improve remotely sensed population estimation. In *RSS 99 earth observation: from data to information.* Cardiff, United Kingdom: University of Wales.

Donnay, J. P., M. J. Barnsley and P. A. Longley. 2001. *Remote sensing and urban analysis.* London: Taylor and Francis.

Dornan, A. 2001. *The essential guide to wireless communications applications*. Upper Saddle River, N.J.: Prentice-Hall.

Douglas, D. H. 1994. Least cost path in GIS using an accumulated cost surface and slope lines. *Cartographica* 31:37-51.

Dutton, W. H., J. G. Blumler and K. Kramer, eds. 1987. *Wired cities: shaping the future of communications*. Boston, Mass.: G. K. Hall.

Egenhofer, M. J. and D. M. Mark. 1995. Naive geography. In *Spatial information theory: a theoretical basis for GIS,* edited by A. U. Frank and W. Kuhn, 1-15. Berlin: Springer.

Emery, W. B. 1991. *Archaic Egypt (second edition)*. Harmondsworth, United Kingdom: Penguin.

Ewing, R. 1994. Characteristics, causes and effects of sprawl: a literature review. *Environmental and Urban Issues FAU/FIU Joint Center.* (Winter):1-15.

Fabrikant, S. I. and B. P. Buttenfield. 2001. Formalizing spaces for information access. *Annals of the Association of American Geographers* 91:263-80.

Feldman, E., F. A. Lehrer and T. L. Ray. 1966. Warehouse locations under continuous economies of scale. *Management Science* 12:670-84.

Fisher, P. 1999. Models of uncertainty in spatial data. In *Geographical information systems: principles, techniques, management and applications,* edited by P. A. Longley, M. F. Goodchild, D. J. Maguire and D. W. Rhind, 191-205. New York: John Wiley & Sons, Inc.

Forte, M. and A. Siliotti, eds. 1997. *Virtual archaeology.* London: Thames and Hudson.

Fotheringham, A. S., C. Brunsdon and M. Charlton. 2001. *Quantitative geography: perspectives on spatial data analysis*. London: Sage Publications.

Fraser, C. S., E. Baltsavias and A. Gruen. 2002. Processing of IKONOS imagery for submitter 3-D positioning and building extraction. *ISPRS Journal of Photogrammetry and Remote Sensing* 1209.

Friedrich, C. J. 1929. *Alfred Weber's theory of the location of industries*. Chicago: University of Chicago Press.

Gabaix, X. 1999. Zipf's law for cities: an explanation. *Quarterly Journal of Economics* 114:739-67.

Gahegan, M. 2001. Visual exploration in geography: analysis with light. In *Geographic data mining and knowledge discovery,* edited by H. J. Miller and J. Han, 260-87. London: Taylor and Francis.

Galster, G., R. Hanson, M. R. Ratcliffe, H. Wolman, S. Coleman and J. Freihage. 2001. Wrestling sprawl to the ground: defining and measuring an elusive concept. *Housing Policy Debate* 12:681-717.

Gamba, P. and B. Houshmand. 2000. Digital surface models and building extraction: a comparison of IFSAR and LiDAR data. *IEEE Transactions on Geoscience and Remote Sensing* 38:1959-68.

Garey, M. R. and D. S. Johnson. 1979. *Computers and intractability: a guide to the theory of NP-completeness.* New York: W. H. Freeman.

Gerke, M., C. Heike and B. M. Straub. 2001. Building extraction from aerial imagery using a generic scene model and invariant geometric moments. IEEE/ISPRS joint workshop on remote sensing and data fusion over urban areas.

Gero, J. S., ed. 1985. *Design optimization.* Orlando, Fla.: Academic Press.

____. 2000. Computational models of innovative and creative design processes. *Technological Forecasting and Social Change* 64:183-96.

Ghosh, A. and G. Rushton. 1987. Progress in location-allocation modeling. In *Spatial analysis and location-allocation models,* edited by A Ghosh and G. Rushton, 1-20. New York: Van Nostrand Reinhold.

Gladwell, M. 2001. *The tipping point: how little things can make a big difference.* New York: Little, Brown and Company.

Glover, F. 1986. Future paths for integer programming and links to artificial intelligence. *Computers and Operations Research* 5:533-49.

Gober, P. 1990. The urban demographic landscape: a geographic perspective. In *Housing demography. Linking demographic and housing markets,* edited by D. Myers, 232-248. Madison, Wisc.: University of Wisconsin Press.

Goldstein, H. 1995. *Multilevel statistical models.* 2d ed. London: Edward Arnold.

Golledge, R. G. 1978. Learning about urban environments. In *Timing space and spacing time I: making sense of time,* edited by T. Carlstein, D. N. Parkers and N. J. Thrift, 76-98. London: Edward Arnold.

Golledge, R. G., V. Dougherty and S. Bell. 1995. Acquiring spatial knowledge: survey versus route-based knowledge in unfamiliar environments. *Annals of the Association of American Geographers* 85:134-58.

Golledge, R. G. and R. J. Stimson. 1987. *Spatial behavior: a geographic perspective.* New York: Guilford Press.

Goodchild, M. F. 1977. An evaluation of lattice solutions to the corridor location problem. *Environment and Planning A* 9:727-38.

Goodchild, M. F. and P. A. Longley. 1999. The future of GIS and spatial analysis. In *Geographical information systems: principles, techniques, management and applications,* edited by P. A. Longley, M. F. Goodchild, D. J. Maguire and D. W. Rhind, 567-80. New York: John Wiley & Sons, Inc.

Goodchild, M. F. and V. Noronha. 1983. *Location-allocation for small computers,* monograph no. 8. Iowa City Department of Geography, University of Iowa.

Graham, S. and S. Marvin. 1999. Planning cyber-cities? Integrating telecommunications into urban planning. *Town Planning Review* 70:98-114.

Grecu, D. L. and D. C. Brown. 1998. Dimensions of machine learning in design. *Artificial Intelligence in Engineering Design, Analysis and Manufacturing* 12:117-21.

Guérois, M. and D. Pumain. 2002. *Urban sprawl in France* (1950-2000). Milan: Franco Angeli.

Haklay, M., M. Thurstain-Goodwin, D. O'Sullivan and T. Schelhorn. 2001. So go downtown: simulating pedestrian movement in town centers. *Environment and Planning B* 28:343-59.

Hall, P. 1983. Decentralization without end? A re-evaluation. In *Essays in honor of Professor Guttmann*, 125-55. London: Academic Press.

____. 1993. Forces shaping urban Europe. *Urban Studies* 30 (8):883-898.

____. 1997. *Cities of tomorrow: an intellectual history of urban planning and design in the twentieth century.* Oxford: Blackwell.

Harper, G. 2002. Using surfaces to inform local policy: modelling deprivation in Brent. Master's thesis, Department of Geomatic Engineering, University College London.

Harris, B. and M. Batty. 1993. Locational models, geographic information and planning support systems. *Journal of Planning Education and Research* 12:184-98.

Harris, R. J. and P. A. Longley. 2002. Creating small area measures of urban deprivation. *Environment and Planning A* 34:1073-93.

Harris, T. and D. Weiner. 1996. GIS and Society—Scientific report of the I-19 meeting. NCGIA technical report 96-7. Santa Barbara, Calif. National Center for Geographic Information and Analysis.

Hart, R. A. and G. T. Moore. 1973. The development of spatial cognition: a review. In *Image and Environment*, edited by R. M. Downs and D. Stea, 246-88. Chicago: Aldine.

Harvey, D. 1969. *Explanation in geography.* London: Edward Arnold.

Harvey, E. O. and W. Clark. 1965. The nature and economics of urban sprawl. *Land Economics* 41:1-9.

Havlin, S. 1995. The distance between Zipf plots. *Physica A* 216:148-50.

Heinrich, E. 1982. *Die Tempel und Heiligtümer im Alten Mesopotamien.* Vol 14. Denkmäler Antiker Architektur. Berlin: de Gruyter.

Helbing, D. 1991. A mathematical model for the behavior of pedestrians. *Behavioral Science* 36:298-310.

Helbing, D., I. Farkas and T. Vicsek. 2000. Simulating dynamical features of escape panic. *Nature* 407:487-90.

Helbing, D., P. Molnar, I. Farkas and K. Bolay. 2001. Self-organizing pedestrian movement. *Environment and Planning B* 28:361-83.

Hensher, D. A. and L. W. Johnson. 1981. *Applied discrete choice modelling*. London: Croom Helm.

Hillsmann, E. L. 1984. The p-median structure as a unified linear model for location-allocation analysis. *Environment and Planning A* 16:305-18.

Holland, J. H. 1995. *Hidden order: how adaptation builds complexity*. Reading, Mass.: Addison-Wesley.

Hopkins, L. D. 2001. *Urban development: the logic of making plans*. Washington, D.C.: Island Press.

Hrycej, T. 1997. *Neurocontrol: towards an industrial control methodology*. New York: John Wiley & Sons, Inc.

Hunter, P. 2001. Location, location, location. *Computer Weekly* 5:64.

Infocities. 1997. About Infocities. www.infocities.eu.int

ISP. 2002. *Carnival public safety project—assessment of route design for the Notting Hill Carnival*. London: Intelligent Space Partnership for the Greater London Authority.

Jacobs, O. L. R. 1993. *Introduction to control theory*. New York: Oxford Science Publications.

Jankowski, P. and T. Nyerges. 2001. *Geographic information systems for group decision making: towards a participatory geographic information science*. London: Taylor and Francis.

Jennings, N. R. 1996. Coordination techniques for distributed artificial intelligence. In *Foundations of distributed artificial intelligence*, edited by G. M. P. O'Hare and N. R. Jennings, 187-210. New York: John Wiley & Sons, Inc.

Johnston, R. A. and T. de la Barra. 2000. Comprehensive regional modelling for long-range planning: linking integrated urban models and geographical information systems. *Transportation Research A* 34:125-36.

Johnston, R. J. 1999. Geography and GIS. In *Geographical information systems: principles, techniques, management and applications*, edited by P. A. Longley, M. F. Goodchild, D. J. Maguire and D. W. Rhind, 39-47. New York: John Wiley & Sons, Inc.

Kecman, V. 2001. *Learning and soft computing: support vector machines, neural networks and fuzzy logic models*. Cambridge, Mass.: MIT Press.

Kemp, B. J. 1989. *Ancient Egypt*. London: Routledge.

Kingston, R. 2002. Web-based PPGIS in the United Kingdom. In *Community participation and geographic information systems*, edited by W. J. Craig, T. M. Harris and D. Weiner, 101-12. London: Taylor and Francis.

Kitazawa, K. 2000. A method of map matching for personal positioning systems. The 21st Asian conference on remote sensing. Taiwan.

Klincewicz, J. G. 1991. Avoiding local optima in the p-hub location problem using tabu search and grasp. *Annals of Operations Research* 14:283-302.

Klosterman, R. 1998. Computer applications in planning. *Environment and Planning B*. Planning and Design Anniversary Issue:32-6.

Knapp, L. 1995. A task analysis approach to the visualization of geographic data. In *Cognitive aspects of human-computer interaction for geographic information systems,* edited by T. L. Nyerges, D. M. Mark, R. Laurini and M. J. Egenhofer, 355-72. Dordrecht, The Netherlands: Kluwer Academic Publishers.

Kozlowski, L. T. and K. J. Bryant. 1977. Sense of direction, spatial orientation and cognitive maps. *Journal of Experimental Psychology: Human Perception and Performance* 4:590-8.

Krugman, P. 1996. *The self-organizing economy.* Cambridge, Mass.: Blackwell.

Krzanowski, R. and J. Raper, eds. 2001. *Spatial evolutionary modeling, spatial information systems.* New York: Oxford University Press.

Kuehn, A. A. and M. J. Hamburger. 1963. A heuristic program for locating warehouses. *Management Science* 9:643-66.

Kuipers, B. 1978. Modeling spatial knowledge. *Cognitive Science* 2:129-53.

Laherrere, J. and D. Sornette. 1998. Stretched exponential distributions in nature and economy: fat tails with characteristic scales. *European Physics Journal* B 2:525-39.

Laurini, R. 2001. *Information systems for urban planning: a hypermedia cooperative approach.* London: Taylor and Francis.

Lautso, K. 2003. The SPARTACUS system for defining and analysing sustainable urban land use and transport policies. In *Planning support systems in practice,* edited by S. Geertman and J. Stillwell, 453-63. Heidelberg, Germany: Springer Verlag.

Lawless, J. 2001. Urban roof morphology extraction using random sampling of LiDAR data. Master's thesis, University College London.

Lessig, L. 2001. *The future of ideas: the fate of the commons in a connected world.* New York: Random House.

Levy, M. and S. Solomon. 1996a. Dynamical explanation for the emergence of power law in a stock market model. *International Journal of Modern Physics* C 7:65-72.

____. 1996b. Power laws are logarithmic Boltzmann laws. *International Journal of Modern Physics* C 7:595-601.

Li, W. 1999. Zipf's Law. Accessed 2/9/2002. linkage.rockefeller.edu/wli/zipf

Lillesand, T. M. and R. W. Kiefer. 2000. *Remote sensing and image interpretation.* 4th ed. New York: John Wiley & Sons, Inc.

Longley, P. A. and M. Batty, eds. 1996. *Spatial analysis: modelling in a GIS environment.* Cambridge, United Kingdom: GeoInformation International.

Longley, P. A. and G. P. Clarke, eds. 1995. *GIS for business and service planning.* New York: John Wiley & Sons, Inc.

Longley, P. A., M. F. Goodchild, D. J. Maguire and D. W. Rhind. 2001. *Geographic information systems and science.* Chichester, United Kingdom: John Wiley & Sons, Inc.

Longley, P. A. and T. V. Mesev. 2002. Measurement of density gradients and space filling in urban systems. *Papers in Regional Science* 81:1-28.

Longley P. A. and C. Tobón. 2002 The scale of urban deprivation. Spatial dependence and heterogeneity in patterns of hardship in urban areas. CASA working paper 59. Centre for Advanced Spatial Analysis, University College London. www.casa.ucl.ac.uk/working_papers.htm#COMPLETE%20LIST

Love, R. F., J. G. Morris and G. O. Wesolowsky. 1988. *Facilities location, models and methods.* New York: North-Holland.

MacEachren, A. M. 1995. *How maps work: representation, visualization and design.* New York: Guilford Press.

Maguire, D. J. 1995. Implementing spatial analysis and GIS applications for business and service planning. In *GIS for business and service planning,* edited by P. A. Longley and G. P. Clarke, 171-91. Cambridge, United Kingdom: GeoInformation International.

____. 2001. Mobile geographic services come of age. *GeoInformatics* 4:6-9.

Maher, M. L. 2000. A model of co-evolutionary design. *Engineering with Computers* 16: 195-208.

Maher, M. L. and P. Pu. 1997. *Issues and applications of case based reasoning in design.* Mahwah, N.J.: Lawrence Erlbaum Associates.

Malacarne, L. C., R. S. Mendes, I. T. Pedron and E. K. Lenzi. 2001. Nonlinear equations for anomalous diffusion: unified power-law and stretched exponential exact solution. *Physical Review* E 63:030101-4.

Malone, T. W. and K. Crowston. 1990. What is coordination theory and how can it help design cooperative work systems? In Proceedings of the Conference on Computer-Supported Cooperative Work. Los Angeles, Calif.

Malpezzi, S. and Guo Wen-Kai. 2001. Measuring 'sprawl': alternative measures of urban form in U.S. metropolitan areas. Working paper. Center for Urban Land Economics Research, University of Wisconsin.

Malthus, T. and C. J. Younger. 1999. Remotely sensing stress in street trees. In *RSS 99 earth observation: from data to information.* Cardiff, United Kingdom: University of Wales.

Mandelbrot, B. B. 1966. Information theory and psycholinguistics: a theory of word frequencies. In *Readings in mathematical social science,* edited by P. Lazarsfeld and N. Henry. Cambridge, Mass.: MIT Press.

Maranzana, F. E. 1964. On the location of supply points to minimize transport costs. *Operational Research Quarterly* 15:261-70.

Mark, D. M. 1999. Spatial representation: a cognitive view. In *Geographical information systems: principles, techniques, management and applications,* edited by P. A. Longley, M. F. Goodchild, D. J. Maguire and D. W. Rhind, 81-9. New York: John Wiley & Sons, Inc.

Martin, D. 1989. Mapping population data from zone centroid locations. *Transactions of the Institute of British Geographers* 14:90-7.

McHarg, I. L. 1969. *Design with nature.* New York: Natural History Press.

Medyckyj-Scott, D. and H. M. Hearnshaw. 1993. *Human factors in geographical information systems.* London: Belhaven Press.

Mino, E. 2000. Experiences of European digital cities. In *Digital cities: technologies, experiences, and future perspectives,* edited by T. Ishida and K. Isbister, 45-56. Berlin: Springer.

Mitasova, H. and L. Mitas. 1998. Process modeling and simulations. Part of *NCGIA core curriculum in GIScience.* www.ncgia.ucsb.edu/giscc/units/u130/u130.html

Mitchell, W. J. 1995. *City of bits: space, place, and the Infobahn.* Cambridge, Mass.: MIT Press.

Montello, D. R. 1995. How significant are cultural differences in spatial cognition? In *Spatial information theory,* edited by A. U. Frank and W. Kuhn, 485-500. Berlin: Springer.

Naville, E. 1910. *The XIth Dynasty temple at Deir el-Bahari II.* London: Egypt Exploration Fund.

Newman, P. and J. Kenworthy. 1999. *Sustainability and cities: overcoming automobile dependence.* Washington, D.C.: Island Press.

Nielsen, J. 1993. *Usability engineering.* San Diego: Morgan Kaufmann.

Nunamaker, J. F., A. R. Dennis, J. S. Valacich, D. Vogel and J. D. George. 1991. Electronic meeting systems. *Communications of the ACM* 34:40-61.

Nyerges, T. L., D. M. Mark, R. Laurini and M. J. Egenhofer, eds. 1995. *Cognitive aspects of human-computer interaction for geographic information systems.* Dordrecht, The Netherlands: Kluwer Academic Publishers.

Office of Government and Commerce. 2001. XML Specification. www.ogc.gov.uk

Office of the Deputy Prime Minister. 2002. Producing boundaries and statistics for town centres. London pilot study summary report. London: The Stationery Office.

Office of the e-Envoy. 2002. Briefings. www.e-envoy.gov.uk/oee/oee.nsf/sections/index/$file/index.htm

Openshaw, S. 1984. *The modifiable areal unit problem.* Concepts and techniques in modern geography 38. Norwich, United Kingdom: GeoBooks.

Ordnance Survey. 2002. OS MasterMap®. Home page. www.ordnancesurvey.co.uk/osmastermap

Orfeuil, J. P. 1996. Urbain et périurbain: qui va où? *Urbanisme* 289:52-7.

Ossowski, S. 1999. *Coordination in artificial agent societies: social structure and its implications for autonomous problem-solving agents.* Berlin and Heidelberg, Germany: Springer-Verlag.

O'Sullivan, D. and P. M. Torrens. 2000. Cellular models of urban systems. In *Theoretical and practical Issues on cellular automata,* edited by S. Bandini and T. Worsch, 108-117. London: Springer-Verlag.

Peeters, D. and I. Thomas. 1997. Distance-Lp et localisations optimales. Simulations sur un semis aléatoire de points. *Cahiers Scientifiques du Transport* 31:55-70.

Peiser, R. 2001. Decomposing urban sprawl. *Town Planning Review* 72:275-98.

Pendall, R. 1999. Do land-use controls cause sprawl? *Environment and Planning B* 26: 555-71.

Pesaresi, M. and A. Bianchin. 2001. Recognising settlement structure using mathematical morphology and image texture. In *Remote sensing and urban analysis,* edited by J. P. Donnay, M. J. Barnsley and P. A. Longley, 55-67. London: Taylor and Francis.

Petrie, W. M. F. 1930. *Antaeopolis, the tombs of Qau.* Publications of the British School of Archaeology in Egypt 51. London: Quaritch.

Petrie, W. M. F. and F. J. E. Quibell. 1896. *Naqada and Ballas.* London: British School of Archaeology in Egypt and Egyptian Research Account.

Piaget, J. and B. Inhelder. 1956. *The child's conception of space.* London: Routledge and Kegan Paul.

Portugali, J. 2000. *Self-organization and the city.* Berlin: Springer.

Power, A. 1998. *Estates on the edge.* London: Palgrave Macmillan.

Preece, J., Y. Rogers, H. Sharp, D. Benyon, S. Holland and T. Carey. 1994. *Human-computer interaction.* Harlow, United Kingdom: Addison-Wesley.

Prud'homme, R. and G. Lee. 1999. Sprawl, speed and the efficiency of cities. *Urban Studies* 36:1849-58.

Razin, E. and M. Rosentraub. 2000. Are fragmentation and sprawl interlinked? North American evidence. *Urban Affairs Review* 35:821-36.

Reed, W. J. 2001. The Pareto, Zipf and other power laws. *Economics Letters* 74:15-19.

Reynolds, C. 1987. Flocks, herds, and schools: a distributed behavioral model. *Computer Graphics* 21 (4):25-34.

Rhind, D. W., ed. 1997. *Framework for the world.* Cambridge, United Kingdom: GeoInformation International.

____. 1999. GIS management issues. In *Geographical information systems: principles, techniques, management and applications,* edited by P. A. Longley, M. F. Goodchild, D. J. Maguire and D. W. Rhind, 583-8. New York: John Wiley & Sons, Inc.

Richardson, H. W. 1973. Theory of the distribution of city sizes: review and prospects. *Regional Studies* 7:239-51.

Ricke, H. 1965. *Das Sonnenheiligtum des Königs Userkaf I,* Beiträge zur ägyptsichen Bauforschung und Altertumskunds 7. Cairo: Schweizerisches Institut für ägyptische Bauforschung und Altertumskunde.

Robinson, W. S. 1950. Ecological correlation and the behaviour of individuals. *American Sociological Review* 67:351-357.

Rosing, K. E. and M. J. Hodgson. 2002. Heuristic concentration for the p-median: an example demonstrating how and why it works. *Computers and Operations Research* 29:1317-30.

Rosing, K. E. and C. S. ReVelle. 1997. Heuristic concentration: two stage solution concentration. *European Journal of Operational Research* 97:75-86.

Rosing, K. E., C. S. ReVelle and D. A. Schilling. 1999. A gamma heuristic for the p-median problem. *European Journal of Operational Research* 117:522-32.

Rossano, M. J., S. O. West, T. J. Robertson, M. C. Wayne and R. B. Chase. 1999. The acquisition of route and survey knowledge from computer models. *Journal of Environmental Psychology* 19:101-15.

Rowe, N. C. and R. S. Ross. 1990. Optimal grid-free path planning across arbitrarily contoured terrain with anisotropic friction and gravity effects. *IEEE Transactions on Robotics and Automation* 6:540-53.

Rubenstein-Montano, B. 2000. A survey of knowledge-based information systems for urban planning: moving towards knowledge management. *Computers, Environment and Urban Systems* 24:155-72.

Ruddle, R. A., S. J. Payne and D. M. Jones. 1997. Navigating buildings in 'desk-top' virtual environments: experimental investigations using extended navigational experience. *Journal of Experimental Psychology: Applied* 3:143-59.

Rydin, Y. 1999. Public participation in planning. In *British planning: 50 years of urban and regional policy,* edited by B. Cullingworth, 184-97. London: Athlone Press.

Schama, S. 2000. *A history of Britain: at the edge of the world?* 3000 B.C.–A.D. 1603. London: BBC Consumer Publishing.

Schelhorn, T., D. O'Sullivan and M. Thurstain-Goodwin. 1999. STREETS: an agent-based pedestrian model. CASA working paper. Centre for Advanced Spatial Analysis, University College London. www.casa.ucl.ac.uk/working_papers

Schilling, D. A., K. E. Rosing and C. S. ReVelle. 2000. Network distance characteristics that affect computational effort in p-median location problems. *European Journal of Operational Research* 127:525-36.

Schroeder, R., A. Huxor and A. Smith. 2001. Activeworlds: geography and social interaction in virtual reality. *Futures* 33:569-87.

Servant, L. 1996. L'automobile dans la ville. *Cahiers du IAURIF* 114.

Shemyakin, F. N. 1962. General problems of orientation in space and space representations. In *Psychological sciences in the USSR*. Vol.1 NTIS report no. TT62-11083, edited by B. G. Anan'yev, 184-225. Washington, D.C.: Office of Technical Services.

Siegel, A. W. and S. H. White. 1975. The development of spatial representations of large-scale environments. In *Advances in child development and behavior,* edited by W. H. Reese, 9-55. New York: Academic Press.

Šiljak, D. D. 1991. *Decentralized control of complex systems.* San Diego, Calif.: Academic Press.

Simon, H. A. 1955. On a class of skew distribution functions. *Biometrika* 42:425-40.

Sleight, P. 1997. *Targeting customers: how to use geodemographic and lifestyle data in your business.* 2d ed. Henley-on-Thames, United Kingdom: NTC Publications.

Smith, A. H. 2001. 30 days in Activeworlds—community, design and terrorism in a virtual world. In *The social life of avatars: presence and interaction in shared virtual environments,* edited by R. Schroeder, 77-89. Berlin: Springer.

Smith, M. J., D. G. Smith, M. Tragheim and D. Holt. 1997. DEMs and orthophotographs from aerial photographs. *Photogrammetric Record* 15:945-50.

Snyder, K. 2002. Tools for community design and decision making. In *Planning support systems in practice,* edited by S. Geertman and J. Stillwell. Berlin: Springer.

Sohn, G. and I. Dowman. In press. Extraction of buildings from high-resolution satellite data. In *Automated extraction of man-made objects from aerial and space images (III),* edited by E. Baltsavias, A. Gruen and I. von Gool. Lisse: Balkema.

Sornnette, D. and R. Cont. 1997. Convergent multiplicative processes repelled from zero: power laws and truncated power laws. *J. Phys. I France* 7:431-44.

Stadelmann, R. 1991. *Die Ägyptischen Pyramiden.* 2d ed. Vom Ziegelbau zum Weltwunder. Mainz: Verlag Philipp von Zabern.

Stanley, H. E., L. A. N. Amaral, S. V. Buldyrev, A. L. Goldberger, S. Havlin, H. Leschhorn, P. Maass, H. A. Makse, C. K. Peng, M. A. Salinger, M. H. R. Stanley and G. M. Viswanthan. 1996. Scaling and universality in animate and inanimate systems. *Physica A* 231:20-48.

Steadman, J. P. 1983. *Architectural morphology: an introduction to the geometry of building plans.* London: Pion Limited.

Steckeweh, H. 1936. *Die Fürstengräber von Qaw.* Leipzig, Germany: Hinrichs.

Stern, E. and D. Leiser. 1988. Levels of spatial knowledge and urban travel modeling. *Geographical Analysis* 20:140-55.

Stewart, J. Q. 1950. The development of social physics. *American Journal of Physics* 18: 239-53.

Still, G. K. 2001. *Crowd dynamics*. Ph.D. diss., University of Warwick. www.crowddynamics.com

Sutton, J. 1997. Gibrat's legacy. *Journal of Economic Literature* 35:40-59.

Swarm Development Group. 2001. Swarm 2.1.1 (software). Santa Fe. www.swarm.org

Teitz, M. and P. Bart. 1968. Heuristic methods for estimating generalized vertex median of a weighted graph. *Operations Research* 16:955-61.

Telecities. 2002. About Telecities. eurocities.poptel.org.uk/telecities/aboutTC/content.htm

Thorndyke, P. W. and B. Hayes-Roth. 1982. Differences in spatial knowledge acquired from maps and navigation. *Cognitive Psychology* 14:560-89.

Thurstain-Goodwin, M. and M. Batty. 2001. The sustainable town centre. In *Planning for a sustainable future,* edited by A. Layard, S. Davoudi and S. Batty, 234-68. London: Routledge.

Thurstain-Goodwin, M. and D. J. Unwin. 2000. Defining and delimiting the central areas of towns for statistical monitoring using continuous surface representations. *Transactions in GIS* 4:305-17.

Tlauka, M. and P. N. Wilson. 1996. Orientation-free representations from navigation through a computer-simulated environment. *Environment and Behavior* 28:647-64.

Tolman, E. C. 1948. Cognitive maps in rats and man. *Psychological Review* 55:189-208.

Tomlinson, R. F. 1998. The Canada geographic information system. In *The history of geographic information systems: perspectives from the pioneers,* edited by T. W. Foresman, 21-32. Upper Saddle River, N.J.: Prentice Hall.

Torrens, P. M. 2001a. Can geocomputation save urban simulation? Throw some agents into the mixture, simmer, and wait. . .. CASA working paper. Centre for Advanced Spatial Analysis, University College London. www.casa.ucl.ac.uk/working_papers

____. 2001b. *New tools simulating housing choices.* C01-006. Berkeley, CA: University of California Institute of Business and Economic Research and Fisher Center for Real Estate and Urban Economics.

____. 2002a. Cellular automata and multi-agent systems as planning support tools. In *Planning support systems in practice,* edited by S. Geertman and J. Stillwell, 205-222. London: Springer-Verlag.

____. 2002b. SprawlSim: modeling sprawling urban growth using automata-based models. In *Agent-based models of land-use/land-cover change,* edited by D. C. Parker, T. Berger, S. M. Manson and W. J. McConnell, 69-76. Louvain-la-Neuve, Belgium: LUCC International Project Office.

Torrens, P. M. and D. O'Sullivan. 2001. Cellular automata and urban simulation: where do we go from here? *Environment and Planning B* 28 (2):163-168.

Transportation Research Board, National Research Council. 1998. *The costs of sprawl—Revisited.* Washington, D.C.: National Academy Press.

Traynor, C. and M. G. Williams. 1995. Why are geographic information systems hard to use? ACM/SIGCHI. CHI 1995 Mosaic of Creativity. Denver, Colo. 288-9.

Trochim, W. M. 2002. Research methods knowledge base. Accessed May 2002. trochim.human.cornell.edu/kb

Trowbridge, C. C. 1913. On fundamental methods of orientation and imaginary maps. Science 38:888-97.

UNCHS. 1982. *Survey of slum and squatter settlements.* Dublin, Ireland: UNCHS and Tycooly International.

____. 1995. Los asentamientos humanos en America Latina y el Caribe. In Reunión Regional de América Latina y el Caribe Preparatoria de la Conferencia de las Naciones Unidas sobre los Asentamientos Humanos (Hábitat II). CEPAL.

____. 1996. *An urbanizing world: global report on human settlements.* Oxford: Oxford University Press.

University of Chicago. 2003. RePast 2.0 (software). Chicago: Social Science Research Computing Program. repast.sourceforge.net

Valladares, L. and M. P. Coelho. 1995. Urban research in Brazil and Venezuela: towards an agenda for the 1990s. In *Urban research in the Developing world,* edited by R. Stren, 123-38. Toronto: Centre for Urban and Community Studies, University of Toronto.

Venables, M. and U. Bilge. 1998. Complex adaptive modelling at J Sainsbury: the SIM-STORE supermarket supply chain experiment. Paper presented at the LSE Strategy and Complexity Seminar. London School of Economics, London. Accessed May 2003. www.lse.ac.uk/LSE/COMPLEX/Seminars/1998/report98mar.htm

Vicsek, T., A. Czirok, E. Ben-Jacob, I. Cohen and O. Shochet. 1995. Novel type of phase transition in a system of self-driven particles. *Physical Review Letters* 75:1226-9.

Vilensky, B. 1996. Can analysis of word frequency distinguish between writings of different authors? *Physica A* 231:705-11.

Watts, D. J. 2003. *Six degrees: the science of a connected age.* New York: Norton.

WDRT. 2001a. Directorate of Community. Woodberry Down Regeneration Team. www.hackney.gov.uk/woodberry/pdf/community.pdf

____. 2001b. Vision, objectives and procurement. 2d ed. Woodberry Down Regeneration Team. www.hackney.gov.uk/woodberry/pdf/vision.pdf

Webber, R. and M. Farr. 2001. MOSAIC: from an area classification system to individual classification. *Journal of Targeting, Measurement and Analysis for Marketing* 10 (1):55-65.

White, M. 2002. Clearer poverty definition 'vital'. *The Guardian* 27 (August):2.

White, R. and G. Engelen. 1997. Cellular automata as the basis of integrated dynamic regional modelling. *Environment and Planning B: Planning and Design* 24:235-46.

Wiener, N. 1948. *Cybernetics: or control and communication in the animal and the machine.* Cambridge, Mass.: Technology Press.

Wilson, P. N. 1997. Use of virtual reality computing in spatial learning research. In *A handbook of spatial research paradigms and methodologies.* Vol. 1, edited by N. Foreman and R. Gillet, 181-206. East Sussex, United Kingdom: Psychology Press.

Witmer, B. G., J. H. Bailey, B. W. Knerr and K. C. Parsons. 1996. Virtual spaces and real world places: transfer of route knowledge. *International Journal of Human-Computer Studies* 45:413-28.

Wolfram, Stephen. 2002. *A new kind of science.* Champaign, Ill.: Wolfram Media.

Wood, J. D., P. F. Fisher, J. A. Dykes, D. J. Unwin and K. Stynes. 1999. The use of the landscape metaphor in understanding population data. *Environment and Planning B* 26:281-95.

Wrigley, N. and M. Lowe. 2002. *Reading retail.* Oxford: Blackwell.

Zhang, J. and M. F. Goodchild. 2002. *Uncertainty in geographical information.* London: Taylor and Francis.

Zhao, H. and R. Shibasaki. 2001. Reconstructing textured CAD model of urban environment using vehicle-borne laser range scanners and line cameras. In *ICVS, LNCS 2095,* edited by B. Schiele and G. Sagerer, 284-97. Berlin: Springer-Verlag.

Zipf, G. K. 1949. *Human behavior and the principle of least effort.* Cambridge, Mass.: Addison-Wesley.

Contributors

(All affiliations are within University College London (UCL) unless otherwise stated. Mailing address for CASA: UCL 1-19 Torrington Place, London WC1E 7HB, United Kingdom.)

Note that materials relating to the book chapters may be found at www.casabook.com.

Katerina Alexiou

Ph.D. Researcher, CASA and the Bartlett School of Graduate Studies

Research interests: computer-aided design and planning; artificial intelligence in design; interactive design models; knowledge-based systems; cooperative design and planning; design and planning support systems

www.casa.ucl.ac.uk/people/Katerina.htm
a.alexiou@ucl.ac.uk

Sinesio P. Alves-Junior

Ph.D. Researcher, CASA and Department of Geography

Research interests: GIS; urban remote sensing; urban models; cellular automata; geodemographic classification systems; urban segregation

www.casa.ucl.ac.uk/people/Sinesio.htm
sinesio.alves@ucl.ac.uk

Joana Barros

Ph.D. Researcher, CASA and the Bartlett School of Graduate Studies

Research interests: complexity theory, urban morphology, agent-based models, cellular automata models, urban growth, and Latin-American cities

www.casa.ucl.ac.uk/people/Joana.htm
j.barros@ucl.ac.uk

Michael Batty

Professor of Spatial Analysis and Director, CASA

Research interests: urban models; complexity theory; visualisation and computer graphics; dynamic modeling; urban morphology; design methods

www.casa.ucl.ac.uk/people/Mike.htm
m.batty@ucl.ac.uk

Susan Batty

Senior Lecturer, CASA and the Bartlett School of Planning

Research interests: comparative planning and formal theories of institutions; urban environmental politics; GIS and urban analysis; comparative planning; public participation; local sustainability

www.casa.ucl.ac.uk/people/Sue.htm
susan.batty@ucl.ac.uk

Elena Besussi

Research Fellow, CASA

Research interests: European spatial planning; qualitative indicators and criteria of spatial differentiation

www.casa.ucl.ac.uk/people/Elena.htm
e.besussi@ucl.ac.uk

Charles Boulton

Consultant, CASA and Cranfield University

Research interests: innovation; complexity theory; management systems

www.nexsus.org/about.shtml
cbboulton@hotmail.com

Nancy Chin

Ph.D. Researcher, CASA and Department of Geography

Research interests: urban growth; urban planning; spatial pattern analysis; urban sprawl

www.casa.ucl.ac.uk/people/Nancy.htm
n.chin@ucl.ac.uk

Paul J. Densham

Reader in Geography, CASA and Department of Geography

Research interests: locational analysis algorithms; spatial decision support systems; parallel algorithms

www.casa.ucl.ac.uk/people/PaulDensham.html
pdensham@geog.ucl.ac.uk

Michael J. de Smith

Ph.D. Researcher, CASA and Department of Geography

Research interests: distance metrics; distance statistics; path measurement; optimal path alignments

www.casa.ucl.ac.uk/people/Mikede.htm
mike@desmith.net

Stephen Evans

Research Fellow, CASA and the Bartlett School of Planning

Research interests: environmental GIS; 3-D GIS; cartography; mapping coral reefs

www.casa.ucl.ac.uk/people/Steve.htm
stephen.evans@ucl.ac.uk

Wolfram Grajetzki

Research Fellow, CASA

Research interests: burial customs and funerary culture; administration and social relations in Middle Bronze Age Egypt; digital recording of ancient artefacts and monuments; publishing using new media

www.casa.ucl.ac.uk/people/Wolfram.htm
w.grajetzki@ucl.ac.uk

Ian Greatbatch

Ph.D. Researcher, CASA and Department of Geography

Research interests: retail systems; geodemographics; GIS and urban remote sensing; qualitative data analysis and uncertainty

www.casa.ucl.ac.uk/people/Ian.htm
i.greatbatch@ucl.ac.uk

Mordechai (Muki) Haklay

Lecturer in Geographic Information Science, and Department of Geomatic Engineering

Research interests: environmental information and its role in decision making; computational geography; usability and GIS; GIS and society

www.casa.ucl.ac.uk/people/Muki.html
m.haklay@ucl.ac.uk

Andrew Hudson-Smith

Research Fellow, CASA

Research interests: 3-D modelling of urban environments; Web-based 3-D model distribution

www.casa.ucl.ac.uk/people/Andy.htm
asmith@geog.ucl.ac.uk

Chao Li

Ph.D. Researcher, CASA and Department of Geography

Research interests: location-based services; spatial data analysis; urban wayfinding

www.casa.ucl.ac.uk/people/Lily.htm
chao.li@ucl.ac.uk

Daryl Lloyd

Ph.D. Researcher, CASA and Department of Geography

Research interests: error and uncertainty in GIS; urban geography; philosophy of uncertainty

www.casa.ucl.ac.uk/people/Daryl.htm
d.lloyd@ucl.ac.uk

Paul A. Longley

Professor of Geographic Information Science, Department of Geography and Deputy Director, CASA

Research interests: quantitative geographical techniques; urban geography; geodemographics; urban remote sensing; GIS and GIScience

www.casa.ucl.ac.uk/people/Paul.htm
p.longley@geog.ucl.ac.uk.

David J. Maguire

Visiting Professor, CASA and Director of Products, Solutions and International, ESRI, USA

Research interests: handheld systems; GIS software architectures; spatial databases; spatial analysis

www.casa.ucl.ac.uk/people/David_Maguire.htm
dmaguire@esri.com

Torsten Schietzelt

Ph.D. Researcher, CASA and Department of Geography

Research interests: spatial decision support systems; locational analysis; GIS for election systems and humanitarian relief

www.casa.ucl.ac.uk/people/Torsten.htm
t.schietzelt@ucl.ac.uk

Narushige Shiode

Research Fellow, CASA and the Bartlett School of Graduate Studies

Research interests: spatial analysis; locational optimisation; cyberspace analysis; virtual reality modeling; visualisation and computer graphics

www.casa.ucl.ac.uk/people/Naru.htm
n.shiode@ucl.ac.uk

Sarah L. Smith

Ph.D. Researcher CASA and Department of Geography, and Research Analyst, Research and Innovation Department, Ordnance Survey (Great Britain)

Research interests: remote sensing; image interpretation; applications of city modelling; understanding error and uncertainty

www.casa.ucl.ac.uk/people/SarahSmith.htm
sarah.smith@ucl.ac.uk

J. Philip Steadman

Professor of Urban and Built Form Studies, CASA and the Bartlett School of Planning

Research interests: energy use in transport and building stock; the geometry of built form; the evolution of building types

www.casa.ucl.ac.uk/people/Phil.htm
j.p.steadman@ucl.ac.uk

Mark Thurstain-Goodwin

Research Fellow, CASA and Consultant, Geofutures

Research interests: urban geography; town centers; property; geodemographics; data mining; sustainable development

www.casa.ucl.ac.uk/people/Mark.htm
mark.thurstain-goodwin@ucl.ac.uk

Carolina Tobón

Ph.D. Researcher, CASA and Department of Geography

Research interests: geovisualisation; usability evaluation; spatial statistics; GIS and GIScience

www.casa.ucl.ac.uk/people/Carolina.htm
c.tobon@ucl.ac.uk

Paul M. Torrens

Ph.D. Researcher, CASA and Department of Geography

Research interests: spatial analysis and modelling; GIS and GIScience; artificial intelligence, artificial life, urban studies

www.casa.ucl.ac.uk/people/PaulTorrens.htm
ptorrens@geog.ucl.ac.uk

Richard Webber

Visiting Professor, CASA and Consultant

Research interests: geodemographic classification systems; comparative studies of urban structure and change; psephology

www.casa.ucl.ac.uk/people/Richard.htm
richardwebber@blueyonder.co.uk

Theodore Zamenopoulos

Ph.D. Researcher, CASA and the Bartlett School of Graduate Studies

Research interests: design science and computing; artificial intelligence in design and planning; computational models of learning and creative search; design and planning support systems; knowledge acquisition and representation

www.casa.ucl.ac.uk/people/Theo.htm
t.zamenopoulos@ucl.ac.uk

Advanced Spatial Analysis: The CASA book of GIS
Editorial assistance by Claudia Naber
Copyediting by Tiffany Wilkerson
Cartographic assistance by Edith M. Punt
Book design, production and image editing by Savitri Brant
Cover design by Savitri Brant
Web design by Sonja Curtis www.casabook.com
Manuscript administration by Sarah Sheppard
Printing coordination by Cliff Crabbe